IET TELECOMMUNICATIONS SERIES 77

Network Design, Modeling, and Performance Evaluation

Other volumes in this series:

Volume 9 **Phase Noise in Signal Sources** W.P. Robins
Volume 12 **Spread Spectrum in Communications** R. Skaug and J.F. Hjelmstad
Volume 13 **Advanced Signal Processing** D.J. Creasey (Editor)
Volume 19 **Telecommunications Traffic, Tariffs and Costs** R.E. Farr
Volume 20 **An Introduction to Satellite Communications** D.I. Dalgleish
Volume 26 **Common-Channel Signalling** R.J. Manterfield
Volume 28 **Very Small Aperture Terminals (VSATs)** J.L. Everett (Editor)
Volume 29 **ATM: The broadband telecommunications solution** L.G. Cuthbert and J.C. Sapanel
Volume 31 **Data Communications and Networks, 3rd Edition** R.L. Brewster (Editor)
Volume 32 **Analogue Optical Fibre Communications** B. Wilson, Z. Ghassemlooy, and I.Z. Darwazeh (Editors)
Volume 33 **Modern Personal Radio Systems** R.C.V. Macario (Editor)
Volume 34 **Digital Broadcasting** P. Dambacher
Volume 35 **Principles of Performance Engineering for Telecommunication and Information Systems** M. Ghanbari, C.J. Hughes, M.C. Sinclair, and J.P. Eade
Volume 36 **Telecommunication Networks, 2nd Edition** J.E. Flood (Editor)
Volume 37 **Optical Communication Receiver Design** S.B. Alexander
Volume 38 **Satellite Communication Systems, 3rd Edition** B.G. Evans (Editor)
Volume 40 **Spread Spectrum in Mobile Communication** O. Berg, T. Berg, J.F. Hjelmstad, S. Haavik, and R. Skaug
Volume 41 **World Telecommunications Economics** J.J. Wheatley
Volume 43 **Telecommunications Signalling** R.J. Manterfield
Volume 44 **Digital Signal Filtering, Analysis and Restoration** J. Jan
Volume 45 **Radio Spectrum Management, 2nd Edition** D.J. Withers
Volume 46 **Intelligent Networks: Principles and applications** J.R. Anderson
Volume 47 **Local Access Network Technologies** P. France
Volume 48 **Telecommunications Quality of Service Management** A.P. Oodan (Editor)
Volume 49 **Standard Codecs: Image compression to advanced video coding** M. Ghanbari
Volume 50 **Telecommunications Regulation** J. Buckley
Volume 51 **Security for Mobility** C. Mitchell (Editor)
Volume 52 **Understanding Telecommunications Networks** A. Valdar
Volume 53 **Video Compression Systems: From first principles to concatenated codecs** A. Bock
Volume 54 **Standard Codecs: Image compression to advanced video coding, 3rd edition** M. Ghanbari
Volume 59 **Dynamic Ad Hoc Networks** H. Rashvand and H. Chao (Editors)
Volume 60 **Understanding Telecommunications Business** A. Valdar and I. Morfett
Volume 65 **Advances in Body-Centric Wireless Communication: Applications and state-of-the-art** Q.H. Abbasi, M.U. Rehman, K. Qaraqe, and A. Alomainy (Editors)
Volume 67 **Managing the Internet of Things: Architectures, theories and applications** J. Huang and K. Hua (Editors)
Volume 68 **Advanced Relay Technologies in Next Generation Wireless Communications** I. Krikidis and G. Zheng
Volume 69 **5G Wireless Technologies** A. Alexiou (Editor)
Volume 70 **Cloud and Fog Computing in 5G Mobile Networks** E. Markakis, G. Mastorakis, C.X. Mavromoustakis, and E. Pallis (Editors)
Volume 71 **Understanding Telecommunications Networks, 2nd edition** A. Valdar
Volume 72 **Introduction to Digital Wireless Communications** Hong-Chuan Yang
Volume 73 **Network as a Service for Next Generation Internet** Q. Duan and S. Wang (Editors)
Volume 74 **Access, Fronthaul and Backhaul Networks for 5G & Beyond** M.A. Imran, S.A.R. Zaidi, and M.Z. Shakir (Editors)
Volume 76 **Trusted Communications with Physical Layer Security for 5G and Beyond** T.Q. Duong, X. Zhou, and H.V. Poor (Editors)
Volume 79 **Satellite Communications in the 5G Era** S.K. Sharma, S. Chatzinotas, and D. Arapoglou
Volume 80 **Transceiver and System Design for Digital Communications 5th Edition** Scott R. Bullock
Volume 905 **ISDN Applications in Education and Training** R. Mason and P.D. Bacsich

Network Design, Modeling, and Performance Evaluation

Quoc-Tuan Vien

The Institution of Engineering and Technology

Published by The Institution of Engineering and Technology, London, United Kingdom

The Institution of Engineering and Technology is registered as a Charity in England & Wales (no. 211014) and Scotland (no. SC038698).

First published 2018

The Institution of Engineering and Technology
Michael Faraday House
Six Hills Way, Stevenage
Herts, SG1 2AY, United Kingdom

www.theiet.org

British Library Cataloguing in Publication Data
A catalogue record for this product is available from the British Library

ISBN 978-1-78561-336-4 (hardback)
ISBN 978-1-78561-337-1 (PDF)

Typeset in India by MPS Limited

I dedicate this book to my parents, Dich-Tung Vien and Thi-Ngoc Ngo, my wife, Tram Pham, and our lovely children, Helen and Harry.

Contents

List of figures xiii
List of tables xvii
Foreword xix
Preface xxi
Abbreviations xxv

1 Internetworking and network fundamentals 1
1.1 Fundamentals of networking 1
1.2 Local area networks 5
1.2.1 LAN devices 5
1.2.2 LAN topologies 7
1.2.3 LAN technologies 9
1.2.4 LAN access methods 12
1.2.5 LAN transmission methods 14
1.3 Wide area networks 16
1.3.1 WAN devices 16
1.3.2 WAN topologies 18
1.3.3 WAN technologies 20
1.4 Open systems interconnection reference model 28
1.4.1 OSI layers 29
1.4.2 Interaction between OSI layers in communications 30
1.4.3 Information exchange and information formats in OSI layers 31
1.5 Internet Protocols 32
1.5.1 TCP/IP suite 33
1.5.2 Internet Protocol 34
1.5.3 Transmission Control Protocol 34
1.5.4 User datagram protocol 34
1.5.5 Other protocols 34
1.6 Internetworks 35
1.6.1 Internetwork hierarchy 36
1.6.2 Internetwork addressing 36
1.6.3 Standards organizations 38
1.6.4 Internetworking devices 38
1.7 Related works 41
1.8 Review questions 41
1.9 Problems 42

2 Routing in computer networks **45**
2.1 Fundamentals of routing 45
2.1.1 Path determination 46
2.1.2 Routing algorithms 46
2.2 Static routing protocols 48
2.3 Dynamic routing protocols 48
2.4 Interior gateway protocols 49
2.4.1 Distance-vector routing protocols 49
2.4.2 Link-state routing protocols 52
2.4.3 Enhanced Interior Gateway Routing Protocol 56
2.5 Border Gateway Protocols 56
2.6 Related works 57
2.7 Review questions 58
2.8 Problems 58

3 Internet Protocol addressing **61**
3.1 An overview of IP address 61
3.1.1 Addressing types 63
3.1.2 Addressing mechanisms 63
3.2 IP version 4 (IPv4) 63
3.2.1 Private IP addresses 64
3.2.2 Classful IP addresses 64
3.3 Subnetting 66
3.4 Variable-length subnetting 69
3.5 Supernetting 70
3.6 IP version 6 71
3.7 Related works 73
3.8 Review questions 73
3.9 Problems 74

4 Network analysis, architecture, and design **79**
4.1 An overview of network analysis, architecture, and design 79
4.2 Network analysis process 81
4.3 Network architecture process 81
4.4 Network design process 82
4.5 Network hierarchical design 83
4.6 Network hierarchical layers 84
4.6.1 Core layer 84
4.6.2 Distribution layer 84
4.6.3 Access layer 85
4.7 Network design approaches and evolution 86
4.8 Network management and security 87
4.8.1 Network design goals 88
4.8.2 Technology evaluation criteria 89
4.8.3 Network-management design 89

4.9 Related works 91
4.10 Review questions 91
4.11 Problems 92

5 Network-requirement analysis 93
5.1 Network services and requirements 93
5.2 Service characteristics 94
5.3 Requirement-analysis process 98
5.4 User-requirement analysis 99
5.5 Application-requirement analysis 100
5.6 Host-requirement analysis 101
5.7 Network-requirement analysis 102
5.8 Requirement-analysis model 103
5.9 Reliability, maintainability, and availability analyses 104
5.10 Related works 109
5.11 Review questions 110
5.12 Problems 111

6 Network flow analysis 115
6.1 Flows 115
6.2 Data sources and sinks 117
6.3 Flow models 117
6.3.1 Peer-to-peer model 118
6.3.2 Client–server model 119
6.3.3 Cooperative computing model 119
6.3.4 Distributed computing model 120
6.4 Flow boundaries 121
6.5 Flow distributions 122
6.6 Flow specification 123
6.7 Flow analysis model 125
6.8 Related works 126
6.9 Review questions 127
6.10 Problems 128

7 Network performance evaluation 129
7.1 Benchmarking: merits and demerits 129
7.2 Simulation: merits and demerits 130
7.3 Analytical modeling: merits and demerits 131
7.4 Systems and system environment 132
7.5 Related works 135
7.6 Review questions 135
7.7 Problems 135

8 Network simulation models 137
8.1 Simulation for emulating and evaluating a system 137
8.2 Simulation models of a system 139

8.3	Steps/phases in a simulation	140
8.4	An example of simulation model	140
8.5	Related works	142
8.6	Review questions	142
8.7	Problems	143
9	**Statistical models in network simulation**	**145**
9.1	Statistical models in queuing systems	145
9.2	Poisson point process	147
9.2.1	Stationary Poisson point process	148
9.2.2	Nonstationary Poisson point process	150
9.2.3	Operations of Poisson point process	151
9.3	Input modeling	152
9.3.1	Data collection	152
9.3.2	Probability distribution identification	153
9.3.3	Parameter estimation	154
9.3.4	Distribution fitness test	155
9.4	Related works	155
9.5	Review questions	156
9.6	Problems	156
10	**Probabilities in performance analysis**	**159**
10.1	Basic concepts and terminology of probability	159
10.2	Axioms and properties of probability	161
10.3	Conditional probability	164
10.4	Independence of events	168
10.5	Related works	170
10.6	Review questions	170
10.7	Problems	171
11	**Random variables in network modeling and simulation**	**175**
11.1	Introduction of random variables	175
11.2	Discrete random variables	176
11.2.1	Probability mass function	177
11.2.2	Cumulative distribution function	178
11.2.3	Expected value	180
11.2.4	Variance and standard deviation	182
11.2.5	Moments	185
11.3	Continuous random variables	186
11.3.1	Cumulative distribution function	187
11.3.2	Probability density function	188
11.3.3	Expected value	190
11.3.4	Variance and standard deviation	190
11.3.5	Moments	191
11.4	Related works	192

11.5 Review questions 192
11.6 Problems 193

12 Discrete random variables and their applications 197
12.1 Review of discrete random variables 197
12.2 Bernoulli random variable 197
12.3 Binomial random variable 200
12.4 Geometric random variable 202
12.5 Pascal random variable 205
12.6 Poisson random variable 208
12.7 Discrete uniform random variable 211
12.8 Related works 214
12.9 Review questions 214
12.10 Problems 215

13 Continuous random variables and their applications 217
13.1 Review of continuous random variables 217
13.2 Continuous uniform random variable 217
13.3 Exponential random variable 222
13.4 Erlang random variable 226
13.5 Normal random variable 230
13.6 Lognormal random variable 235
13.7 Related works 238
13.8 Review questions 238
13.9 Problems 239

14 Random variable generation in network simulation 241
14.1 Inverse transform technique 241
14.1.1 Bernoulli random variable generation 242
14.1.2 Geometric random variable generation 243
14.1.3 Discrete uniform random variable generation 245
14.1.4 Continuous uniform random variable generation 246
14.1.5 Exponential random variable generation 247
14.2 Transformation techniques 248
14.2.1 Binomial random variable generation 248
14.2.2 Pascal random variable generation 249
14.2.3 Erlang random variable generation 250
14.2.4 Lognormal random variable generation 250
14.3 Related works 251
14.4 Review questions 252
14.5 Problems 253

15 Queuing theory for network modeling and performance evaluation 255
15.1 An introduction to queuing theory 255
15.2 Queuing systems in computer and communication networks 256

15.3 Poisson point process for queuing systems 257
15.4 Performance measures in queuing systems 258
15.4.1 Interarrival and service time in Poisson point process 259
15.4.2 Waiting and response time 259
15.4.3 The number of packets in the queue and in the system 260
15.4.4 Traffic intensity and server utilization 260
15.4.5 Little's law for queuing systems 261
15.5 Kendall's notation for queuing systems 262
15.6 Related works 263
15.7 Review questions 264
15.8 Problems 264

16 Single-server queues—network behaviors and analysis 267
16.1 $M/M/1$ Queue 267
16.2 $M/M/1/K$ Queue 271
16.3 M/M/1/K/K Queue 274
16.4 Related works 278
16.5 Review questions 279
16.6 Problems 279

17 Multi-server queues—network behaviors and analysis 281
17.1 $M/M/c$ Queue 281
17.2 $M/M/c/K$ Queue 286
17.3 $M/M/c/K/K$ Queue 291
17.4 Related works 296
17.5 Review questions 296
17.6 Problems 297

References 299
Index 315

List of figures

Figure P.1 Dependence graph of chapters xxiii
Figure 1.1 Local area network 2
Figure 1.2 Wide area network 3
Figure 1.3 Campus area network 3
Figure 1.4 Metropolitan area network 4
Figure 1.5 Home area network 4
Figure 1.6 Personal area network 5
Figure 1.7 Hub and its network symbol 6
Figure 1.8 Switch and its network symbol 6
Figure 1.9 Router and its network symbol 7
Figure 1.10 Bus topology 7
Figure 1.11 Star topology 8
Figure 1.12 Ring topology 8
Figure 1.13 Extended star topology 9
Figure 1.14 Unshielded twisted-pair (UTP) 10
Figure 1.15 Shielded twisted-pair (STP) 10
Figure 1.16 Optical fiber 10
Figure 1.17 Carrier sense multiple access with collision detection (CSMA/CD) 12
Figure 1.18 Carrier sense multiple access with collision avoidance (CSMA/CA) 13
Figure 1.19 Control token technique 14
Figure 1.20 Unicast transmission 15
Figure 1.21 Broadcast transmission 15
Figure 1.22 Multicast transmission 16
Figure 1.23 A WAN switch 17
Figure 1.24 Modems in WAN 17
Figure 1.25 A channel service unit/digital service unit (CSU/DSU) in WAN 18
Figure 1.26 Access server in WAN 18
Figure 1.27 A multiplexer in WAN 19
Figure 1.28 Full mesh topology 19
Figure 1.29 Partial mesh topology 20
Figure 1.30 Circuit switching 21
Figure 1.31 Packet switching 21
Figure 1.32 Virtual circuit switching 22
Figure 1.33 Integrated Services Digital Network (ISDN) devices 23

Figure 1.34 X.25 devices 24
Figure 1.35 Frame relay devices 25
Figure 1.36 Asynchronous transfer mode (ATM) devices 26
Figure 1.37 Switched multimegabit data service (SMDS) devices 27
Figure 1.38 Digital subscriber line (DSL) 27
Figure 1.39 Open systems interconnection (OSI) layers 28
Figure 1.40 Upper and lower OSI layers 31
Figure 1.41 Protocol data unit (PDU) in OSI model 32
Figure 1.42 Transmission Control Protocol (TCP)/Internet Protocol (IP) 33
Figure 1.43 Internetwork hierarchy 36
Figure 1.44 MAC address 37
Figure 2.1 Interior gateway protocol (IGP) versus exterior gateway protocol (EGP) 48
Figure 2.2 Routing loop in Example 2.1 52
Figure 2.3 Open Shortest Path First (OSPF) routing protocol 54
Figure 2.4 OSPF over backbone area 54
Figure 2.5 A graph in Example 2.2 where the value on the edge represents weight or path cost 55
Figure 2.6 Border Gateway Protocols (BGPs) with Internal BGP (IBGP) and External BGP (EBGP) 57
Figure 2.7 Figure for Problems 4 and 5 59
Figure 2.8 Figure for Problem 6 59
Figure 3.1 IP address with network mask in Example 3.1 62
Figure 3.2 IP address with subnet mask in Example 3.1 62
Figure 3.3 An IPv4 address 64
Figure 3.4 IP address classes 65
Figure 3.5 Subnetting 66
Figure 3.6 Variable-length subnet mask in Example 3.5 69
Figure 3.7 Supernetting in Example 3.8 71
Figure 3.8 IPv6 address in Example 3.9 72
Figure 3.9 Figure for Problem 10 75
Figure 3.10 Figure for Problem 11 76
Figure 4.1 Network analysis, architecture, and design process 80
Figure 4.2 Network hierarchical design with interconnectivity 83
Figure 4.3 Network hierarchical layers 84
Figure 4.4 One-layer design 85
Figure 4.5 Two-layer design 86
Figure 4.6 Network hierarchy over content-delivery network (CDN) 87
Figure 5.1 System components 94
Figure 5.2 Packet error probability (PER) versus the number of bits 97
Figure 5.3 Channel capacity versus signal-to-noise ratio (SNR) 98
Figure 5.4 Requirement-analysis process between two end nodes 99
Figure 5.5 Requirement-analysis model 103
Figure 5.6 Time between failure (TBF) and time to repair (TTR) of a system 105
Figure 5.7 Devices connected in series 106

Figure 5.8 Devices connected in parallel 107
Figure 5.9 Figure for Problem 5. 111
Figure 5.10 Figure for Problem 7. 112
Figure 5.11 Figure for Problem 8. 113
Figure 6.1 Flow types 116
Figure 6.2 Peer-to-peer model 118
Figure 6.3 Client–server model 119
Figure 6.4 Cooperative computing model 120
Figure 6.5 Distributed computing model 121
Figure 6.6 Flow boundaries 122
Figure 6.7 Flow distribution 123
Figure 6.8 Flow analysis model 125
Figure 7.1 System components in a simulation model 132
Figure 7.2 Data flow between two end devices 133
Figure 7.3 Components in a banking system 134
Figure 7.4 Components in a single-server queuing system 134
Figure 8.1 Simulation phases 141
Figure 8.2 Simulation phases in a queuing system 142
Figure 9.1 Interarrival time in Poisson point process $N(t)$ 149
Figure 9.2 Input modeling in network simulation 152
Figure 10.1 Event in an experiment 161
Figure 11.1 Random variable for representing a random event 175
Figure 11.2 pmf of random variable in Example 11.2 177
Figure 11.3 cdf of random variable in Example 11.2 179
Figure 11.4 cdf of a continuous uniform random variable in $[1, 3]$ 187
Figure 11.5 pdf of a continuous uniform random variable in $[1, 3]$ 188
Figure 12.1 pmf of $X \sim Ber(0.6)$ 199
Figure 12.2 pmf of $X \sim Bin(20, 0.6)$ 201
Figure 12.3 pmf of $X \sim Geo(0.6)$ 203
Figure 12.4 pmf of $X \sim Pas(4, 0.6)$ 206
Figure 12.5 pmf of $X \sim Poi(5)$ 209
Figure 12.6 pmf of $X \sim U(1, 5)$ 211
Figure 13.1 pdf of $X \sim U(1, 5)$ 219
Figure 13.2 cdf of $X \sim U(1, 5)$ 220
Figure 13.3 pdf of $X \sim Exp(5)$ 223
Figure 13.4 cdf of $X \sim Exp(5)$ 223
Figure 13.5 pdf of $X \sim Erl(3, 5)$ 227
Figure 13.6 cdf of $X \sim Erl(3, 5)$ 228
Figure 13.7 pdf of $X \sim N(1, 9)$ 232
Figure 13.8 cdf of $X \sim N(1, 9)$ 232
Figure 13.9 pdf of $X \sim LN(1, 0.25)$ 236
Figure 13.10 cdf of $X \sim LN(1, 0.25)$ 236
Figure 14.1 Samples of $X \sim Ber(0.6)$ 243
Figure 14.2 Samples of $X \sim Geo(0.6)$ 244
Figure 14.3 Samples of $X \sim U(1, 5), X \in \mathbb{Z}$ 245

Figure 14.4 Samples of $X \sim U(1, 5)$ 246
Figure 14.5 Samples of $X \sim Exp(5)$ 247
Figure 14.6 Samples of $X \sim Bin(20, 0.6)$ 248
Figure 14.7 Samples of $X \sim Pas(4, 0.6)$ 249
Figure 14.8 Samples of $X \sim Erl(3, 5)$ 250
Figure 14.9 Samples of $Y \sim N(1, 0.25)$ 251
Figure 14.10 Samples of $Y \sim LN(1, 0.25)$ 252
Figure 15.1 Components in a queuing system 257
Figure 15.2 Performance measures in a queuing system 262
Figure 16.1 $M/M/1$ queue 267
Figure 16.2 $M/M/1/K$ queue 271
Figure 16.3 $M/M/1/K/K$ queue 275
Figure 17.1 $M/M/c$ Queue 282
Figure 17.2 $M/M/c/K$ Queue 287
Figure 17.3 $M/M/c/K/K$ Queue 292

List of tables

Table 2.1	Dijkstra's algorithm for finding the shortest distance in Example 2.2	56
Table 3.1	Private IP address range	64
Table 3.2	IP address classes	65
Table 3.3	Subnetting in Example 3.3	67
Table 3.4	Laboratories in Example 3.4	68
Table 3.5	Subnetting in Example 3.4	68
Table 3.6	Rooms in Example 3.6	70
Table 3.7	Variable-length subnetting in Example 3.6	70
Table 3.8	IPv6 packet format	73
Table 3.9	Subnetting of Class C address	74
Table 3.10	Subnetting of Class B address	75
Table 3.11	List of workgroups in Problem 12	76
Table 3.12	List of departments in Problem 14	77
Table 5.1	Table for Problem 5	112
Table 12.1	Formulas of discrete random variables	198
Table 13.1	Formulas of continuous random variables	218

Foreword

About the book

This book introduces a methodological approach to network design that enables readers to evaluate a network situation and identify the most important aspects to be monitored and analyzed. It shows how to go from the analysis of initial network requirements to the network architecture design, modeling, simulation, and evaluation. Practice exercises are given for selected chapters, and case studies take the reader through the whole network-design process. It is ideal for practitioners and researchers working in all aspects of network services.

About the author

Quoc-Tuan Vien is currently a senior lecturer in computing and communications engineering within the Faculty of Science and Technology, Middlesex University, London, UK. Prior to joining Middlesex University, he worked as postdoctoral research assistant with the School of Science and Technology, Nottingham Trent University, Nottingham, UK. He received his PhD degree from Glasgow Caledonian University, UK, in 2012.

Dr Vien is a senior member of the IEEE, a member of the IET, and a fellow of the HEA. He has authored/coauthored two books, five book chapters, and over 70 publications in major conference proceedings and ISI journals in which he is a leading author of 43 papers. His current research interests include physical layer security, network coding, non-orthogonal multiple access, energy harvesting, spectrum sensing, device-to-device communications, relay networks, cognitive radio networks, heterogeneous networks, wireless network-on-chip, public safety networks, and cloud radio access networks.

He was a recipient of the Best Paper Award at the IEEE/IFIP 14th International Conference on Embedded and Ubiquitous Computing in 2016. He has been an editor of *International Journal of Digital Multimedia Broadcasting*, a guest editor of the EAI Endorsed Transactions on Industrial Networks and Intelligent Systems, a program cochair for the EAI International Conference on Industrial Networks and Intelligent Systems (INISCOM 2018), a technical symposium cochair for the International Conference on Recent Advances in Signal Processing, Telecommunications and Computing (SigTelCom 2017, 2018), and a TPC member of over 100 conferences such as the IEEE PIMRC, ICNC, VTC, ISWCS, ATC, etc. He has also served as a session chair at IEEE flagship conferences including the IEEE WCNC, VTC, and ISWCS. He was honored as an exemplary reviewer of the IEEE Communications Letters in 2017.

Preface

About the book

Over the last five decades, internetworking has gained significant interest in connecting people with people and machines with machines through computer networks. Internetworking provides sustainable links between various types of network topology connecting from a small local area network to a wide area network. As a remarkable example of internetworking, the Internet has played an important role in our daily life with apparent benefits. By surfing the web, we are facing the outside world with up-to-date news, connecting with old friends, and knowing new friends that geographical distance does not allow us to meet face-to-face. Imagine you are planning to travel and what happens without the Internet? In fact, with the Internet, your travel would be much simpler and enjoyable when everything can be organized according to a routine, such as booking airplane tickets, reserving hotel rooms, tracking maps, checking the weather forecast, researching the culture of the new area, etc. All of these conveniences depend on the network-architecture design and modeling, which require not only a background knowledge but also extra effort and investment on the learning of various models and techniques to analyze their effectiveness in practical situations.

Intended audience

This book is intentionally designed for undergraduate and graduate programs in networking, communications, and computer engineering, as well as communications, IT, and networking professionals, researchers and scientists involved in the planning, design, development, testing, and operation of network services. This book is also suitable for self-study to gain a basic knowledge of computer-network design to be able to develop a simple network model for further analysis at advanced level.

Developed as a textbook, this book can benefit lecturers and students in networking, communications, and computer engineering. The chapters are particularly designed as a series of independent modules that can be combined in a number of ways for designing the courses.

Plan of the book

This book will help you to understand how to evaluate a network situation and to identify the most important network aspects that need to be monitored and analyzed. This will be done through the introduction of a methodological approach to network design. This book will also introduce the concepts of network modeling, analysis and simulation, and examine the techniques facilitating such work. It will provide you with the appreciation of the design and development simulation software as appropriate models to evaluate the pure performance and availability, as well as performability of computer networks.

This book covers the following topics:

- Internetworking and internetworking units
- Systematic approach to network design:
  - Requirements and flow analysis
  - Logical design
  - Addressing and routing
  - Internetworking strategies: switching, routing, hierarchy, and redundancy
- Probability and statistics used in modeling networks:
  - Probabilities
  - Random variables and their families
  - Random-variate generation
  - Application of statistical models in network simulation
- Input modeling techniques and output data analysis
- Queueing theory and analytical modeling of nodes and networks:
  - Queueing models and their variations
  - Analysis of queueing models and their applications in network design
- Performance evaluation of computer networks
  - Network simulation
  - Modeling for analytical solutions and simulation
  - Modeling and simulation of single-and multi-server queueing systems
  - Modeling of complex systems
  - Building simulation models and analytical models

In this book, I will provide detailed steps from the analysis of initial network requirements to the network architecture design, modeling, simulation, and evaluation. In particular, there are more focuses on the statistical and queueing models to show that they are particularly useful for the network design and analysis. This book is directed at readers with little or no background in networking. For the readers with great interest, related works are provided in a separate section of every chapter for further study. Also, this book contains a number of review questions and problems at the end of each chapter for the readers to practice.

Usage of the book

This book is specifically designed for undergraduate and graduate programs in networking, communications, and computer engineering, as well as communications, IT, and networking professionals, researchers and scientists involved in the planning, design, development, testing, and operation of network services.

For best use of the book, the students of undergraduate programs in networking, communications, and computer engineering are suggested to follow the whole structure of the book. The students of postgraduate programs who are supposed to have good background knowledge in computer networks and used to work on network design/development can have a quick review of Chapters 1–3.

The book consists of several homework problems that are organized to assist both instructors and students with reference to the relevant sections.

Dependence graph

The dependence between chapters in the book is summarized in Figure P.1 where some chapters can be skipped depending on the background of the students and also on the preferences of the instructor.

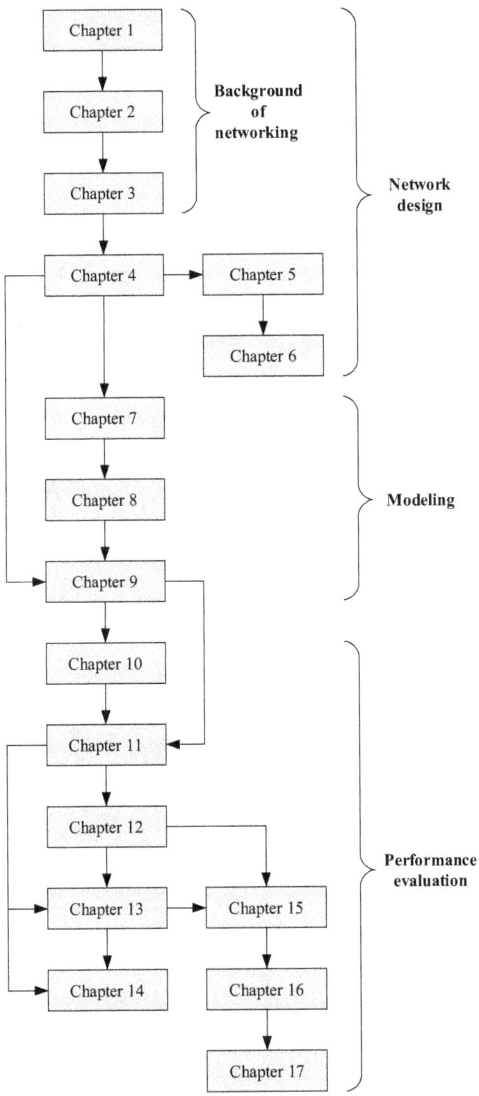

Figure P.1 Dependence graph of chapters

Abbreviations

ABR	area border router
ADSL	asymmetric digital subscriber line
ANSI	American National Standards Institute
AP	access point
ARP	address resolution protocol
AS	autonomous system
ASCII	American Standard Code for Information Interchange
ATM	asynchronous transfer mode
BER	bit error rate
BGP	Border Gateway Protocol
bps	bits per second
BRI	Basic Rate Interface
CAN	campus-area network
CAPEX	CAPital EXpenditures
CCITT	Consultative Committee for International Telephony and Telegraphy
cdf	cumulative distribution function
CDN	content delivery network
CIDR	classless inter-domain routing
CIR	committed information rate
CoS	classes of service
CPE	customer premises equipment
CSMA/CA	carrier sense multiple access with collision avoidance
CSMA/CD	carrier sense multiple access with collision detection
CSU	channel service unit
CTS	clear-to-send
DARPA	Defense Advanced Research Projects Agency
DCE	data circuit-terminating equipment
DDoS	distributed denial-of-service
DHCP	Dynamic Host Configuration Protocol
DNS	domain name system
DSL	digital subscriber line
DSU	digital service unit
DTE	data terminal equipment
DUAL	diffusing update algorithm
DVRP	distance-vector routing protocol

EBCDIC	Extended Binary Coded Decimal Interchange Code
EBGP	External Border Gateway Protocol
ECMA	European Computer Manufacturing Association
EGP	Exterior Gateway Protocol
EIGRP	Enhanced Interior Gateway Routing Protocol
ES	end system
FCFS	first-come first-served
FIFO	first-in first-out
FIRO	first-in random-out
FTP	File Transfer Protocol
Gbps	gigabits per second
GHz	gigahertz
GIF	Graphics Interchange Format
GIGO	garbage-in–garbage-out
HAN	home-area network
HDSL	high bit-rate digital subscriber line
HTTP	HyperText Transfer Protocol
Hz	hertz
IAB	Internet Activities Board
IANA	Internet Assigned Numbers Authority
IBGP	Internal Border Gateway Protocol
ICMP	Internet Control Message Protocol
IEEE	Institute of Electrical and Electronics Engineers
IETF	Internet Engineering Task Force
IGP	interior gateway protocol
IGRP	Interior Gateway Routing Protocol
IP	Internet Protocol
IPv4	Internet Protocol version 4
IPv6	Internet Protocol version 6
IS	intermediate system
IS-IS	intermediate system to intermediate system
ISDN	Integrated Services Digital Network
ISO	International Organisation for Standardization
ITU	International Telecommunication Union
ITU-T	ITU-Telecommunication Standardization Sector
IWU	internetworking unit
JPEG	Joint Photographic Experts Group
Kbps	kilobits per second
LAN	local-area network
LIFO	last-in first-out
LLC	Logical Link Control
LSRP	link-state routing protocol
MAC	media access control
MAN	metropolitan-area network

Mbps	megabits per second
MIP	management information base
MPEG	Motion Picture Experts Group
MTBF	mean time between failures
MTTR	mean time to repair
MTU	maximum transmission unit
MUX	multiplexer
NAT	network address translator
NIC	network interface card
NPPP	Nonstationary Poisson Point Process
NT	network termination
OPEX	operating expenses
OSI	open systems interconnection
OSPF	Open Shortest Path First
P2P	peer-to-peer
PAN	personal-area network
PC	personal computer
pdf	probability density function
PDN	public data network
PDU	protocol data unit
PER	packet error rate
pmf	probability mass function
PPP	Poisson point process
PRI	Primary Rate Interface
PSE	packet-switching exchange
PSN	packet-switched network
PSTN	public switched telephone network
PVC	permanent virtual circuit
Q–Q	quantile–quantile
QoS	quality of service
RARP	Reverse Address Resolution Protocol
RFC	Request For Comments
RIP	Routing Information Protocol
RMA	reliability, maintainability and availability
RTS	request-to-send
SDSL	symmetric digital subscriber line
SDU	service data unit
SIP	SMDS Interface Protocol
SMDS	switched multimegabit data service
SMTP	Simple Mail Transfer Protocol
SNI	Subscriber Network Interface
SNMP	Simple Network-Management Protocol
SNR	signal-to-noise ratio
SPPP	stationary Poisson point process

STP	shielded twisted pair
SVC	switched virtual circuit
TA	terminal adapter
TCP	Transmission Control Protocol
TE	terminal equipment
TFTP	Trivial File Transfer Protocol
TIFF	Tagged Image File Format
ToS	types of service
UDP	User Datagram Protocol
UTP	unshielded twisted pair
VC	virtual circuit
VDSL	very-high-data-rate digital subscriber line
VLSM	Variable-Length Subnet Mask
WAN	wide-area network
WLAN	wireless local-area network
WWW	World Wide Web

Chapter 1

Internetworking and network fundamentals

It has been decades since the first computer was produced to do a very basic calculation. When first developed, the computer was simply a standalone machine to help people in either solving problems that require high-complexity computational algorithms or storing data or providing amusement with a variety of entertainment and was therefore named as a personal computer (PC). Over the time, such PC has been keeping changing to meet the users' non-stopping requirements in daily life. Along with a number of social activities of high interactive demands between users of distant areas, a two-way flow of information between the PC and the user is shown to be insufficient and hence necessitate the networking concept to link all the PCs in a proper manner that allows them to understand each other even though they have different configurations and may be located at different continents.

Aiming at showing an overview picture of computer networks, in this chapter, I will first outline the fundamentals of networking in Section 1.1 with different network models. Starting from small-size networks, Section 1.2 will discuss local area network (LAN) for limited geographic area with variant topologies, technologies and access mechanisms. Section 1.3 will then consider wide area network (WAN) for a large geographic area, where I will highlight some well-known WAN devices and topologies along with relevant technologies over network media. In order to model the communications between different devices in different networks, a reference model or framework is required. Section 1.4 will provide a brief explanation of seven layers in open systems interconnection (OSI) framework. As the most widely used protocol over the Internet, Transmission Control Protocol/Internet Protocol (TCP/IP) suite with a variety of communication protocols will be presented in Section 1.5. A brief account of internetworks and internetworking units will be explained in Section 1.6 where I will show the functionalities of intermediary devices for computer networking, including repeaters/hubs, bridges/switches, and routers/gateways.

1.1 Fundamentals of networking

In the early days of computer networks, only data was shared between connected PCs of the same types. A simplest network is when we connect two PCs via either wire or wireless to share data. With the advance in communication technology, more computer systems can be linked together to make a complex computer network. The evolved technology also allowed sharing graphic, voice and even video streams over various types of computer systems. In the 1990s, the Internet was emerged as a novel concept to form a global community as a global communication network where people at any location or time zone can simply surf the World Wide Web (WWW) to exchange the information. Since then, the Internet has become a part of our daily life. However, behind such high-technology communication network, there are a number of interconnections between electronic devices and computer systems that we are not aware of their presence.

Depending on the size, service, and functionality of a network, there are many types of computer networks as follows:

- LAN (cf. Figure 1.1): connects the devices which are geographically close to each other. As an extension of wired LAN, wireless LAN (WLAN) enables the over-the-air connection of wireless enabled devices.
- WAN (cf. Figure 1.2): consists of two or more LANs connecting the devices which are geographically far apart.
- Campus area network (CAN) (cf. Figure 1.3): connects LANs within either a school campus or an enterprise campus.
- Metropolitan area network (MAN) (cf. Figure 1.4): connects LANs within a town or a city but has a smaller coverage area compared to WAN.
- Home area network (HAN) (cf. Figure 1.5): connects devices within an individual home.
- Personal area network (PAN) (cf. Figure 1.6): is the smallest network connecting devices around an individual person, which could be PCs, telephones, mobile phones, tablets, personal digital assistants (PDAs), printers, etc.

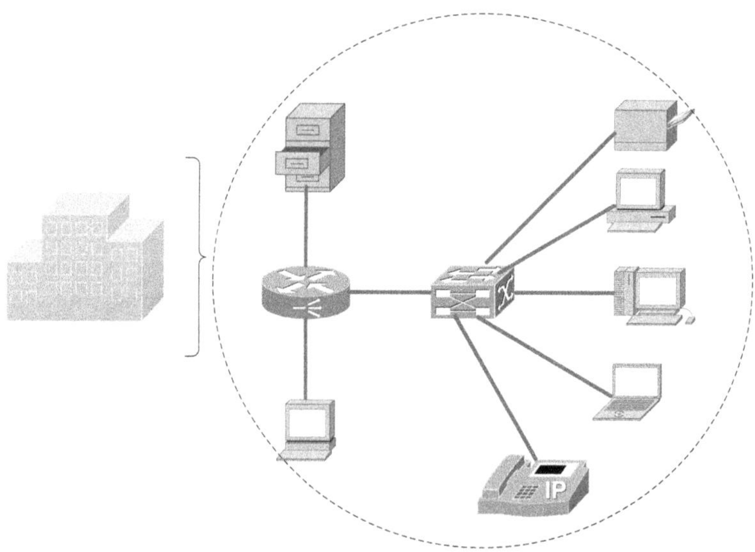

Figure 1.1 Local area network

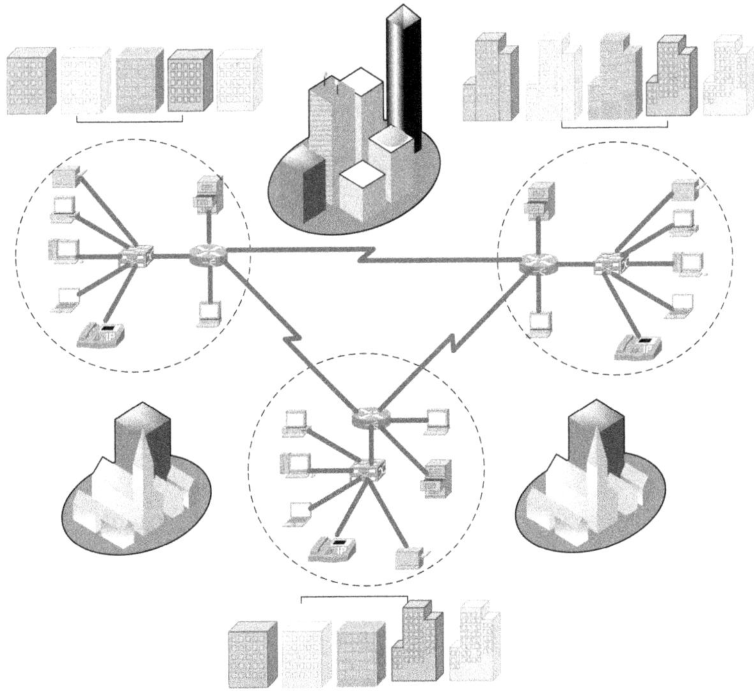

Figure 1.2 Wide area network

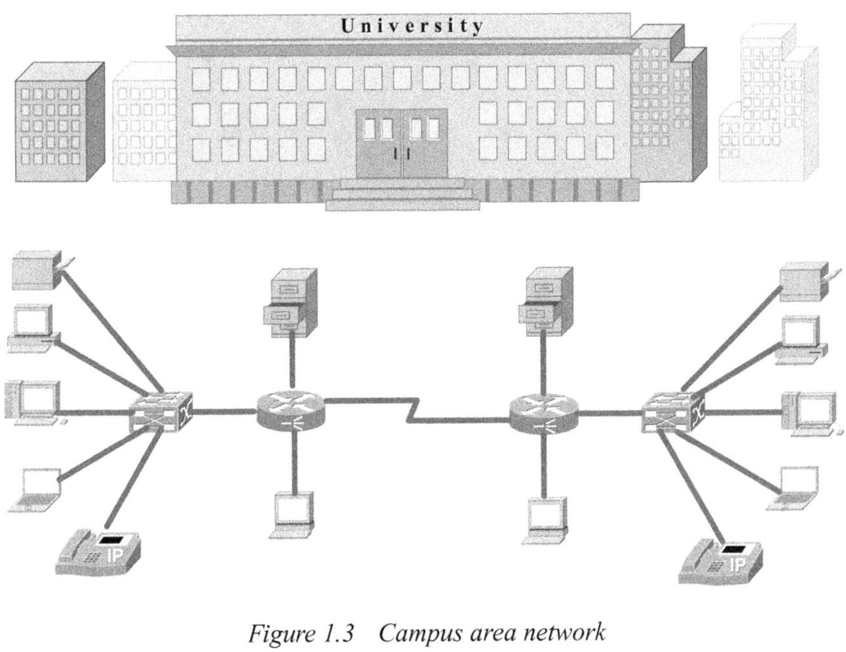

Figure 1.3 Campus area network

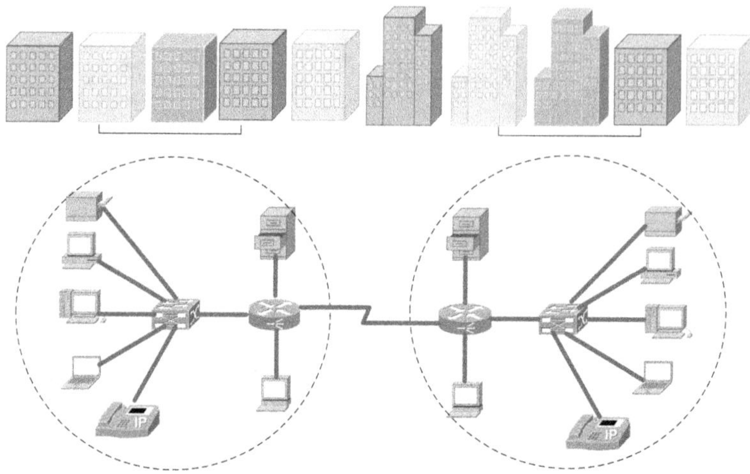

Figure 1.4 Metropolitan area network

Figure 1.5 Home area network

As listed above, the size of the geographical area is exploited to name distinct groups of computer networks. In the scope of this book, I will scrutinize the LANs and WANs in most parts, while the CANs, MANs, PANs, and HANs can be referred to as special models with specific group of users and geographic boundaries.

Figure 1.6 Personal area network

1.2 Local area networks

A LAN is first defined as a group of PCs connected within a limited geographic area, e.g., in one floor of a building (cf. Figure 1.1). With the development of technology, the LAN can be referred to a larger group of devices in different floors of the building.

Over the LANs, given a short communication distance, it is expected that the devices can communicate with each other at a high data rate. In other words, a high bandwidth (cf. Definition 1.1) can be allocated in the LANs. The LANs also allow multiple access with full-time connectivity of different devices which can be locally managed at the access point (AP) in a centralized manner or distributed manner without the AP.

Definition 1.1. *Bandwidth, in computing, is defined as the maximum data rate or information capacity having units of bits per second (bps).*[1]

1.2.1 LAN devices

In order to connect devices within a LAN, appropriate devices should be selected. There are three well-known LAN devices including hubs, switches, and routers. In this subsection, a brief introduction of these devices is provided for an overview of their functionality, while I will discuss about them in detail in Section 1.6.4.

1.2.1.1 Hubs

Hubs are the basic intermediate network devices to link end network devices to a shared medium in a small LAN with low throughput requirements (cf. Definition 1.2).

[1]Note that the bandwidth is defined differently in digital communications, which is used to represent the operation frequency range measured in hertz (Hz). In the rest of this book, bandwidth and capacity are used interchangeably, unless otherwise indicated.

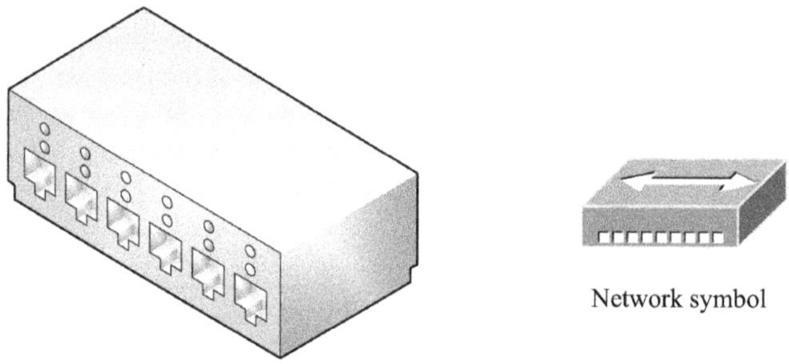

Figure 1.7 Hub and its network symbol

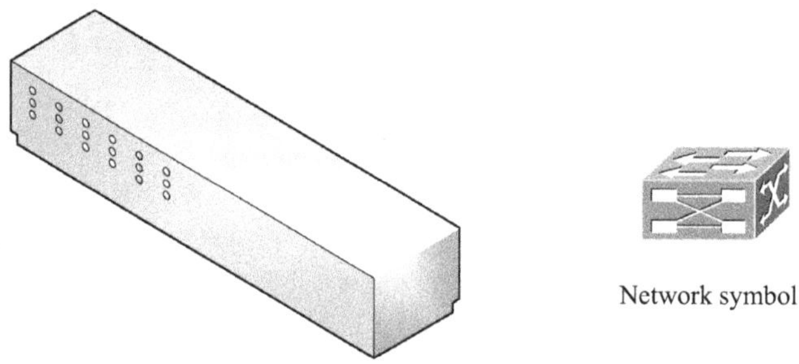

Figure 1.8 Switch and its network symbol

A simplest form of the hubs is a repeater with only two physical incoming and outgoing ports, which used to extend the network coverage. The hubs are generally regarded as multi-port repeaters to regenerate the received signals and send them over all ports (cf. Figure 1.7).

Definition 1.2. *Throughput is defined as the actual data transfer rate in bps.*

1.2.1.2 Switches

Switches are normally used in a large network to connect devices to a LAN by regenerating the received data frame and sending it to a certain destination port (cf. Figure 1.8). A predecessor of switches is a bridge that simply connects a LAN to another LAN via two physical ports. The switches typically have multiple ports and thus can be regarded as multi-port bridges.

1.2.1.3 Routers

Routers are intermediary network devices used to connect LANs and WANs (cf. Figure 1.9). The routers receive and forward data packets based on addresses. Instead of using hardware to forward the packets as in switches, the routers use software to manage the packet forwarding with different protocols and also support different network technologies.

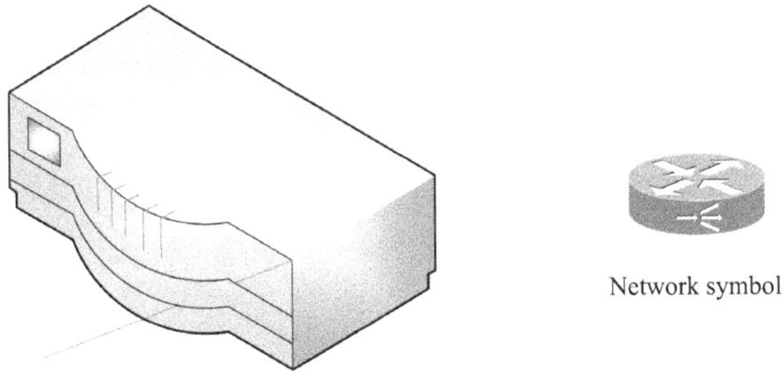

Figure 1.9 Router and its network symbol

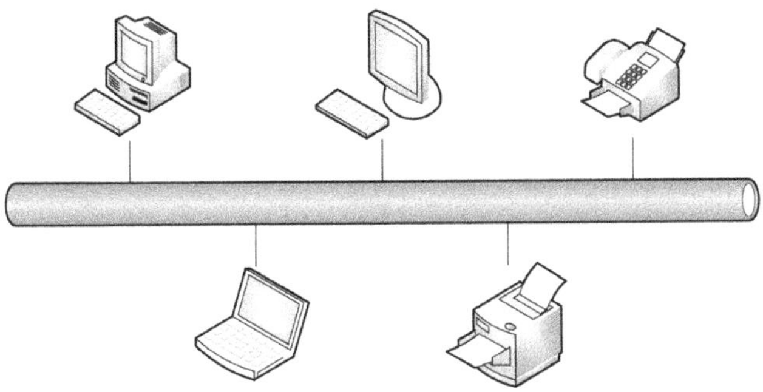

Figure 1.10 Bus topology

1.2.2 LAN topologies

In a LAN, devices are connected via physical links following different geometric shapes. Apart from the simplest point-to-point connection, electronic devices and LAN intermediary devices can be connected via various physical topologies, which can be listed in the following sections.

1.2.2.1 Bus

Bus topology has been deployed for connecting PCs since the early implementation of historic Ethernet where all devices are connected to a single central cable, namely, the bus, as the backbone of the network and they use a shared medium (cf. Figure 1.10).

1.2.2.2 Star

In a star topology, devices are connected to a central point of the network, e.g., hubs, over the same shared medium as in bus topology (cf. Figure 1.11).

1.2.2.3 Ring

In a ring topology, devices are linked together in a closed loop with no central point. Each device sequentially receives and passes the data frames to the next device until reaching the destination (cf. Figure 1.12).

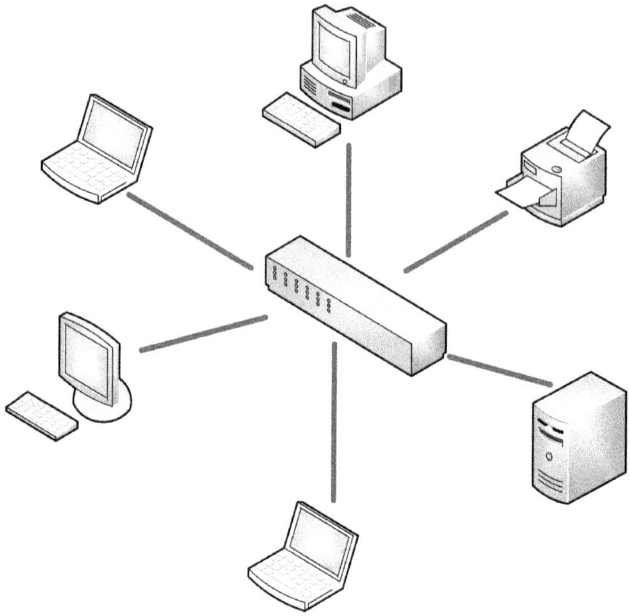

Figure 1.11 Star topology

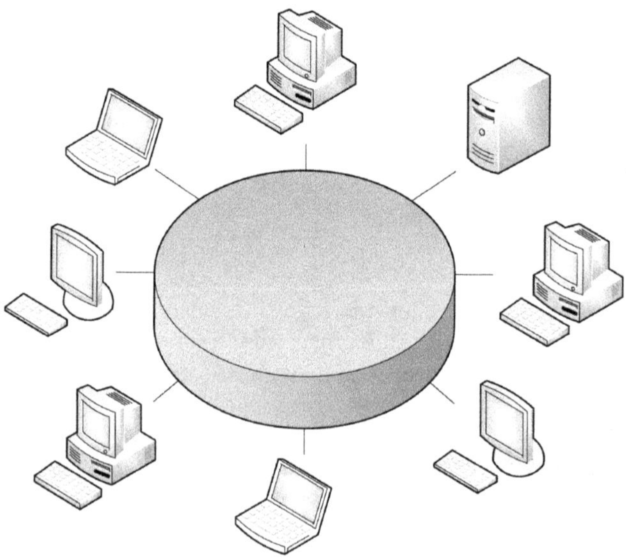

Figure 1.12 Ring topology

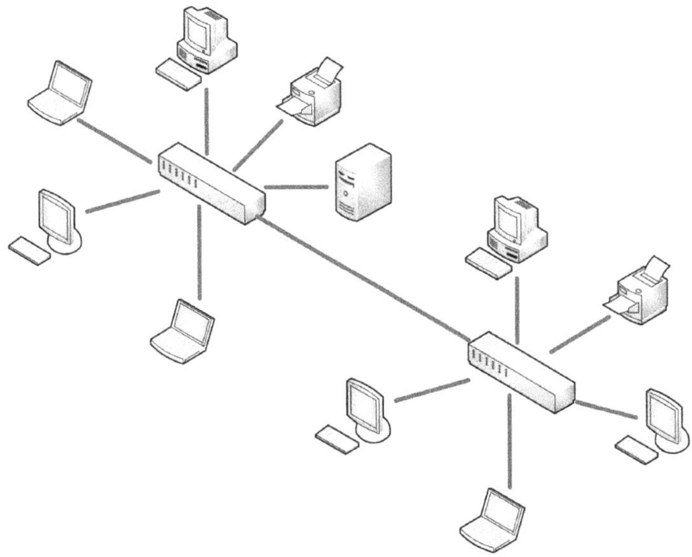

Figure 1.13 Extended star topology

1.2.2.4 Extended star

Extended star topology is used to enlarge the coverage area of the star topology by using additional repeaters/hubs (cf. Figure 1.13).

1.2.3 LAN technologies

In order to assist the communication between workstations, PCs, printers, servers, etc., a combination of both software and hardware is required with technologies to maintain the traffic in the shared network. Basically, there are two LAN technologies including Ethernet and token ring.

As a family of LAN products, Ethernet is the most widely used technology defined by the Institute of Electrical and Electronics Engineers (IEEE). The Ethernet has sustained for over 35 years since the first Ethernet standard with the first products were developed in 1980. The Ethernet has been a fast and reliable network solution for bus and star topology (cf. Sections 1.2.2.1 and 1.2.2.2) with evolved technologies that can provide data speed in the range of 10 Mbps to 10 Gbp. There are different types of the Ethernet standard depending on cabling and transmission media, which can be over either copper cable with unshielded twisted-pair (UTP) (cf. Figure 1.14) or shielded twisted-pair (STP) (cf. Figure 1.15), or optical fiber (cf. Figure 1.16) or wireless. The current Ethernet media options include two general types of copper cable: UTP and STP, plus several types of optical fiber cable.

In the following, I will describe some well-known Ethernet covered by the IEEE 802.3 and IEEE 802.11 standards, followed by an overview of token ring technology and IEEE 802.5 standard for ring topology.

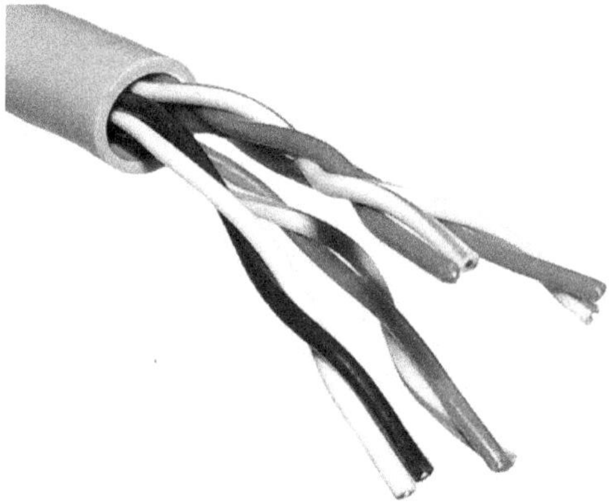

Figure 1.14 Unshielded twisted-pair (UTP)

Figure 1.15 Shielded twisted-pair (STP)

Figure 1.16 Optical fiber

1.2.3.1 10BaseT Ethernet

10BaseT or 10-Mbps Ethernet is a standard IEEE 802.3 Ethernet for connecting devices in a LAN. It uses two pairs of UTP Category-3 cables to carry 10 Mbps half-duplex or full-duplex communications in a star topology over a distance of up to 100 m in length. In a half-duplex system, the communication between two devices is realized separately but not simultaneously, whereas the devices in a full-duplex system can communicate with each other simultaneously at the same time.

1.2.3.2 100BaseTX and 100BaseFX fast Ethernet

Known as fast Ethernet, 100BaseTX uses two pairs of UTP Category-5 or Type-1 STP cables, while 100BaseFX employs fiber-optic cables to transmit data at a rate of 100 Mbps for either half-duplex or full-duplex communications. The fast Ethernet uses the star topology and also supports the 10BaseT standard Ethernet networks.

1.2.3.3 1000BaseT and 1000BaseX gigabit Ethernet

1000BaseT Gigabit Ethernet provides 1 Gbps full-duplex transmission over a distance of up to 100 m by employing four pairs of UTP Category-5 cable. Instead of using UTP cables, another Gigabit Ethernet standard, namely, 1000BaseX, was also proposed to use fiber-optic cables exploiting the advantages of optical fiber in data transmission for high reliability and long distance as well as its extendibility to a much higher data rate in later version of Ethernet standards.

1.2.3.4 10GBaseSR and 10GBaseLR 10Gbps Ethernet

One of the latest developments of Ethernet standards to meet the high-bandwidth requirements is the usage of fiber-optic cables to achieve a data rate of up to 10 Gbps, which is also known as 10 Gigabit Ethernet or 10GbE. Specifically, there are 10GBaseSR and 10GBaseLR standards, where the 10GBaseSR employs multi-mode fiber-optic cables for a short-range transmission of up to 300 m in a LAN or a MAN, whereas 10GBaseLR makes use of single-mode fiber-optic cables to provide a transmission range of up to 40 km and thus can be deployed in different LANs, MANs, and even WANs.

1.2.3.5 Wireless Ethernet

Commonly referred to as Wi-Fi or WiFi, the IEEE 802.11 is a WLAN technology to establish a wireless connection between devices for LAN extension, cross-building access, nomadic access and ad hoc networks. In order to enable and maintain the wireless connectivity function, the WLAN requires a wireless AP and a wireless network interface card (NIC) adapter to provide wireless communication capability to each device. The WLANs have emerged with various IEEE 802.11 standards in use with different specifications to operate at different frequency bands, e.g., 2.4 GHz (802.11b/g/n), 3.65 GHz (802.11y), 4.9 GHz (802.11j), 5 GHz (802.11a/n/h/j/ax/ac), 5.9 GHz (802.11p), and 60 GHz (802.11ad/aj/ay). Another technology for short-range wireless communication is Bluetooth which is normally used in PANs (cf. Section 1.1). The Bluetooth follows the IEEE 802.15 standards.

1.2.3.6 Token ring

Originally developed by IBM in the 1970s, Token Ring network has been known as IBM's LAN technology. After IBM, the IEEE also proposed a related IEEE 802.5 standard which is regarded as a shadow of the token ring. The IEEE 802.5 and the token ring are interchangeable and compatible with each other. Both standards are employed in token-passing networks where a small frame, namely, token, is moved around the network and only the device keeping the token has the right to transmit.

1.2.4 LAN access methods

Over the shared media of a LAN with a number of devices, contention is likely to take place when these devices all want to send data at the same time. It is vital to have methods to manage the media access of the devices in the network. In order to answer the question how the devices in a LAN access the network and share the medium, in this subsection, I will discuss three access control methods, including carrier sense multiple access with collision detection (CSMA/CD), carrier sense multiple access with collision avoidance (CSMA/CA), and control token.

1.2.4.1 Carrier sense multiple access with collision detection

CSMA/CD is currently used in IEEE 802.3 Ethernet to deal with the contention when transferring data in a LAN. The CSMA/CD detects collisions in a bus topology (cf. Section 1.2.2.1) of the wired networks. When a device wants to transmit information, it first listens to see if any other device(s) is currently transmitting data over the network (cf. Figure 1.17). If the medium is free with no data transmission in progress, the device

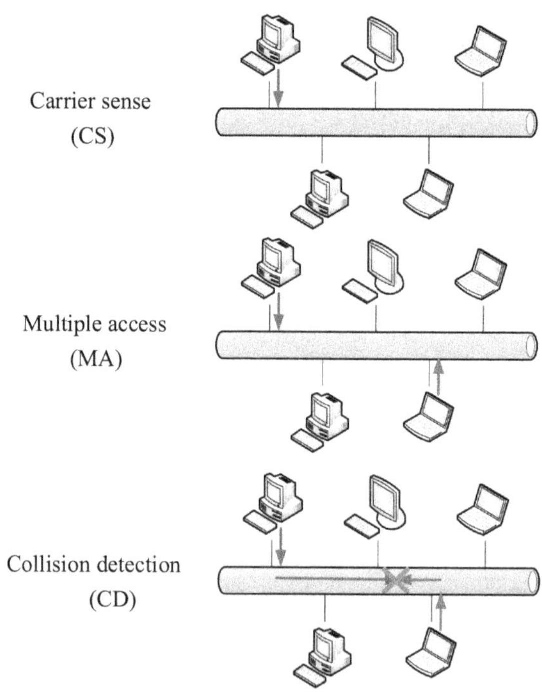

Figure 1.17 Carrier sense multiple access with collision detection (CSMA/CD)

can transmit its data. Otherwise, the device has to wait until the occupied transmission finishes prior to transmitting its data. However, there is a likely scenario that collision may happen when two devices in the network transmit data simultaneously. In that case, both devices back off, and each device will wait a random time period before retransmitting its data. Such collision indeed occurs quite often in a network with many devices and thus causes performance degradation of the Ethernet in a busy network.

1.2.4.2 Carrier sense multiple access with collision avoidance

CSMA/CA is especially developed for the wireless medium in IEEE 802.11 WLAN (cf. Section 1.2.3.5). The CSMA/CA is basically similar to the CSMA/CD for wired LAN in carrier sensing, but devices in the WLAN have to first request for the right to send to avoid collisions. In order to send data, the sender transmits a request-to-send (RTS) frame to the receiver indicating the time required for data transmission and the receiver sends a clear-to-send (CTS) frame to let the sender know that the channel is reserved for the data transmission (cf. Figure 1.18). The RTS and CTS frames are therefore of great importance in the WLAN with CSMA/CA to avoid the collisions.

1.2.4.3 Control token

Control token is a media access technique used mostly in a ring topology (cf. Section 1.2.2.3) with token ring technology and the IEEE 802.5 standard (cf. Section 1.2.3.6). Each device sends its data only if it has a token which is a small frame used to manage the token passing in the network (cf. Figure 1.19). The token is moved around the

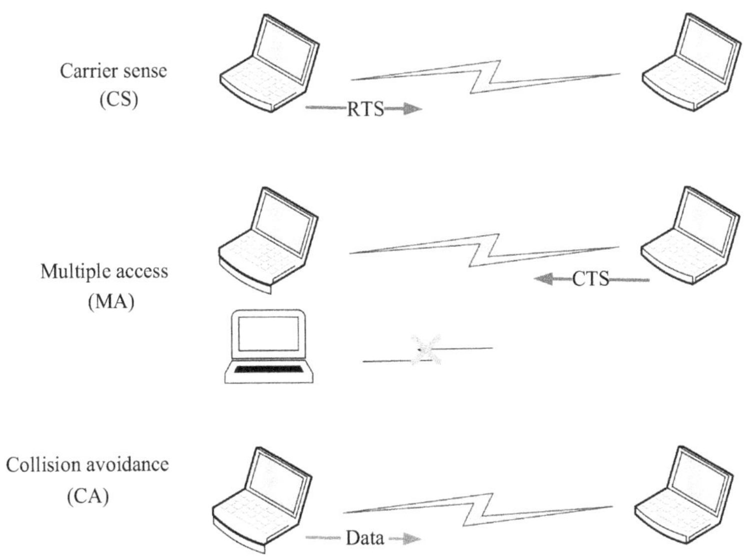

Figure 1.18 Carrier sense multiple access with collision avoidance (CSMA/CA)

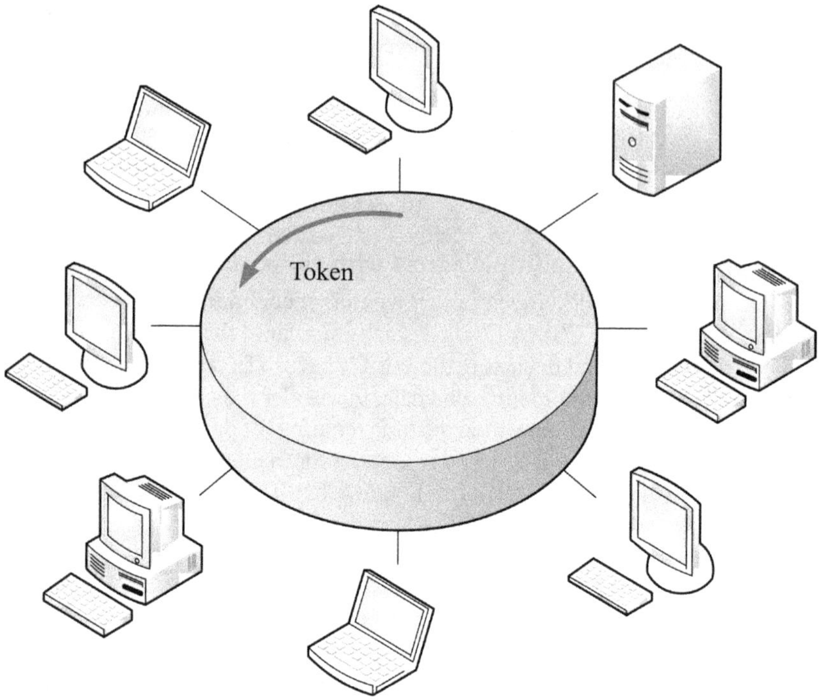

Figure 1.19 Control token technique

network. If a device receives the token, it is granted the right to transmit data. However, if the device holding the token does not want to send any data then it will pass the token to the next device. Such token passing helps prevent the collisions in the network. The data frame circulates the ring until it reaches the intended receiver. The receiver copies the data frame for further processing and forward it to continue circulating the ring until it reaches the sender for verifying if has been received by the receiver. Although the collisions are unlikely to occur with control token, the delay caused by circulating the token around the network is substantial in a large network with a number of devices.

1.2.5 LAN transmission methods

Data transmission within a LAN can be classified into three essential types, including unicast, multicast, and broadcast, which can be defined as follows.

1.2.5.1 Unicast transmission

Unicast transmission is used to express the data communication between one device and another device (cf. Figure 1.20). There are only one sender and one receiver. The

Figure 1.20 Unicast transmission

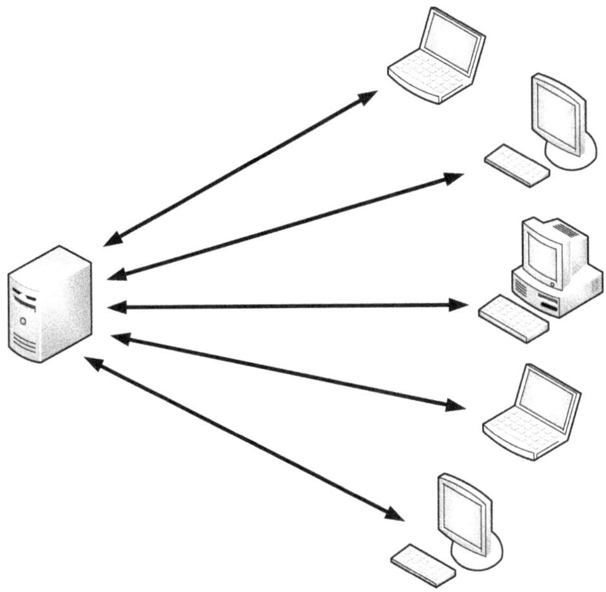

Figure 1.21 Broadcast transmission

sender sends the data packet across the network to a specified receiver based on the destination address. The unicast transmission is still of greatest importance in LAN.

1.2.5.2 Broadcast transmission

Broadcast transmission represents the data communication where a device sends data to all other devices in the network (cf. Figure 1.21). There is only one sender, while all other devices are receivers interested in receiving data. In the broadcast transmission, the data packet is sent with a broadcast address and its copies will be received by all the receivers.

1.2.5.3 Multicast transmission

Multicast transmission indicates the data communication where one or more devices send data to a specific set of other devices (cf. Figure 1.22). There are one set of senders and another set of receivers. The destinations of the data packet are determined with

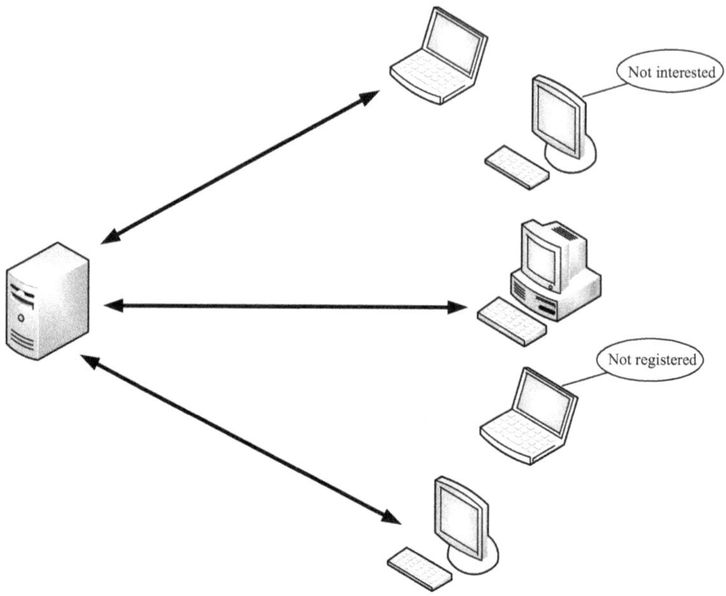

Figure 1.22 Multicast transmission

a multicast address and a copy of the packet will be sent to each receiver in the set of the multicast destination address.

1.3 Wide area networks

A WAN can be simply understood as a large data communication network connecting two or more LANs to cover a broad geographical area (cf. Figure 1.2). The WAN can be spread across multiple cities and countries and even in different continents. Devices in the WAN are typically connected via common carrier circuits of service providers, i.e., telephone/mobile phone companies, with serial links allowing them to connect to public networks, i.e., the Internet.

1.3.1 WAN devices

There are different types of devices using in a WAN depending on the deployed technologies. In the following, I will describe some common WAN devices, including WAN switches, access servers, modems, channel service unit (CSU)/digital service unit (DSU), and multiplexers (MUXs). Other specific devices with respect to each WAN technology will be discussed in Section 1.3.3.

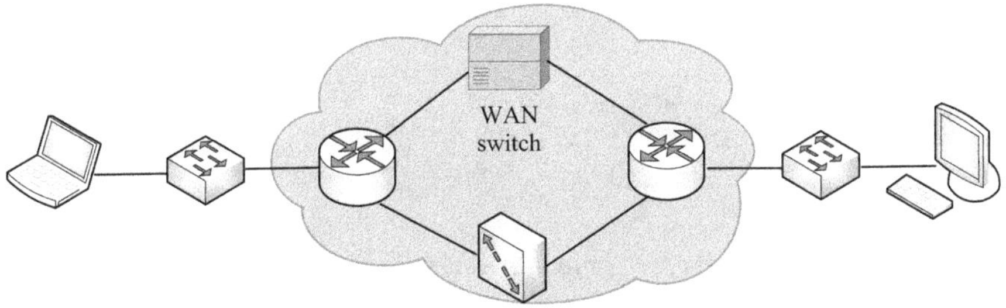

Figure 1.23 A WAN switch

Figure 1.24 Modems in WAN

1.3.1.1 WAN switches

Similar to a LAN switch (cf. Section 1.2.1.2), a WAN switch is a multi-port intermediary device having the same function of device connection (cf. Figure 1.23). However, instead of connecting devices in the LAN, the WAN switches are used in carrier networks to direct traffic of different WAN technologies, e.g., frame relay, X.25, and switched multimegabit data service (SMDS) which will be discussed in Section 1.3.3.

1.3.1.2 Modems

A modem is basically a device that has the function of both modulator and demodulator to convey analog signals over the medium, e.g., telephone lines. A modem at the sender converts the digital data to analog signals for transmission and another modem at the receiver will do the reverse job by changing the received analog signals into the original digital data (cf. Figure 1.24). Depending on the modulation/demodulation, the modems may have different data rates.

1.3.1.3 Channel service unit/digital service unit

A CSU/DSU is a device used to convert a digital data frame from a router of a LAN into a frame suitable for digital circuits used in a WAN and vice versa from a WAN to a LAN (cf. Figure 1.25). The CSU/DSU processes the signals transmitted to or received from the WAN, manages line control, and also provides signal timing for the communication between devices in the LAN and the WAN.

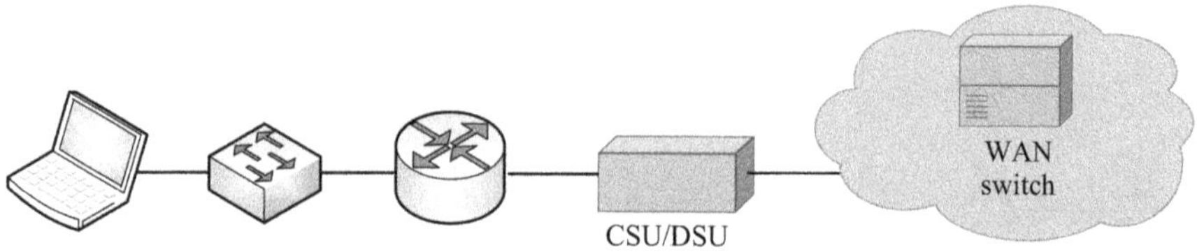

Figure 1.25 A channel service unit/digital service unit (CSU/DSU) in WAN

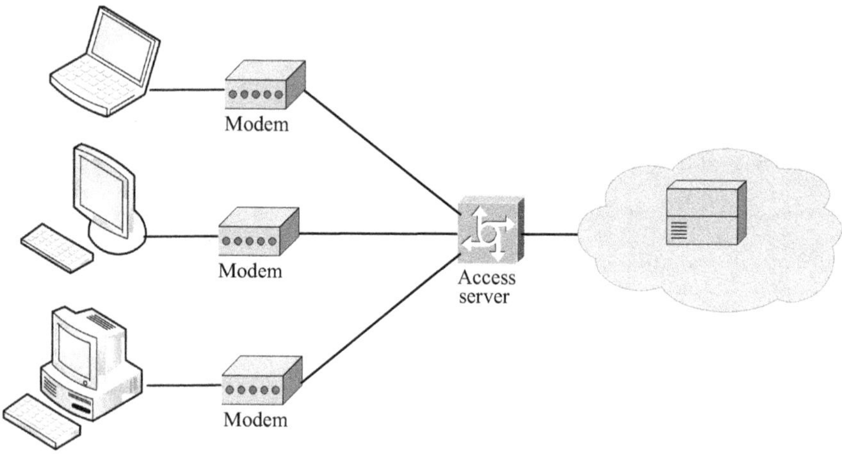

Figure 1.26 Access server in WAN

1.3.1.4 Access server

An access server is a center managing dial-up calls from the devices or hosts requesting to login an Internet service or dial-out connections into a WAN, and vice versa with dial-in connections (cf. Figure 1.26).

1.3.1.5 Multiplexers

A MUX is a device used to combine data channels from different WAN technologies into a single data stream which can be then transmitted over a common line with an integrated CSU/DSU (cf. Figure 1.27). The MUX also allows us to combine both data and voice over the same line connecting to the WAN. The MUX can therefore eliminate a number of individual connections for each channel saving significantly operation and implementation costs.

1.3.2 WAN topologies

A WAN is used to connect multiple remote LANs of different sites using the intermediary devices as described in Section 1.3.1. There are numerous approaches to connect these WAN devices with different topologies depending on their locations. Apart from the basic point-to-point, star, and ring topologies as in the LAN (cf. Sections 1.2.2.2 and 1.2.2.3), there are hybrid and mesh topologies. As the name says, a

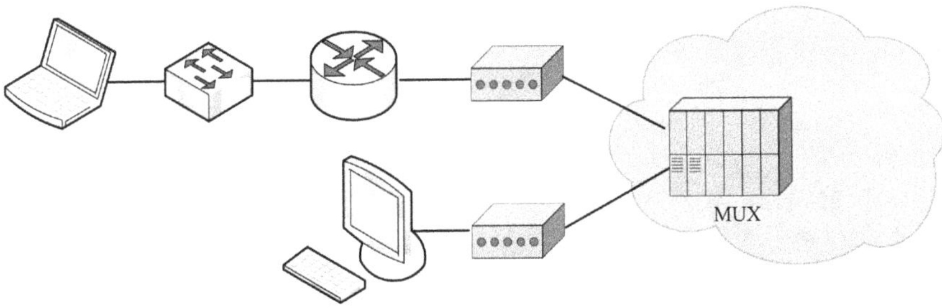

Figure 1.27 A multiplexer in WAN

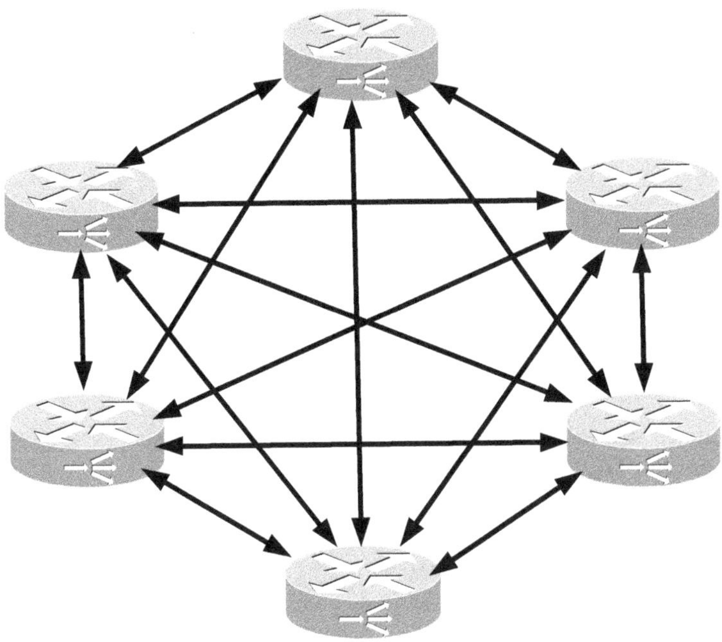

Figure 1.28 Full mesh topology

hybrid topology is simply a combination of two or more basic topologies, while, in a mesh network, the devices are connected to each other over different links. In the following, I will discuss two basic types of mesh topology including full mesh and half mesh topologies.

1.3.2.1 Full mesh

In a full-mesh WAN (cf. Figure 1.28), each device has a connection to all other devices in the network. With a number of redundant links, this topology provides the best performance with the lowest possibility of failure as a single broken link does not cause much affect on the data transmission. However, the full mesh connection

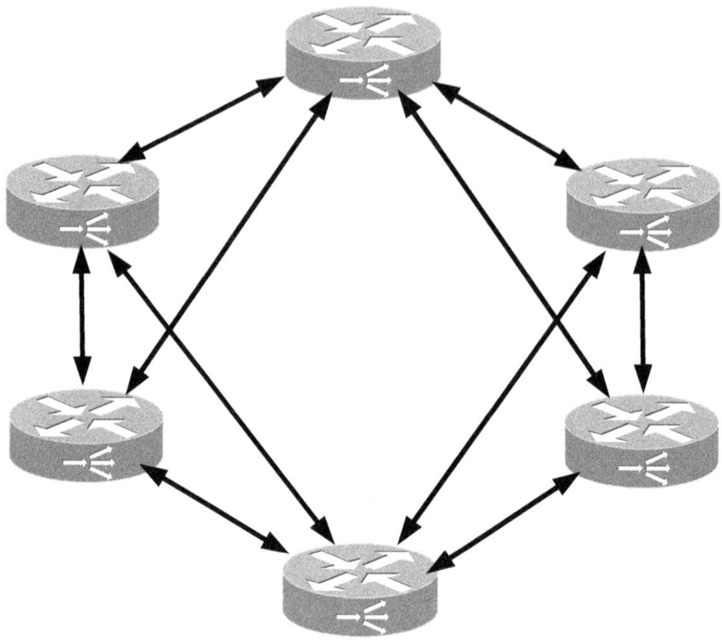

Figure 1.29 Partial mesh topology

can be applied only in a small network, whereas it is impractical in a large WAN with a large number of devices.

1.3.2.2 Partial mesh

A partial mesh topology is deployed to partially connect devices with each other in a WAN so that there are at least two devices connected via two or more links to other devices in the network (cf. Figure 1.29). Such topology helps to reduce the cost of the redundancy implementation in the network compared to a full mesh topology.

1.3.3 WAN technologies

In a WAN, to connect all devices around the world, simply using physical devices and cables is insufficient, but that is a complicated process which requires numerous technologies to sustain and maintain the communications. In this subsection, I will first discuss two foundational technologies for switching, including circuit switching and packet switching, along with connectionless and connection-oriented services with virtual circuits (VCs). Then, I will provide an overview of various underlying technologies and protocols associated with WANs, e.g., Integrated Services Digital Network (ISDN), X.25, frame relay, asynchronous transfer mode (ATM), SMDS, and digital subscriber line (DSL).

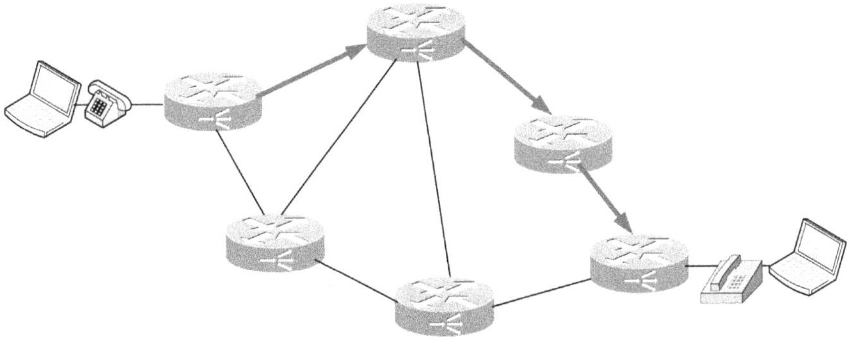

Figure 1.30 Circuit switching

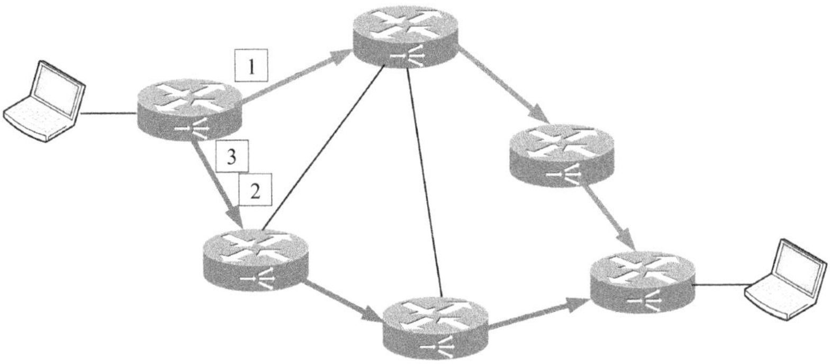

Figure 1.31 Packet switching

1.3.3.1 Circuit switching

Circuit switching is a simple way to send data from a sender to a receiver by physically connecting them when a data communication is required between them (cf. Figure 1.30). Such connection has to be established prior to transmitting data at the sender and will last until the communication finishes. For this reason, the circuit-switched communications are also referred to as connection-oriented communications. The switched circuits operate as in a public switched telephone network for voice communication. A typical instance of switched circuits can also be found in ISDN that will be described in detail in Section 1.3.3.4.

1.3.3.2 Packet switching

Packet switching is a fundamental WAN technology for digital communications where all devices transmit data in the form of packets over the shared medium with common carrier resources (cf. Figure 1.31). Since the packets can be delivered in either connectionless or connection-oriented manners, there are two types of packet switching, including connectionless packet switching, also known as datagram packet switching, and connection-oriented packet switching, also known as VC packet switching.

This accordingly increases network efficiency and saves significantly the cost of the customers in leasing the point-to-point lines from a carrier network provider. In packet-switched networks (PSNs) with VCs, connections are virtually set up between the WANs and the carrier network allowing the customers in different WANs share the same carrier network which is referred to as a cloud. Some examples of the PSNs include X.25, frame relay, ATM, and SMDS, which will be sequentially discussed in Sections 1.3.3.5–1.3.3.8.

1.3.3.3 Virtual circuits

In PSNs, the carrier network provider set up a virtual connection between customer sites in different WANs to deliver data packets between them. The virtual connection is actually realized via a number of logical circuits within the shared carrier network and is therefore named as VCs (cf. Figure 1.32). Depending on the frequency of usage of the VC, following are the two types of VC:

- *Switched VC (SVC)*: consists of circuit establishment, data transfer, and circuit termination. The SVCs are dynamically set up when required for data transmission, so they are used for sporadic communication which occurs at irregular intervals or only in a few places. Such dynamic process, although requires a high bandwidth use and resource for establishment and termination, nevertheless saves the cost of constantly setting up VC.
- *Permanent VC (PVC)*: consists of only data transfer mode which is realized using a permanently established VC. The PVCs are used in the situation when constant data transmission is required between two remote customer sites. The PVCs help save bandwidth use with the preset VCs, however result a considerably high cost for constantly maintaining the availability of the VCs.

1.3.3.4 Integrated Services Digital Network

ISDN supports both digital telephone and data transport services through the usage of existing telephone wires provided by telephone companies. Basically, in the ISDN,

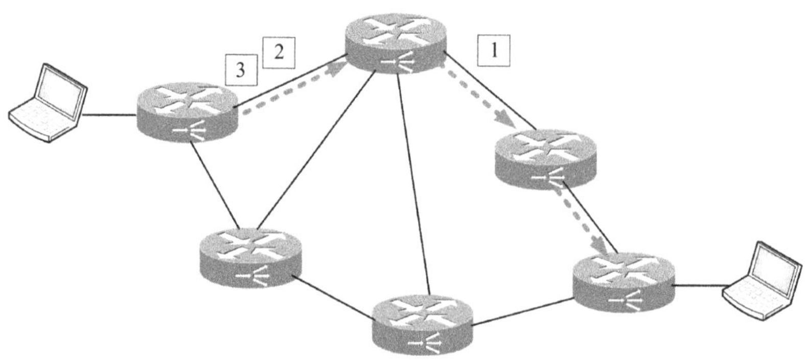

Figure 1.32 Virtual circuit switching

when a router has voice, text, data, etc. to transmit to another router at a remote site, an ISDN circuit is firstly initiated to set up a voice call to the remote network with a telephone number. The data communication between the two devices will be carried out when the connection between them is authentically established. As in circuit switching, the connection will be terminated when the data communication is complete.

Applications of ISDN involve providing additional telephone lines in homes for tasks requiring integrated voice and data, high-speed file transfer, resource sharing, and video-conferencing.

ISDN devices include the following (cf. Figure 1.33):

- *Terminals*: There are two types of terminal equipment (TE). TE type 1, i.e., TE1, and TE type 2, i.e., TE2. The TE1s are specialized ISDN terminals to connect to the ISDN network through four-wire, twisted-pair digital links, while the TE2s connect to the ISDN network through a terminal adapter (TA).
- *TAs*: An ISDN TA can be either a standalone device or an integrated board within TE2, which is basically used to connect the ISDN to other interfaces, e.g., serial interface on a router.
- *Network-termination (NT) devices*: also include two NT types, i.e., NT1 and NT2, corresponding to TE1 and TE2 to connect the four-wire subscriber wiring of TE to the conventional two-wire local loop at customer sites.
- *Line-termination equipment*: is a device physically connected to the phone company switch.
- *Exchange-termination equipment*: is a device used to connect between the switches within the phone network.

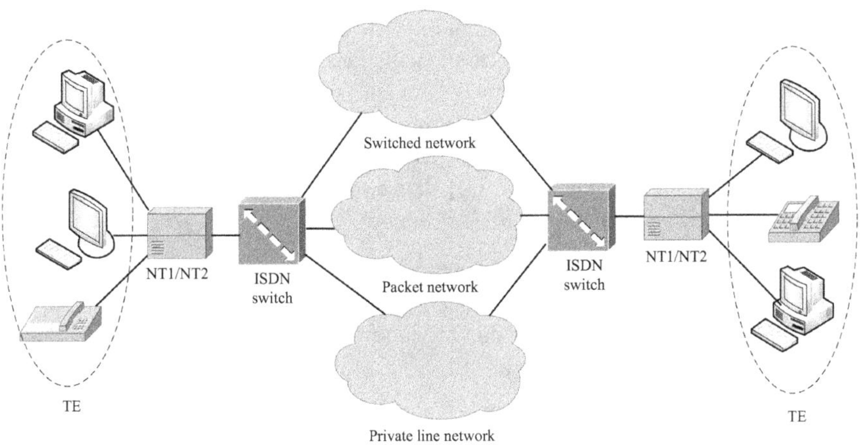

Figure 1.33 Integrated Services Digital Network (ISDN) devices

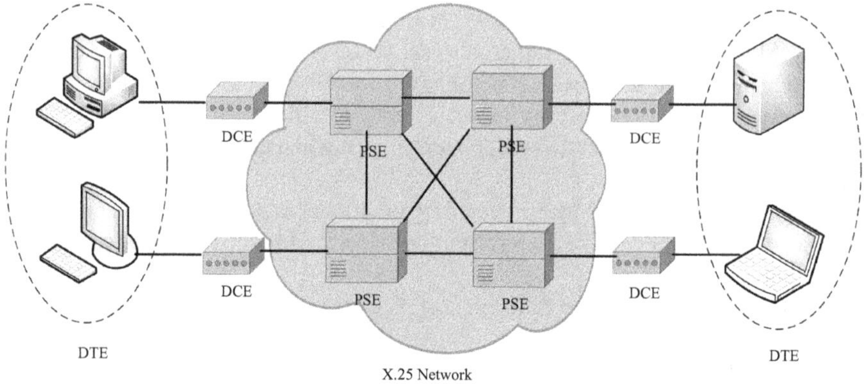

Figure 1.34 X.25 devices

ISDN provides two following types of services:

- *Basic Rate Interface (BRI)*: offers two bearer (B) channels and one delta (D) channel (2B + D), where the B channel carries data at the rate of 64 kbps and the D channel is for signaling at lower rate of 16 kbps.
- *Primary Rate Interface (PRI)*: offers more B channels with higher signaling rate of 64 kbps to support large applications. For instance, the PRI services in North America and Japan have 23 B channels and 1 D channel, while other parts of the world have PRI services with 30 B channels and 1 D channel.

1.3.3.5 X.25

X.25 is an International Telecommunication Union-Telecommunication Standardization Sector (ITU-T) standard protocol in WAN that describes how to connect user devices and network devices in PSNs along with their maintenance and operation procedures. The X.25 can operate effectively and independently of the type of systems in the network. In order to ensure reliable communication among network devices, the X.25 contains both SVCs and PVCs (cf. Section 1.3.3.3) within the physical circuit.

X.25 devices include data TE (DTE), data circuit-terminating equipment (DCE), and packet-switching exchange (PSE) (cf. Figure 1.34).

- DTEs are simply end devices, e.g., terminals, PCs, or network hosts.
- DCEs are communication devices, e.g., modems and switches, that act as an interface to link the DTEs at the ESs and the PSE at the carrier provider to transfer data across the network.

1.3.3.6 Frame relay

Frame relay is a typical WAN protocol used in PSNs with packet-switching technology. The frame relay was originally designed to work together with ISDN to enable end devices to dynamically share network resources. The frame relay was standardized

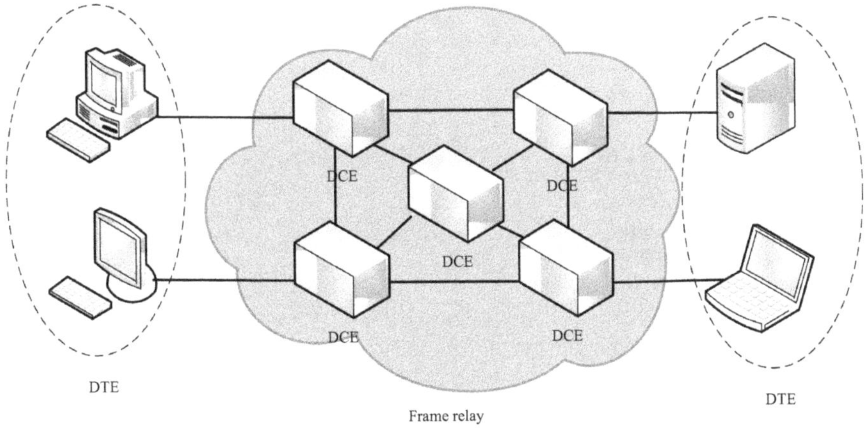

Figure 1.35 Frame relay devices

by the ITU-T and is an American National Standards Institute (ANSI) standard in the United States (US).

Similar to X.25, frame relay devices basically include DTEs and DCEs (cf. Figure 1.35). The data communication between devices in PSN is realized via either SVCs or PVCs along with two following techniques:

- *Variable-length packets*: The packets of different lengths are switched across the VCs until reaching the destination and thus enhance the network efficiency.
- *Statistical multiplexing*: The statistical multiplexing is used to control the network access of devices to efficiently allocate bandwidth in frame relay PSN.

Due to its similarity with X.25, frame relay is also regarded as a streamlined version of X.25. However, the frame relay typically operates over WAN facilities and consequently achieves a higher performance with more reliable connection services when compared to X.25.

1.3.3.7 Asynchronous transfer mode

ATM is an ITU-T standard for delivering not only data but also voice and video simultaneously over the same medium. Similar to X.25 and frame relay, the ATM is based on packet-switching technology and the data communication in ATM is realized via either SVCs or PVCs. However, the ATM makes further use of circuit switching to develop a cell-switching multiplexing technology. Such asynchronous combination technique makes the ATM more efficient than the traditional synchronous technologies.

ATM is based on cell relay which is a high-speed transmission method to transfer information in terms of cells which are fixed-size units. Each cell has a size of 53 bytes (or octets) containing cell-header information in 5 bytes and payload in 48 bytes. Given such small size, the cells can travel faster across the network without

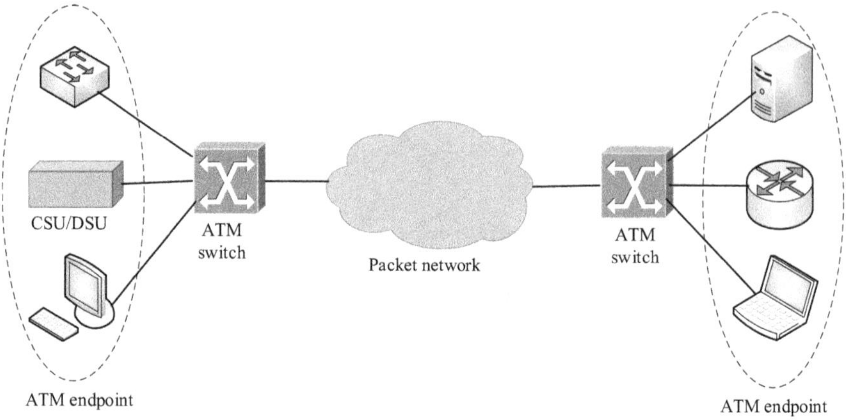

Figure 1.36 Asynchronous transfer mode (ATM) devices

suffering the latency caused by waiting the large data packet. The ATM therefore provides high-speed WANs and LANs.

ATM devices (cf. Figure 1.36) consist of the following:

- *ATM switch*: has similar role as the switches used in different technology but specifically designed for cell relay in an ATM network.
- *ATM endpoint*: is basically an ES which could be routers, LAN switches, CSU/DSU, and workstations.

1.3.3.8 Switched multimegabit data service

SMDS is a high-speed, packet-switched, datagram-based WAN technology used for sending and receiving data over public networks, namely, public data networks (PDNs). By using fiber-optic or copper cables, the SMDS can support high-speed data transmission.

SMDS devices consist of the following components (cf. Figure 1.37):

- *Customer premises equipment (CPE)*: is a TE at the customer site, which could include end devices, e.g., terminals and PCs, and intermediary devices, e.g., modems, multiplexers, and routers.
- *Carrier equipment*: generally includes high-speed WAN switches complying with network specifications and standards for carrier networks.
- *Subscriber network interface (SNI)*: is the interface between CPE and carrier equipment to provide service transparently to the customer by using SMDS interface protocol.

1.3.3.9 Digital subscriber line

DSL technology is a WAN technology that exploits existing twisted-pair telephone lines to convey high-bandwidth multimedia data across the network to customer sites (cf. Figure 1.38). Depending on the employed technologies, provided services, and

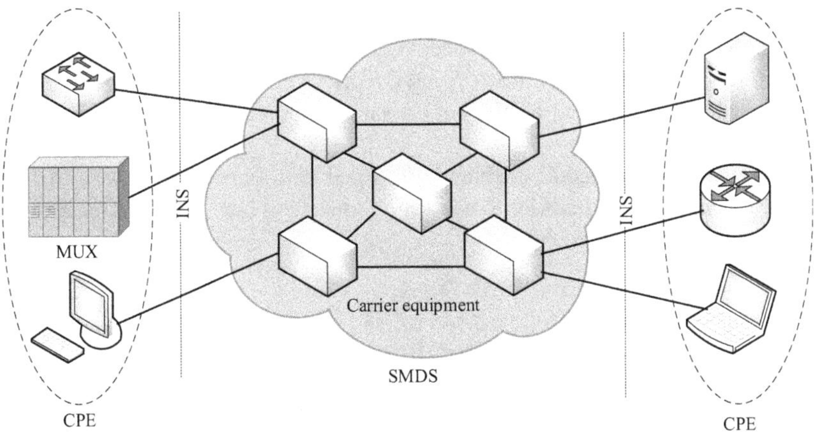

Figure 1.37 Switched multimegabit data service (SMDS) devices

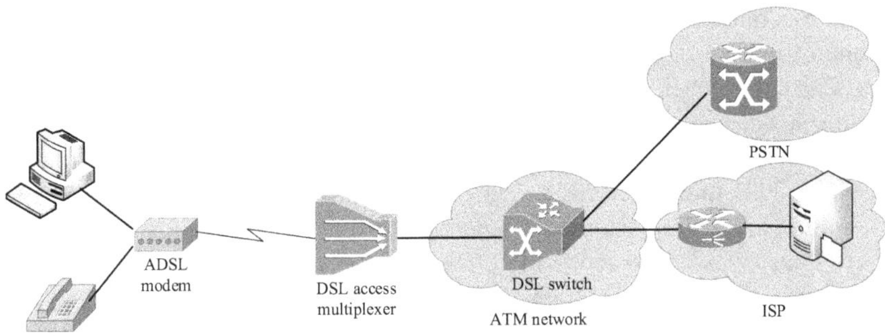

Figure 1.38 Digital subscriber line (DSL)

supported data rate, there are various xDSL forms, of which can be typically listed as follows:

- *Asymmetric DSL (ADSL)*: is the most popular associated with DSL technology. In the ADSL, more bandwidth is allocated for downlink compared to uplink, and thus suitable for users who prefer to download data, surf internet, watch online movie, etc., rather than uploading or sharing data online. With the asymmetry in ADSL circuit design dividing the bandwidth of a telephone line into multiple channels for high-speed downstream, medium-speed duplex, and basic telephone service, the ADSL can enable data flow for both downstream and upstream, while guaranteeing uninterrupted basic telephone service. The ADSL is therefore playing a crucial role and will keep promoting in telephony markets.
- *Symmetric DSL (SDSL)*: provides symmetrically high-speed data communication of variable rate up to 1.54 Mbps. The SDSL, though supports only data yet does

not allow analog signal as in ADSL, is easy to deploy and is thus beneficial for the services required only high-speed data transmission over a long distance.

- *High-bit-rate DSL*: is a symmetric version of DSL that can offer full-rate service of 1.5 Mbps over two-wire pairs but does not support adaptive-rate services as in SDSL.
- *Very high-data-rate DSL*: enables high-speed data transmissions of nearly ten times than that of the ADSL over short distances using twisted-pair copper telephone lines as in ADSL.

1.4 Open systems interconnection reference model

In order to convey data between devices across different network infrastructures over a long distance, intermediary hardware is certainly vital to connect them together. However, how can an application in one device communicate with another application in another device?

In 1984, the International Organization for Standardization (ISO) decided to establish a common reference model, namely, OSI, as a conceptual framework for communication between PCs in the network. With the OSI reference model, the vendors and developers can design digital communication products and software programs that can interoperate and compatible with each other.

The OSI reference model divides the process of communication between two end devices, i.e., PCs, into seven layers, each of which specifies distinct network functions and tasks and can be implemented independently (cf. Figure 1.39). For instance, a PC wants to send a message to another PC. The message will be sent from an application of the source PC go down through all the layers across the network and then go up through the same layers at the destination PC. In this section, I will first describe seven

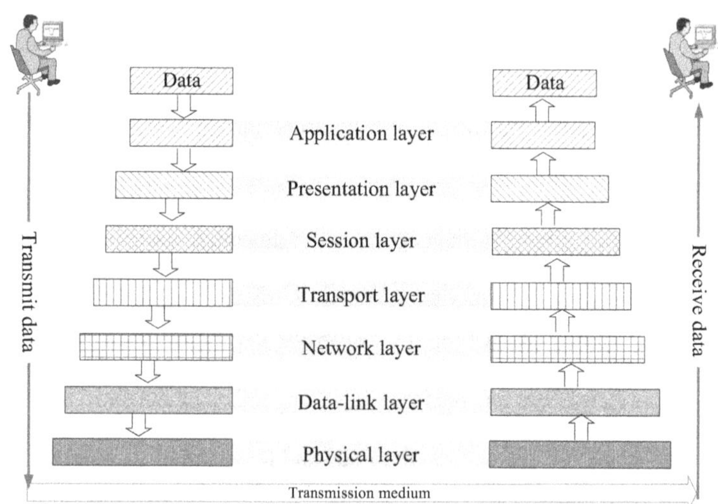

Figure 1.39 Open systems interconnection (OSI) layers

layers of the OSI model, followed by the introduction of the interactions between these layers in communications as well as the information format associated with each layer.

1.4.1 OSI layers

Starting from the layer nearest to the user running various applications, the seven OSI layers in Figure 1.39 can be described in the following sections.

1.4.1.1 Layer 7—application layer

Application layer is the OSI layer 7 at which the end users interact directly with the application. This layer is not an application but provides services for an application program to guarantee effective communication between applications in the network. At the application layer, the functions and services provided involve identifying and authenticating communication partners, assessing network resources, maintaining synchronization of communications, and determining protocol for the software application. Some examples of protocols at the physical layer include File Transfer Protocol (FTP), Telnet, Simple Mail Transfer Protocol (SMTP), and Simple Network Management Protocol (SNMP), which will be discussed in detail in Section 1.5.5.

1.4.1.2 Layer 6—presentation layer

Presentation layer is normally a part of an operating system (OS) to provide a variety of coding and conversion functions on the data at the application layer. The presentation layer basically turn the data received from the application layer into a presentation format that can be converted back at the other end using common data representation formats, common data compression schemes, and common data encryption schemes. Some well-known standards for text include Extended Binary Coded Decimal Interchange Code (EBCDIC) and American Standard Code for Information Interchange (ASCII), those for video include QuickTime and Motion Picture Experts Group (MPEG), those for graphic image include Graphics Interchange Format (GIF), Joint Photographic Experts Group (JPEG), and Tagged Image File Format (TIFF).

1.4.1.3 Layer 5—session layer

Session layer establishes, manages, coordinates, and terminates communication sessions, including service requests and responses between applications at two end devices. The services at the session layer also involve authentication and reconnection of the communication sessions when interrupted.

1.4.1.4 Layer 4—transport layer

Transport layer segments the data received from the session layer into packets for transporting across the network. The services at the transport layer includes error checking of the received data making sure that the data is delivered reliably in a proper sequence with flow control. Multiplexing and VCs are also set up at this layer to enable data from various applications to be transmitted between the devices on a single physical link. On the Internet, the transport protocols are TCP and user datagram protocol (UDP), which will be described in Sections 1.5.3 and 1.5.4, respectively.

1.4.1.5 Layer 3—network layer

Network layer manages the addressing and routing of the data packets received from the transport layer to the right destination based on the network address. A typical example of an network layer implementation is the IP on the Internet. Detailed description of the IP will be provided in Section 1.5.2.

1.4.1.6 Layer 2—data-link layer

Data-link layer establishes reliable data transmission across the physical link in the form of network frames. Functions at the data-link layer include physical addressing, network topology formation, error notification, sequencing of frames, and flow control. The data-link layer has the following two sub-layers:

- *Logical link control (LLC) sub-layer*: manages communications between devices over a shared physical data link. The LLC is defined in the IEEE 802.2 specification.
- *Media access control (MAC) sub-layer*: manages protocol access to the physical network medium via a MAC address which is a unique identifier defined by the IEEE for a NIC of a manufacturer.

1.4.1.7 Layer 1—physical layer

Physical layer determines various mechanisms and specifications for conveying bit streams over the network. The physical layer specifications are defined depending on the hardware facilities, which could include voltage levels, bandwidth, maximum transmission distance, etc.

1.4.2 Interaction between OSI layers in communications

The communications between two applications in two end devices have to go through all seven OSI layers as defined in the above subsection, but in reverse directions. Assume that an application in PC A wants to send information to another application in PC B. The information flow from the application program in PC A will be passed from the application layer sequentially through all layers to the physical layer of PC A. In the form of bit streams, the information will be sent across the transmission medium to PC B where the information will go through all layers from the physical layer to the application layer. Then, the received information will be processed with an application program in PC B, which complete the communication process.

In general, a layer in the OSI model communicates with the layer directly above it (except the application layer), the layer directly below it (except the physical layer), and its peer layer in the remote PC making use of the services provided by adjacent layers. For instance, the data-link layer in PC A interacts with its adjacent network layer and physical layer in addition to the data-link layer in PC B.

Considering the characteristics and interaction of the seven layers, the OSI model, for simplicity, can be divided into two sets (cf. Figure 1.40):

- *Upper OSI layers*: include the application, presentation, and session layers to manage the information process related to the software.
- *Lower OSI layers*: include the transport, network, data-link, and physical layers to control the data delivery until reaching the transmission medium.

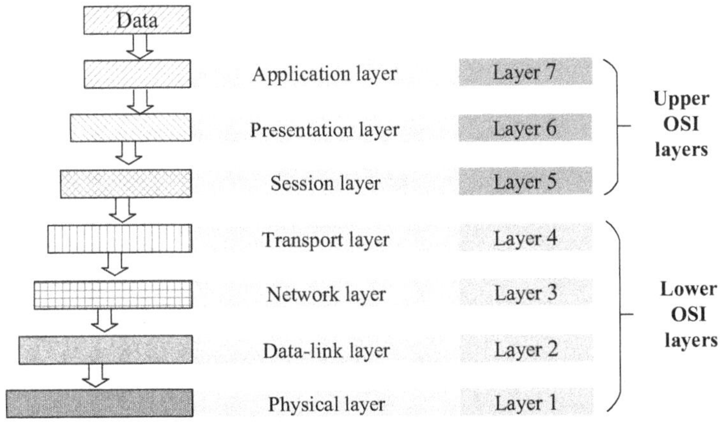

Figure 1.40 Upper and lower OSI layers

1.4.3 Information exchange and information formats in OSI layers

For information exchange between PCs in different systems or networks, control information is required in the OSI layers (cf. Section 1.4.1). The control information is defined in variety of forms to exchange between peer layers. Specifically, either headers or trailers or both are prepended to the data information when it passes down from upper layers. Such process is called encapsulation where the data is enclosed with the control information to allow its readability or decodability at the other end device. In reverse, decapsulation is performed at the destination PC to sequentially remove the appended headers and/or trailers.

Through the encapsulation process, there are various information formats, in which the common ones are as follows:

- *Frame*: is an information unit at data-link layer, consisting of the data-link layer header and/or trailer and data encapsulated at the network layer.
- *Packet*: is an information unit at network layer consisting of the network layer header and/or trailer and data encapsulated at the transport layer.
- *Segment*: is an information unit at transport layer employing TCP protocols with packet segmentation. A segment consists of the transport layer header and/or trailer and data encapsulated at the session layer.
- *Datagram*: has the same format as segment at transport layer but associated with UDP protocols.
- *Message*: is an information unit often at application layer.
- *Cell*: is a fixed-size information unit at data-link layer but is specially used in PSNs with ATM (cf. Section 1.3.3.7) and PDNs with SMDS (cf. Section 1.3.3.8).

Besides the above formats, there exist a general term, namely, data unit, to represent information units at different OSI layers. Common data units consist of service data unit (SDU) and protocol data unit (PDU). The SDU is a unit of data passed down from an upper layer to a lower layer that has not been encapsulated, while

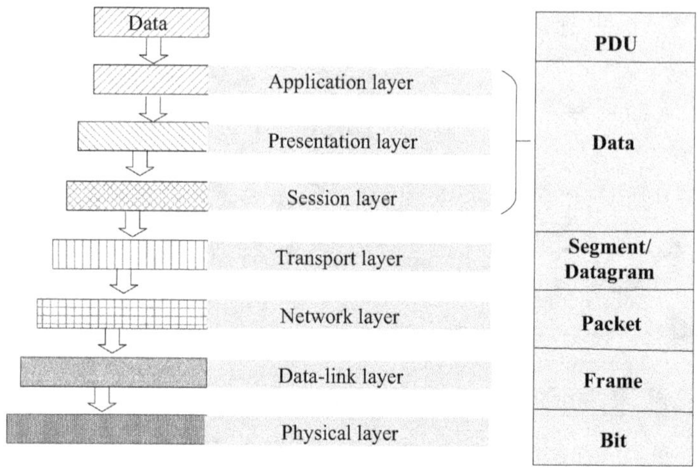

Figure 1.41 Protocol data unit (PDU) in OSI model

the PDU refers to a unit of data after encapsulating the SDU with control information. The PDUs are defined according to the OSI model layers (cf. Figure 1.41). In the upper OSI layers, i.e., application, presentation, and session layers, PDU is referred to as data. In the lower OSI layers, there are four PDUs as follows:

- *Layer 4—transport layer*: PDU is a segment (TCP)/datagram (UDP).
- *Layer 3—network layer*: PDU is a packet.
- *Layer 2—data-link layer*: PDU is a frame.
- *Layer 1—physical layer*: PDU is a bit.

1.5 Internet Protocols

Although the OSI reference model is well described through seven layers (e.g., Section 1.4), it is not a communication method, yet a conceptual framework for communication between two end devices in the network. An actual data transmission or communication is only enabled and activated through a communication protocol which is a set of rules that regulate the exchange information and make the communication more efficient over a network medium. There are a variety of communication protocols operating at different layers in LANs and WANs. This section will discuss the most widely used protocol suite used by the Internet called IPs, of which the most communication protocols are the TCP and the IP. In order to find an appropriate way for the packets to travel from the source to the destination, it is crucial to have routing protocols. Various routing protocols will be explained in detail in Chapter 2, while the rest of this section will outline other protocols, including UDP, Internet Control Message Protocol (ICMP), Address Resolution Protocol (ARP), and

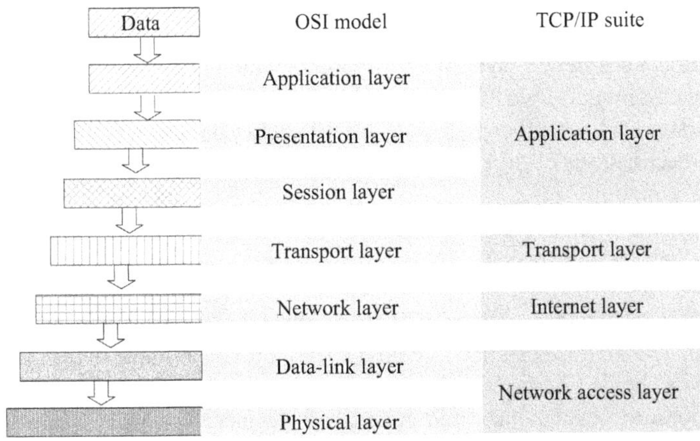

Figure 1.42 Transmission Control Protocol (TCP)/Internet Protocol (IP)

application-layer protocols, such as FTP, Trivial FTP, HyperText Transfer Protocol (HTTP), Telnet, SNMP, SMTP, and domain name system (DNS).

1.5.1 TCP/IP suite

TCP/IP suite[2] is a dominant commercial protocol architecture for the Internet and the WWW. Invented by Vinton G. Cerf and Robert E. Kahn in 1974, the TCP/IP was indeed specified and extensively used 10 years before the proposal of the OSI framework by the ISO. The TCP/IP has been developed by the US Defense Advanced Research Projects Agency and has been regarded as the heart of the Internet providing fundamental communication protocols until now.

Although there is no official model for the TCP/IP suite, it is really a working one with the following layers (cf. Figure 1.42):

- *Application layer*: provides network services to applications with many application protocols, e.g., HTTP, FTP, Telnet, SNMP, SMTP, DNS, NFS, XDR, RPC.
- *Host-to-host or transport layer*: deals with the transmission of the data with two main protocols including TCP and UDP.
- *Internet layer*: incorporates how the packet is constructed for transmission, represented by the IP.
- *Network access layer*: encapsulates IP packets into frames for transmission, map IP address to physical hardware address, i.e., MAC address, and transmit data physically over the medium.

The above layers of the IPs spread across all seven layers of the OSI reference model with some similarities yet simplicity with only four layers.

[2]TCP/IP suite can also be referred to as the Internet protocol suite.

1.5.2 Internet Protocol

IP is a protocol at Internet layer of TCP/IP suite, which is corresponding to network layer in the OSI model. The IP contains packets which are blocks of data to be directed over networks using routing protocols. An IP packet basically include a source IP address, a destination IP address and the actual data to be conveyed along with some control information.

1.5.3 Transmission Control Protocol

TCP is a protocol at transport layer of TCP/IP suite, which has the same name as in the OSI model. The TCP is designed to help the IP to provide reliable data transmission. The TCP is a connection-oriented communication protocol supplying a VC between end-user applications at source and destination devices. The TCP divides outgoing messages into segments and then passes them to the IP for routing to the destination where the messages will be reassembled from the incoming segments. In order to provide reliable communications, the TCP also allows the devices to resend the lost packets and erroneously received packets through an acknowledgment mechanism.

Several well-known application-layer protocols that use TCP include FTP, HTTP, Telnet, and SMTP.

1.5.4 User datagram protocol

UDP is also a protocol at transport layer of TCP/IP suite as in the TCP. However, the UDP is designed for connectionless communications. The UDP is considered as an unreliable protocol since there is no acknowledgment mechanism for retransmission of the lost packets. The UDP also provides no software checking for the segment delivery and thus cannot guarantee the packet delivery. Although the UDP seems to have no benefit over the TCP, such lack of control information for simple communication makes the UDP much faster than the TCP. The UDP is therefore more appropriate for time-sensitive and real-time applications, e.g., multimedia, video streaming, and broadcasting, where the latency of the packet delivery is a major concern.

Several well-known application-layer protocols that use UDP include DNS, TFTP, and SNMP.

1.5.5 Other protocols

There are a variety of protocols corresponding to applications and services at different layers, which can be listed as follows:

- *ICMP*: is a network-layer IP that is designed to report errors of the message transmission.
- *ARP*: is a communication protocol that is used to map IP network addresses to hardware addresses, MAC addresses, corresponding to data-link layer in the OSI model. Inversely, the IP addresses can be obtained from the MAC addresses using Reverse ARP.

- *Application-layer protocols*: include some widely known protocols at different port numbers as follows (sorted in ascending order of the port number):
  - *FTP*: is a TCP used to handle data transfer on port 21.
  - *Telnet*: is a TCP used to manage network devices at the command prompt level on port 23.
  - *SMTP*: is a TCP used to provide electronic mail services on port 25.
  - *DNS*: is a UDP used to translates domain names of the network into IP addresses on port 53.
  - *Trivial FTP (TFTP)*: is a UDP used to convey data on port 69 in connectionless communication with no requirement of session establishment and release.
  - *HTTP*: is a TCP used by web browsers on port 80.
  - *SNMP*: is a TCP/UDP (port 161/162) used by network administrators to monitor and report abnormal network conditions, configure, and control network devices.

1.6 Internetworks

As shown in previous sections, there exists a variety of network models with different sizes and functionalities. It is vital to connect them together in a large network, namely, internetwork, by using intermediary devices.

When considering a large group with a number of individual elements, one of the questions that can be raised is how to share a limited resource available in the group to every element in an efficient way. This concern indeed applies to the internetworks which also require an efficient resource sharing and effective communication capability between all individual networks. In other words, network management plays an important role in internetworks to cope with the incompatibility of different devices as well as the inconsistency of different network models.

With the evolved technologies, many manufacturers, vendors, and operators are continuing to create new machines with different configurations, OSs with different programs, and also different transmission media with advanced communication mechanisms. Internetworking therefore has to enable the compatibility of them allowing the cooperation of PCs from different vendors with different OSs using different transmission media on different networks in different continents.

Internetworking is not simply providing a connection between devices to link them together. Such process sounds simple, but it in fact requires complicated tasks with challenges to be faced. There are a number of concerns regarding the connectivity, compatibility, flexibility, reliability, and security in the internetworks. For instance, one of the critical issues is how to provide a reliable and secured communication between different devices over different media in different systems given a number of advanced technologies evolved along with cyberattacks developed in parallel, e.g., distributed denial-of-service attack and recently WannaCry ransomware attack have affected computers worldwide.

In order to easily manage such a large network, internetworks require a network hierarchy for organizing devices in systems and an addressing mechanism to identify

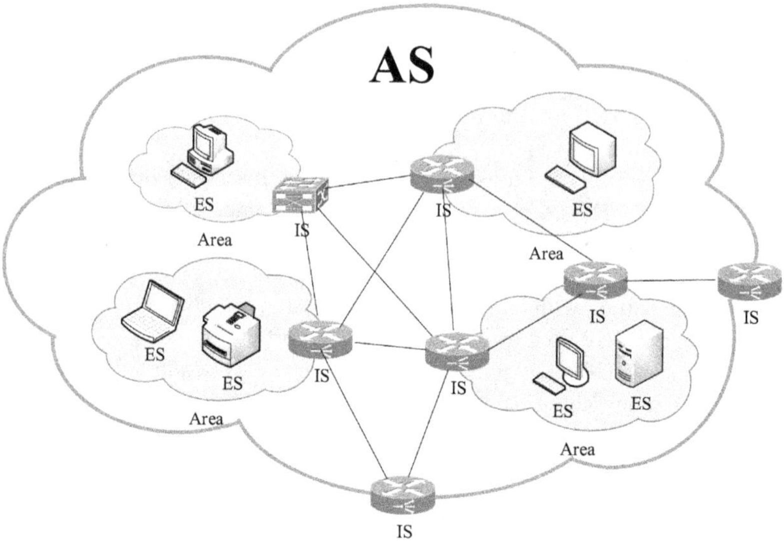

Figure 1.43 Internetwork hierarchy

these devices following international standards. In the following, internetwork hierarchy and addressing will be described, followed by the introduction of some well-known standards organizations and intermediary devices used in the internetworks.

1.6.1 Internetwork hierarchy

For ease of management, the ISO has defined key terminologies for different systems as follows (cf. Figure 1.43):

- *ES*: consists of PCs, printers, terminals, etc. at the end users who only perform tasks with applications.
- *Intermediate system (IS)*: consists of repeaters/hubs, switches/bridges, and routers, which are used to connect PCs in either the same or different networks.
- *Autonomous system (AS)*: is a "large" network consisting of a number of networks interconnected via routers and controlled by a common administrative entity. Due to its large size, the AS is normally divided into subnetworks or areas depending on their location and routing policies in the network.

1.6.2 Internetwork addressing

To identify devices in the internetworks, there are basically two internetwork address types, including

1.6.2.1 MAC addresses

Each LAN interface has a unique MAC address assigned by the manufacturer. A PC in the ES has only a LAN interface with only a MAC address, while internetworking

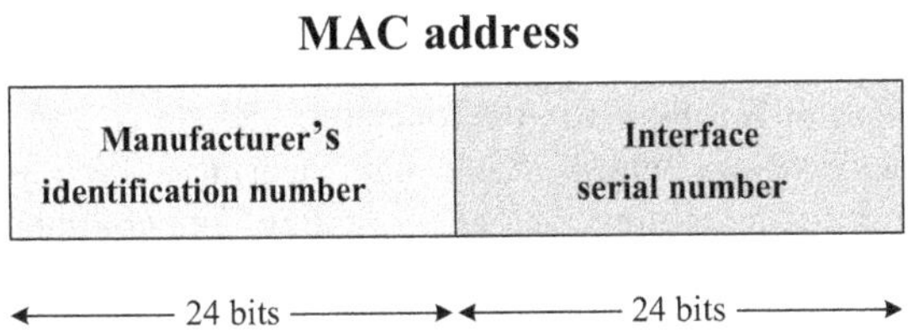

Figure 1.44 MAC address

devices in IS have multiple LAN interfaces for interconnection and thus should have multiple MAC addresses. Since MAC is a sublayer of data-link layer in the OSI framework (cf. Section 1.4.1.6), the MAC addresses are also referred to as a subset of data-link layer addresses.

A MAC address consists of 48 bits written in hexadecimal (or base-16) numbers (cf. Figure 1.44). For simple representation, there are two formats of MAC addresses, i.e., MM-MM-MM-SS-SS-SS or MMMM.MMSS.SSSS. The first 24 bits are identification number of the manufacturer and the remaining 24 bits are the interface serial number assigned by the manufacturer.

Example 1.1. *A MAC address 00000ABC1982 can be formatted as*

- *00-00-0A-BC-19-82*
- *0000.0ABC.1982*

1.6.2.2 Network layer addresses

A network layer address is used at the network layer in the OSI model (cf. Section 1.4.1.5) as virtual or logical addresses. The network layer address is required for network layer protocols, i.e., routing protocols, to determine the location of devices following the internetwork hierarchy model (cf. Section 1.6.1). Depending on the number of physical network connections at the internetworking devices, they may have one or multiple network layer addresses.

Over the Internet with IP routing protocol, each device has a logical address called IP addresses. There are two versions of IP addresses, namely, IP version 4 (IPv4) and IP version 6 (IPv6). An IPv4 address consists of 32 bits represented in dot-decimal format as A.B.C.D where A, B, C, and D are in decimal representing eight bits called an octet. An IPv6 address is much longer including 128 bits, i.e., 16 octets, represented in hexadecimal numbers and separated by colons for every two octets. The IP addressing will be discussed in detail in Chapter 3.

Example 1.2. *IPv4 and IPv6 addresses:*

- *IPv4: 172.10.10.82*
- *IPv6: 1982:BA10:C12:7:10:14:12:30*

1.6.3 Standards organizations

Standards are required to allow for interoperability between devices in different networks. The standards are helpful in not only ensuring a large market for equipment from different manufacturers, but also enabling the communications between the products from different vendors. However, the standards may cause freezing technologies, and there may have multiple standards for the same things.

For the standard development, prior to putting the standards into formal specifications, organizations provide forums and online societies for discussion, e.g., Internet Society, ATM Forum, Internet Activities Board with Request For Comments documents. Some well-known standards organizations can be listed as follows:

- ISO
- International Telecommunication Union—Telecommunication Standardization Sector (ITU-T)—formerly Consultative Committee for International Telephony and Telegraphy (CCITT)
- European Computer Manufacturing Association (ECMA)
- Electrical Industries Association (EIA)
- Telecommunication Industries Association (TIA)
- American National Standards Institute (ANSI)

1.6.4 Internetworking devices

To enable data communications between networks or areas within an AS, internetworking devices are required to connect:

- LAN-to-LAN
- LAN-to-WAN
- WAN-to-WAN

There are four primary types of devices, including:

- Repeaters/hubs
- Bridges/switches
- Routers
- Gateways

A quick glance on the above list is that some of them except gateways have the same names as those discussed in Section 1.2.1 when I briefly introduced intermediary LAN devices. Indeed, those devices, i.e., repeaters, hubs, bridges, switches, and routers, are also internetworking devices having the same functions as in LAN. With reference to the OSI layers (cf. Section 1.4.1), the internetworking devices can be classified as

- *Repeaters/hubs*: OSI layer-1 devices at the physical layer.
- *Bridges/switches*: OSI layer-2 devices at the data-link layer.
- *Routers*: OSI layer-3 devices at the network layer.
- *Gateways*: operate at or above OSI layer 4, i.e., from the transport to application layer.

This subsection will discuss more details of the devices in the internetworks.

1.6.4.1 Computers

Along with the intermediary devices in the internetworks, there is a must-have device that every computer system cannot run without it. That is computer or PC or terminal.

Every PC has a unique way to identify itself, which is based on a physical address. A PC has a physical address regardless it is connected to internetwork or runs on its own. The physical address can be found on the NIC which is a circuit board or a chip allowing a PC to communicate with other PCs in the network.

The physical address is programmed into a chip on the NIC and is assigned by the hardware manufacturer before it leaves the factory. Since the physical address is also referred to as MAC address (cf. Section 1.6.2.1), the NIC is a layer-2 device at the data-link layer. The address is unique on every NIC card, so if the NIC were replaced on a PC, the physical address of the PC would change to that of the new NIC.

1.6.4.2 Repeaters/Hubs

A repeat, as the name says, is basically an intermediary device used to extend the coverage of LANs. It has two ports and the data flow can travel in both ways, i.e., two-way communications. When receiving the data from one port, the repeaters may clean up the noise, amplify, and pass the data at bit level to another port; hence, the repeaters function at the physical layer of the OSI framework. The repeaters therefore depend on the transmission media, but they are independent of the communication protocols. Affected by the transmission media, the repeaters are normally confined to a house or a single building.

The repeaters provide no traffic isolation of all data flows from the computers attached to them in the network. The monitoring of the operation at the repeaters is limited, and thus using the repeaters should be aware of the connection links, especially those which are likely to fail. The number of repeaters in the network is also restricted due to their limited transmission distance.

Following are the two common types of repeaters:

- *Simple repeaters*: are simply used to get around limitations in cable length caused by signal loss or timing dispersion. They pass on individual data bits without performing any processing at the packet level. The collisions are also passed along with the data with no restriction. Therefore, the number of the repeaters that can be used in the network is limited and the basic Ethernet design requires that signals must be able to get from one end of the network to the other within a specified amount of time and a maximum allowable length.
- *Buffered repeaters*: operate at the level of the whole data packets. Rather than passing bit by bit as in the simple repeaters, the buffered repeater processes an entire packet received from a network, stores the packet in an internal buffer, and then retransmits it at the other end port to another network. The collisions are not passed and repeated, and thus, there is no restriction of the number of the buffered repeaters that can be used in the network. There is only a requirement that the packets received from two networks should have the same format, while the requirement that both the networks should be of the same type can be relaxed.

As an extension of repeaters, a hub is simply regarded as a multi-port repeater. The hub is a very common internetworking device and is used as the center of a star topology network.

1.6.4.3 Bridges/Switches

A bridge, like a repeater, is used to store and transmit packets and also dependent on the transmission media. However, the bridge functions at the data-link layer of the OSI model. The bridge cannot recognize the protocols used at the layers above the data-link layer, but it simply passes all protocols onto the network.

One of the advantages of the bridge is that it can isolate traffic by dividing the network to reduce traffic load on the network. Therefore, more repeaters can be used. Operating at the data-link layer, the bridges can determine whether the destination MAC address carried by data is not part of the same network segment as its source; however, it cannot determine what network segment the data should be sent to. In that case, the bridges pass randomly data across all segments of the network. Such broadcasting may cause broadcast storms, which is a major drawback of the bridge.

With multiple ports, a multi-port bridge is named as a switch and is popularly used in internetworks.

1.6.4.4 Routers

Considering a large network including a number of network segments or subnetworks using different protocols, hubs and even switches are shown to be not able to support all of them. In such a complex network, it is required to have an intermediary device that can find the best path for the data transmission between the segments based on their address. A router was therefore designed.

The routers operate at the network layer of the OSI framework. Different from the hubs and switches, the routers are independent of the transmission media. The routers are generally much more complex and expensive than hubs and switches, and thus they are used only when routing is required to connect networks having dissimilar protocols. The routers make their decisions on the path routing based on network layer address (cf. Section 1.6.2.2) rather than the destination MAC address (cf. Section 1.6.2.1). In Chapter 2, routing protocols will be discussed in detail.

1.6.4.5 Gateways

Basically, a gateway can be viewed as a general version of a router, which also has a basic function to connect different networks. However, the gateways operate at the transport layer or above layers in the OSI model. The gateway can convert data format from the application program, languages, architecture, and communication protocols at a system to make them conform with others at the destination system. The gateway is therefore more complex than the router in enabling the communication and understanding between different architectures and environments.

1.7 Related works

For more information, the readers are suggested to refer to the following selected references:

Data Networking Spohn *et al.* in 2002 [1], Stallings in 2006 [2], and Dye *et al.* in 2011 [3].

LAN Fraser in 1983 [4], Chu and Gans in 1987 [5], Pahlavan *et al.* in 1995 [6], Crow *et al.* in 1997 [7], Woodcock in 1999 [8], Clark and Hamilton in 1999 [9], Bray and Sturman in 2000 [10], Stallings in 2001 [11], and Roshan and Leary in 2003 [12].

WAN Decina and Scace in 1986 [13], Lechleider in 1991 [14], Thorpe and Ross in 1992 [15], McDysan and Spohn in 1994 [16], Brown and Malis in 1998 [17], Ginsburg in 1999 [18], and Bellamy in 2000 [19].

OSI Model ISO/IEC standard 7498-1 in 1994 [20].

TCP/IP Cerf and Khan in 1974 [21] and Rodriguez *et al.* in 2001 [22].

Internetworks Boggs *et al.* in 1980 [23], Teare in 1999 [24], and Stallings in 2003 [25].

1.8 Review questions

[Q1.1] List popular computer network models and explain how to differentiate between them.

[Q1.2] Differentiate between hubs, switches, and routers in a LAN.

[Q1.3] How are the devices physically connected in a LAN? Discuss advantages and disadvantages of each topology.

[Q1.4] Compare different technologies used in a LAN.

[Q1.5] What are the differences between a WLAN and a LAN, between Wi-Fi and Bluetooth, and between token ring and Ethernet.

[Q1.6] Which access methods are employed in the IEEE 802.3, IEEE 802.5, and IEEE 802.11? Discuss and explain the reasoning for the selected methods in each standard.

[Q1.7] Name three transmission methods in a LAN and give an example for each of them.

[Q1.8] List some common devices used in a WAN and discuss their usage.

[Q1.9] Differentiate between full mesh and partial mesh topologies. Which one is feasible in a large network?

[Q1.10] What are the differences between circuit switching, datagram packet switching and VC packet switching? Discuss the advantages and disadvantages of them for data transmission.

[Q1.11] Differentiate between SVC and PVC in terms of resource requirement.

[Q1.12] What services are provided by ISDN and what data rates are offered by each service?

[Q1.13] What are the similarities and differences between X.25, Frame relay, ATM, and SMDS?

[Q1.14] Describe some typical forms of DSL and their applications.

[Q1.15] List seven layers of the OSI reference model in order and the corresponding PDU at each layer.

[Q1.16] Describe TCP/IP suite with its layers and discuss the necessity of layering with the TCP/IP.

[Q1.17] Differentiate between TCP and UDP.

[Q1.18] List and briefly describe some application-layer protocols using TCP and/or UDP.

[Q1.19] What are internetworks?

[Q1.20] How to identify devices in the internetworks?

[Q1.21] Describe briefly intermediary devices used in internetworks and discuss how to differentiate between them.

1.9 Problems

[P1.1] Consider a network consisting of N devices. How many cable links are required in total and how many ports are required in each device if they are connected in a ring, bus, star, and full mesh topology? Illustrate the answer with five devices.

[P1.2] Sketch a hybrid topology to connect 40 devices with a star backbone and four ring networks. Modify the topology using a ring backbone and two bus networks.

[P1.3] Given six devices with ten cables, design a partial mesh network using all cables.

[P1.4] Consider a PCN employing VC to connect node A and node B via three nodes 1, 2, and 3. The length of the message to be sent is 40 octets and every packet contains a header of 3 octets for control information. Assume that it takes 1 ms to send an octet, and there is no delay caused by the processing at the nodes.

- Determine the transmission time to send a message from A to B.
- Suppose that the message is split into two packets. Find the new transmission time from A to B. Discuss the findings when the message is split further into four or five or ten packets.
- Let N, K, L, and H denote the message length in octets, header length in octets per packet, the number of packets in a message, and the number of hops between two end devices, respectively. Assume that N is divisible by L. Derive the expression of the transmission time between these two end nodes, which is also known as end-to-end delay.
- Find the optimal L, as a function of K, L, and H, so as to minimize the end-to-end delay in VC-based PCN.

[P1.5] Derive the general expression of end-to-end delay in a network employing either circuit switching or datagram packet switching or VC packet switching as a function of the following parameters:

- N: message length in octets
- K: header length in octets per packet
- H: number of hops between two end devices
- L: number of packets per message
- R: link data rate in bps
- T_s: call setup time, in seconds, for circuit switching and VC
- T_p: propagation delay over a hop in seconds per hop

Find the optimal L, as a function of the other parameters, to minimize the end-to-end delay in datagram packet switching.

Chapter 2
Routing in computer networks

In an internetwork with a number of network segments using a variety of protocols and network topologies, the communications between devices in these subnetworks require a powerful device to connect them in an efficient way. It is therefore vital to have a router as an intermediary device that not only plays the role of connecting the subnetworks but also routes the data via an optimal path toward the destination node.

Routing is basically finding the path between the source and destination nodes based on their addresses. The work of the router sounds like that of switches; however, the routers operate at the network layer of the OSI framework, while the switches function at the data-link layer using MAC addresses. The routers are independent of transmission media and are generally much more complex requiring high investment expenditure compared to other intermediary devices at lower layers. In order to route the messages between networks, a routing protocol is crucial, of which path determination is the most complicated process.

In this chapter, for a first view of routing in computer networks, I will first discuss in Section 2.1 the fundamentals of routing along with an overview of algorithms for path determination. Static versus dynamic routing protocols will be described in Sections 2.2 and 2.3, respectively. Considering the routing protocols in different network infrastructures, interior gateway protocols (IGPs) and exterior gateway protocols (EGPs) will be sequentially presented in Sections 2.4 and 2.5 with relevant protocols, such as Routing Information Protocol (RIP), Interior Gateway Routing Protocol (IGRP), Open Shortest Path First (OSPF), Enhanced IGRP (EIGRP), and Border Gateway Protocol (BGP).

2.1 Fundamentals of routing

As briefly outlined in Section 1.6.4.4 of Chapter 1, routers can make their decisions on the path from a sender to a receiver based on the network addresses, but how do they work?

The routers need a routing protocol which has enough network layer addressing information for user traffic to be directed from one network to another one. There are basically two activities involved in the routing process, including path determination and packet switching. While the packet switching requires a simple algorithm to examine the address of the next hop for the packets to be forwarded, the path determination, which will be explained in the following subsection, is rather complicated with different routing protocols.

2.1.1 Path determination

In fact, a routing protocol is not simply finding a path, but it should be in an efficient way to carry messages in terms of packets between networks given a number of available paths. The routing protocols define the format of some fields within a packet as a preliminary information for the path determination. With the routing protocols, it should be noticed that they are used to exchange information between routers and do not take into account any user data and traffic.

In order to determine which path for the packets to travel, routing protocols define route metrics which are the costs of reaching a destination. The costs here could be the number of hops or other measurement of communications, such as delay and bandwidth. For path determination, there are routing tables containing route information in the network which can be regarded as an abstract of the real network topology. The routing tables are managed by routers and are used in the routing algorithms to calculate the optimal path to a destination. Depending on the routing algorithms, there may have different route information stored in these tables. When receiving a packet, the routers check its destination address and route it to the next hop following the optimal path. The routers are therefore required to communicate between them to keep updating their routing tables.

2.1.2 Routing algorithms

There exist various routing algorithms depending on network resources, router capacity, and route metrics as well as practical requirements. The design of all these routing algorithms however has the same desired outcome which is the best or optimal route(s) from one point to another point. These are required to be stable but robust to be able to confront the hardware failures in unpredictable circumstance that may cause load unbalancing, overburden, and even interruption. Moreover, the routing algorithms should be designed in an efficient way with low complexity and low overhead to allow them function in the condition of limited physical resources. The convergence of the algorithms implemented at the routers is also a critical design goal to guarantee that the best route can be determined after a limited number of routing update messages, and thus the routing loops or network outages can be refrained from the slow network convergence. With a number of existing network infrastructures with evolved technologies, the routing algorithms should also be adaptable to different network models and settings along with internetworking devices and terminals with different configurations.

With different design goals, the routing algorithms are classified based on different criteria, such as network infrastructures, hierarchies, complexity, convergence, adaptability, and flexibility. They may be contrary in meaning and functionality but can supplement each other if appropriately used. For instance, some can be listed as follows:

- *Flat routing algorithms* where all nodes are regarded as peers of the same level versus *hierarchical routing algorithms* which are based on the network hierarchies as modeled in an organization (cf. Section 1.6.1).
- *Single-path routing algorithms* for finding only one best path from source to destination versus *multipath routing algorithms* which support multiple paths to the same destination for load sharing.
- *Distance-vector routing algorithms* which simply use the distances between neighboring nodes with low complexity versus *link-state routing algorithms* which rely on the routing updates of the entire network and thus require high-complexity processing at the routers but allow fast convergence.
- *Static routing algorithms* of which the routes are manually preset by a network administrator irrespective of possible succeeding network changes versus *dynamic routing algorithms* which automatically find alternative routes as the network topology or traffic changes.

Basically, with respect to a specific router in a network, there are two kinds of network connection which are either directly or remotely connected. The associated networks of the router can be therefore categorized as follows:

- *Directly connected networks*: are networks connected directly with the router interfaces via directly connected routes.
- *Remote networks*: are networks connected through other routers by using routing protocols with either static or dynamic routes. Details of the static and dynamic routing protocols will be presented in Sections 2.2 and 2.3, respectively.

The routing algorithms can be either implemented within an enterprise or across the Internet. The routing protocols can be accordingly grouped into following two groups (cf. Figure 2.1):

- *IGP*: is used to describe a routing protocol employed within an autonomous system (AS) in the internetwork hierarchy. Depending on the employed routing algorithms, there are different forms of the IGP, which will be discussed in Section 2.4.
- *EGP*: is a routing protocol used for exchanging routing information among ASs on the Internet. The EGP is however obsolete after its third version and has been replaced by a new standardized protocol called BGP. Two forms of the BGP, including Internal BGP (IBGP) and External BGP (EBGP), will be described in Section 2.5.

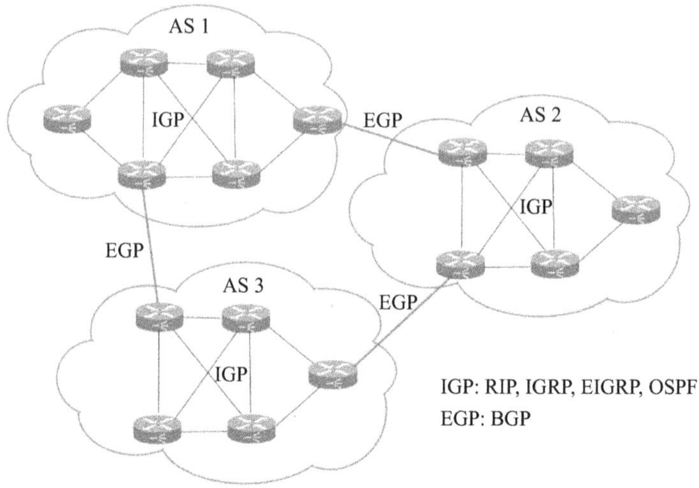

Figure 2.1 Interior gateway protocol (IGP) versus exterior gateway protocol (EGP)

2.2 Static routing protocols

A static routing protocol can be identified as bus routes based on which the bus uses the same route everyday to travel between two end stations through the same number of stops. In the context of networking, the bus can be regarded as a packet, whereas the intermediary stops are routers in the network.

The static routes are manually defined by system administrators and are fixed regardless of network changes or reconfiguration. These routes are easy to configure and maintain in a small network, but it will be time-consuming with a considerably increased complexity as the network size increases. The static routing is therefore suitable for only a network with simple topologies, a small number of nodes, and when the links are unlikely to change. Since the administrators are required to configure the routes, their intervention is indispensable whenever the network topology changes. The static routing protocols are however shown to provide more secure communications with no extra resources for traffic reduction and also with minimal processing.

2.3 Dynamic routing protocols

Unlike static routing, dynamic routing protocols allow the packets to find a new path to the destination in an adaptive manner, and thus the dynamic routing is also known as adaptive routing. This fact is related to the circumstance when driving a car with a number of alternative routes depending on different factors, such as traffic conditions, weather, etc.

In computer networks, the adaptation property of dynamic routing protocols helps find substitute routes as the network topology or traffic changes, such as the loss of nodes and the interruption of connection links. The packets can be automatically rerouted around the broken links and the failure nodes via the next optimal path when the previous one is not available. The dynamic routing is independent of the network size, and thus more scalable and suitable for complex network topologies. However, there are some disadvantages of such dynamics when the network is less secure and substantial resources are required to manage the routing process.

2.4 Interior gateway protocols

An IGP is a general term to describe a routing protocol employed within an AS for exchanging routing information between gateways. Depending on the routing mechanisms and metrics, the IGPs can be divided into two groups, including distance-vector routing protocols (DVRPs) and link-state routing protocols (LSRPs), which will be described in the following subsections. In addition, there exist hybrid routing protocols having the features of both the DVRPs and LSRPs, of which a typical example is EIGRP designed by Cisco.

2.4.1 Distance-vector routing protocols

DVRPs are protocols based on the distance and direction between nodes in the network. The DVRPs use either the Bellman–Ford algorithm to compute the shortest paths from a source vertex to all other vertexes in a weighted digraph or Ford–Fulkerson algorithm to compute the maximum flow or diffusing update algorithm full-state machine (DUAL FSM) specially developed for EIGRP.

The DVRP requires that a router informs its neighbors of topology changes periodically by passing the routing table to neighbor nodes in the network. The cost of reaching a destination is quantified using route metrics which are stored in a distance vector in the routing table. Employing various algorithms, a router can calculate the direction and distance to other routers.

Each router does not have any knowledge about the full network topology. A router advertises its distance vector to other routers and also receives the advertisement of the distance vectors from the other routers. These advertisements are then accumulated in the routing table at each router.

Without the requirement of an actual map of the network topology with hierarchical design, the routing update process in the DVRP is simple and does not require much knowledge of the administrators, but its main disadvantage is slow convergence and may cause routing loops in some cases.

Two typical examples of DVRPs are as follows:

- *RIP*, *RIP Version 2 (RIPv2)*, and *RIP next generation (RIPng)*: where the route metric is the hop count.
- *IGRP*: which is based on a composite metric including bandwidth, delay, load, reliability, and maximum transmission unit (MTU).

2.4.1.1 Routing Information Protocol

As one of the oldest and simplest DVRP, RIP was originally specified in the RFC 1058 published in 1988 and has been most widely used on the Internet for exchanging routing information among gateways.

In the RIP, the hop count is used as the metric for path selection. The maximum allowable hop count is 15 hops and the routing updates are broadcast every 30 s by default setting in the update timer of the RIP. Apart from the update timer, the RIP has other specific timers including invalid timer, flush timer, and hold-down timer.

Although the RIP is extensively used in different networks due to its low complexity, it has some limitations as the number of hops is restricted and slow convergence. The routes in the RIP will be dropped if the number of hops exceeds 15 hops. In other words, the destination is considered as unreachable when the metric value in the routing table is 16. This limitation however helps prevent routing loops from counting to infinity, which is also known as count-to-infinity problem. The routing loops will be explained in detail in Section 2.4.1.3.

The RIP has three versions as follows:

- *RIP version 1 (RIPv1)*: is the original version of the RIP which was defined in RFC 1058. The RIPv1 supports only subnets of the same class having the same length or fixed length, while it does not support different-sized subnets with variable length subnetting since there is no subnet information carried in the routing updates. The RIPv1 is thus also known as a classful routing protocol. In particular, there is no router authentication in the RIPv1, which makes it exposed to the possibility of being attacked.
- *RIP version 2 (RIPv2)*: was originally described in RFC 1058 in 1993 and was later standardized in RFC 2453 in 1998. The RIPv2 allows more information to be included in the RIP packets. Specifically, an RIPv2 packet contains additional subnet and authentication information, and thus the RIPv2, as a classless routing protocol, can support variable-length subnets and authentication mechanism. The hop count is also limited by 15 hops and the RIPv2 maintains backward compatibility with RIPv1. In order to reduce unnecessary load on those hosts which are not involved in routing, instead of broadcasting the routing table as in the RIPv1, the RIPv2 multicasts it to all adjacent routers using an IP multicast address 224.0.0.9. Additionally, the RIPv2 packet includes route tags to distinguish between internal routes learned with the RIP and those learned with other protocols.
- *RIP next generation (RIPng)*: was defined in RFC 2080 in 1997 as an extension of the RIPv2 to support the next generation of IP, i.e., IPv6 (cf. Section 1.6.2.2).

2.4.1.2 Interior Gateway Routing Protocol

IGRP is a DVRP developed in the mid-1980s by Cisco to exchange the routing information between Cisco gateways or routers within an AS. The IGRP overcomes the limitations of the RIP which has only a single routing metric, i.e., hop count, limited by 15 hops.

Similar to the RIP, the IGRP employs distance-vector routing algorithm to determine the direction and distance to any link in the internetwork and also has four timers consisting of update timer, an invalid timer, a hold-down timer, and a flush timer. However, the default configurable hop count of the IGRP-routed packets is 100, which can be configured up to 255, and the routing updates are advertised over longer intervals of 90 s. Therefore, the IGRP can handle complex topologies in large networks, providing scalability and flexibility for segments having different bandwidth and delay characteristics.

In particular, the IGRP uses a combination of variables consisting of bandwidth, delay, load, reliability, and MTU to determine a composite metric which can be adjusted through the use of preset constants. Here, the MTU is the maximum packet size in bytes that a particular interface can handle and the composite metric is a function of the bandwidth, delay, load, and reliability.

Let B, D, L, and R denote the bandwidth, delay, load, reliability, respectively. The composite metric for the IGRP, i.e., M_{IGRP}, is computed by

$$M_{\text{IGRP}} = \left(K_1 B + \frac{K_2 B}{256 - L} + K_3 D\right) \frac{K_5}{R + K_4}, \tag{2.1}$$

where K_1, K_2, K_3, and K_4 are constant metric weights used to define with the router configuration command.

By default, $K_1 = K_3 = 1$ and $K_2 = K_4 = K_5 = 0$, and thus from (2.1), $M_{\text{IGRP}} = B + D$. This means that the IGRP chooses a route based on the bandwidth and the delay by default.

Although the IGRP has been shown to be one of the most successful routing protocols of Cisco, it lacks of support for variable-length subnetting as in the RIPv1. Instead of designing an IGRP version 2 as the RIPv2 with that additional capability, Cisco decided to develop an EIGRP as an hybrid approach of both the DVRP and LSRP, which will be discussed in detail in Section 2.4.3.

2.4.1.3 Routing loops

Routing loops are common problems in the routing update process in DVRP caused by slow convergence. The routing loops are basically formed when there is a misunderstanding between the routers during the routing update and thus cause an endless loop between the routers. This means that a packet is continuously transmitted through the routers but could not reach its intended destination.

Example 2.1. *Consider a simple network with only three routers R1, R2, and R3 (cf. Figure 2.2). A packet at R1 network is supposed be sent to R3 network through R2. The link between R2 and R3 is down but R2 has not yet informed R1 about this breakage. A routing loop can be formed as follows:*

- *R1 assumes that there is a path to R3 through R2 and transmits the data to R2.*
- *Since R2 knows of the broken link between R2 and R3, it tries to reach R3 via R1 and thus sends the original data back to R1.*
- *R1 receives the same original data from R2 and then consults its routing table. However, since R1 has not been informed about the broken link between R2 and*

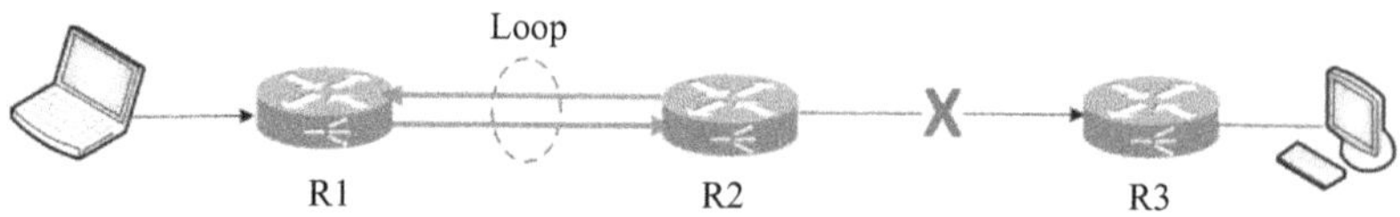

Figure 2.2 Routing loop in Example 2.1

R3, its routing table also shows that the packet can reach R3 via R2, and hence R1 resends its data to R2; and so on.

The packet will encircle endlessly between routers R1 and R2.

With an infinite number of traffics looping to and fro between routers, the routing loop causes a significant waste of spectrum and burdened load for the routers' CPUs. The routing update process will also be affected in future with more routing loops, especially in a large network. Therefore, it is critically important to have a mechanism to get rid of such routing loops.

There are a variety of approaches to prevent routing loops from happening, which can be listed as follows:

- *Count-to-infinity condition*: is the condition when the routing updates increase the metric value to "infinity." Different routing protocols have different "infinity" values. For example, RIP defines that the maximum allowable hop count is 15 hops, which means "infinity" is 16 hops.
- *Hold-down timers*: are used to eliminate regular update messages from improperly restoring a bad route. The hold-down timers only allow the routers to hold any network changes for a specified period of time. If a route is identified as likely to be down, then any other information for that route will be ignored for a hold-down period depending on the routing protocol. This hold-down period helps all routers have enough time to acquire the status of the unreachable network. For example, the hold-down timer in RIP is set for 180 s by default.
- *Split horizon rule*: states that a router should not advertise its routing table back to the interface where it receives the update. The split horizon rule can be applied together with either route poisoning to prevent a router from sending packets through an invalid route or poison reverse to explicitly mark the route as unreachable in the routing update. For example, the RIP has a poisoned route with a metric of 16.

2.4.2 Link-state routing protocols

LSRP, in contrast to DVRP (cf. Section 2.4.1), is performed by every router in the network given the fact that the full picture or complete network topology is available at all routers.

Specifically, every router constructs a map of the connectivity or topology of the network in the form of a graph showing which routers are connected to which other routers. In order to form the routing table, every router calculates the best next hop to reach all other routers in the network rather than sharing the routing table with only its neighbors as in the DVRP. Here, the cost of reaching a destination is also calculated based on route metrics and only connectivity-related information, not the whole routing table, is passed between the routers.

Given the full knowledge of the network topology, the LSRP only requires update when there is a network change instead of performing periodic updates. The LSRP is particularly helpful for large networks which require fast convergence; however, it requires more CPU power and memory for processing at the routers compared to the DVRP, which make the LSRP more expensive to implement. Also, details of the network hierarchy are required and the administrators should have a good knowledge of how to design a network with the LSRP.

Two typical examples of LSRPs are as follows:

- *OSPF*: which was designed by the Internet Engineering Task Force (IETF) working group.
- *Intermediate System to Intermediate System (IS–IS)*: which was designed by the ISO for the OSI framework, not the TCP/IP protocol suite.

Although the OSPF and IS–IS were designed by different groups with different standards, they both have many similarities in providing the necessary routing functionality with link-state routing algorithm. So, only the OSPF is selected to discuss in the following subsection.

2.4.2.1 Open Shortest Path First

OSPF is a typical LSRP in the group of IGPs operating within an AS. Designed by the IETF working group since its establishment in 1988, the OSPF was created to overcome the shortcoming of the RIP which was shown to be incapable of handling large internetworks. The OSPF has been then the most widely used IGP in heterogeneous networks.

The OSPF has the following two versions:

- *OSPF version 2 (OSPFv2)*: was defined in RFC 2328 in 1998 for IPv4 networks.
- *OSPF version 3 (OSPFv3)*: was first defined in RFC 2740 for IPv6 networks and then updated in RFC 5340 in 2008.

As the name stated, the OSPF is an open protocol having specification available in the public and it is based on Shortest Path First algorithm. Specifically, Dijkstra's algorithm, which will be discussed in detail in Section 2.4.2.2, is employed to find the shortest paths between the routers in the network topology.

The OSPF recreates the exact topology of the entire internetwork as in the LSRP through sharing the connectivity-related information, namely, link-state advertisements (LSAs), among routers. The LSAs consist of information about the interfaces at the routers, metrics used, and other variables. Then, based on the LSAs, the OSPF can determine the shortest paths between the routers. Here, the routing metric in the OSPF is the path cost which can be simply computed as inversely proportional to the bandwidth or link capacity. For example, a link with a capacity of 100 Mbps is assumed to have a cost of one, then the 10 and 20 Mbps links will have costs of ten and five, respectively. In order to balance the traffic load dynamically, the link metric can be more complicated with the introduction of cost factors depending on the link availability, reliability, and latency in addition to data throughput.

With the same concept of the LSRP, the OSPF operates following network hierarchy and can quickly detect the network changes allowing fast convergence to the new topology. As shown in Section 1.6.1, the AS is the largest entity consisting of a number of networks which can be grouped into different areas. These areas are connected via area border routers (ABRs) (cf. Figure 2.3). The LSAs from all the routers in the same area, namely, internal routers, are collected to build a topological database for that area, which is then managed and shared between the ABRs to have an overall picture of the internetwork. Hence, the topological databases should be identical at all the internal routers within the same area.

Depending on the location of the source and destination within the same area or in different areas, there are two corresponding kinds of OSPF routing, namely, intra-area and inter-area routing. In order to distribute or share routing information between areas, the OSPF requires a backbone area which consists of all ABRs, other attached routers, and also other non-backbone areas (cf. Figure 2.4). All the routers having an interface connected to the backbone area are thus named as backbone routers. These backbone routers employ inter-area routing over the backbone topology, while the intra-area routing is performed at internal routers given only the knowledge of their individual area topology.

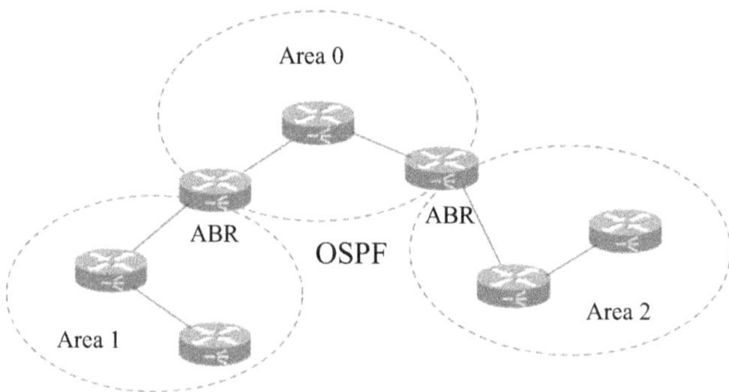

Figure 2.3 Open Shortest Path First (OSPF) routing protocol

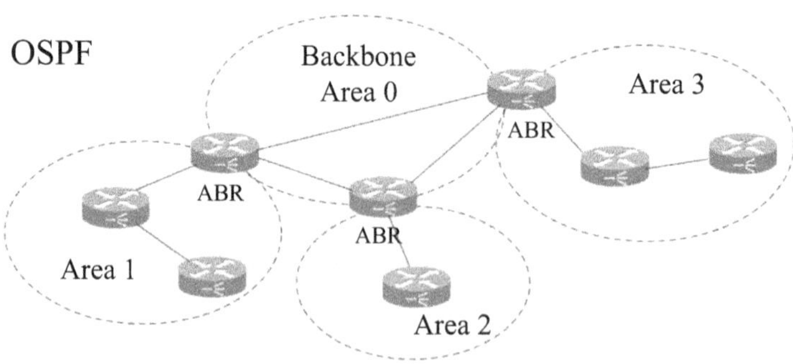

Figure 2.4 OSPF over backbone area

2.4.2.2 Dijkstra's algorithm

One of the fundamental problems in connection with graphs is to find the path having the shortest length between nodes in a graph. In 1956, Edsger W. Dijkstra, a computer scientist, came up with an idea on how to formulate and solve this routing problem, which was then published 3 years later and well known as Dijkstra's algorithm.

The Dijkstra's algorithm has a number of variants applied in different areas. In particular, it has been widely used in the OSPF routing protocol to find the shortest routes between a node and all other nodes to build a shortest path tree.

Given a set of nodes connected in a graph with a nonnegative weight value, i.e., link metric or path cost, on every edge, the Dijkstra's algorithm assigns initial distance values which will be then updated as in the following steps:

- *Step 1*: Assign a provisional distance value to every node: zero for the initial node and infinity (∞) for all other nodes.
- *Step 2*: Mark all nodes as unvisited and set the initial node as the current node. The unvisited set consists of all the unvisited nodes.
- *Step 3*: For the current node, consider all of its unvisited neighbors and calculate the tentative distances from this current node to the neighboring nodes. If this new distance is less than the previously assigned distance, then overwrite with the new one; otherwise, keep the old one.
- *Step 4*: After considering all the neighboring nodes of the current node, mark the current node as visited and remove it from the unvisited set. This visited node will not be checked ever again and its recorded distance is now final and should be minimal.
- *Step 5*: Set the unvisited node with the smallest distance (from the initial node) as the next "current node" and continue from Step 3 until the destination node is marked as visited.

Example 2.2. *Find the shortest path between two nodes A and J in Figure 2.5.*
The shortest path from A to J can be determined following Dijkstra's algorithm as in Table 2.1 where the bold texts in each column for each node are used to trace back the shortest route from the initial node to that node and the bold values are the corresponding distance value. In this example, the shortest route from A to J can be traced back as J <- I <- G <- F <- D <- C <- A having the shortest distance of 28.

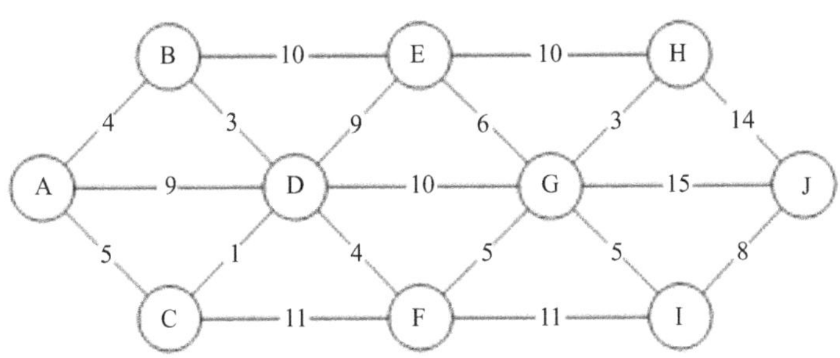

Figure 2.5 A graph in Example 2.2 where the value on the edge represents weight or path cost

Table 2.1 Dijkstra's algorithm for finding the shortest distance in Example 2.2

A	B	C	D	E	F	G	H	I	J
(0)	(∞)	(∞)	(∞)	(∞)	(∞)	(∞)	(∞)	(∞)	(∞)
	$(4,A)$	$(5,A)$	$(9,A)$	(∞)	(∞)	(∞)	(∞)	(∞)	(∞)
	(4,A)	**(5,A)**	$(6,C)$	$(14,B)$	(∞)	(∞)	(∞)	(∞)	(∞)
			(6,C)	$(14,B)$	$(10,D)$	$(16,D)$	(∞)	(∞)	(∞)
				(14,B)	**(10,D)**	$(15,F)$	(∞)	$(21,F)$	(∞)
						(15,F)	$(18,G)$	$(20,G)$	$(30,G)$
							(18,G)	**(20,G)**	**(28,I)**

2.4.3 Enhanced Interior Gateway Routing Protocol

EIGRP is an advanced version of IGRP (cf. Section 2.4.1.2). The EIGRP was also designed by Cisco and was officially described in RFC 7868 in May 2016 after nearly three decades from the development of the predecessor IGRP.

The EIGRP was evolved from the demands of the large-scale internetworks to support classless IP addresses with variable-length subnetting. Although Cisco classifies the EIGRP as a DVRP (cf. Section 2.4.1), it functions as a balanced hybrid routing protocol integrating the capabilities of both the DVRP and LSRP (cf. Section 2.4.2).

Furthermore, the EIGRP employs DUAL to improve its operational efficiency in finding the best path as well as alternative paths with only loop-free routes. The DUAL was first developed in 1998 by J.J. Garcia-Luna-Aceves at SRI International to improve the Dijkstra algorithm (cf. Section 2.4.2.2). Similar to the IGRP, the EIGRP determines the path using a composite metric depending on bandwidth, load, delay, reliability, and MTU (cf. (2.1)). Due to this fact, the EIGRP was considered as a DVRP. However, in addition to the routing table, the EIGRP uses neighbor table and topology table via DUAL, which allow the routers only send routing table data when there is an actual change in the internetwork rather than sending the routing table periodically. With this functionality, the EIGRP also operates as an LSRP, and thus it is generally viewed as a hybrid routing protocol.

2.5 Border Gateway Protocols

BGP is a routing protocol employed as an EGP to exchange routing information between ASs, for instance over the Internet with different Internet service providers (ISPs). After a number of drafts, the current version of the BGP is version 4 (BGP-4) which was described in RFC 4271 published in 2006.

The BGP is used to advertise network addresses between ISPs and ASs. Generally, a customer network employ an IGP, e.g., RIP, OSPF, and IGRP or EIGRP, for exchanging routing information between routers within the network. These customer networks are connected to the ISPs which then use the BGP to exchange routing information between them.

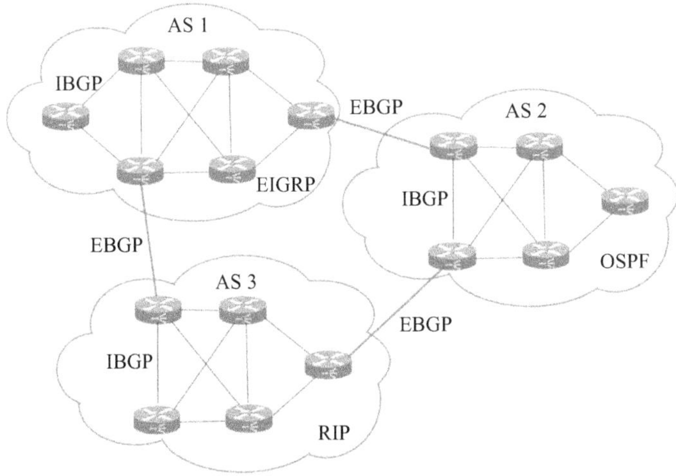

Figure 2.6 Border Gateway Protocols (BGPs) with Internal BGP (IBGP) and External BGP (EBGP)

Although BGP is normally applied between different ASs and ISPs on the Internet, it can also be used for routing within an AS to coordinate and synchronize routing policy of routers in the AS. The BGP therefore supports both session types which are referred to as EBGP and IBGP (cf. Figure 2.6).

The BGP is based on paths and network policies configured by network administrator to make core routing decision and thus functions as a DVRP. In order to propagate and manage the routes, there is a mechanism for route mapping which includes a set of rules describing different actions. However, considering a full mesh configuration of all routers within an AS employing the IBGP, each router is required to connect to all other routers, which causes a quadratic grow of network connections between the routers. Therefore, as a means to alleviate this scaling problem, the BGP makes use of a variety of route parameters, namely, attributes, to define different route policies for route selection process, as well as implementing route reflection (cf. RFC 4456) and AS confederation (cf. RFC 5065). These additional features enable the BGP to meet the scalability as well as the continuing changes of the Internet infrastructure.

2.6 Related works

For more information, the readers are suggested to refer to the following selected references:

Routing protocols Sportack and Fairweather in 1999 [26], Doyle and Carroll in 2005 [27], and Graziani and Allan in 2007 [28].

DVRP Ford in 1956 [29], Ford and Fulkerson in 1956 [30], and Bellman in 1958 [31].

RIP Hedrick in 1988 [32], Malkin and Minnear in 1997 [33], and Malkin in 1998 [34].
IGRP Rutgers in 1991 [35] and technotes of Cisco in 2005 [36].
LSRP Clausen and Jacquet in 2003 [37].
OSPF Moy in 1998 [38,39] and Coltun *et al.* in 2008 [40].
Dijkstra's algorithm Dijkstra in 1959 [41].
EIGRP Garcia-Lunes-Aceves in 1993 [42], Pepelnjak in 1999 [43], and Savage *et al.* in 2016 [44].
BGP Rekhter *et al.* in 1995 and 2006 [45,46], Halab and McPhersoni in 2000 [47], White *et al.* in 2004 [48], Bates *et al.* in 2006 [49], and Traina *et al.* in 2007 [50].

2.7 Review questions

[Q2.1] Describe the routing process of packets in the internetwork.
[Q2.2] Differentiate between routing algorithms.
[Q2.3] Explain the metrics used in routing protocols.
[Q2.4] What are two basic groups of IGPs? Describe the fundamental differences between them.
[Q2.5] In RIP, which metric is used and how to determine if the destination is unreachable?
[Q2.6] Explain the purpose of update timer in RIP.
[Q2.7] What is the limitation of the RIP?
[Q2.8] Describe three RIP versions and their applications.
[Q2.9] What is composite metric in IGRP? Explain how to determine and choose appropriate parameters for this metric subject to specific routing demands.
[Q2.10] Explain when the routing loops happen and approaches to prevent them.
[Q2.11] Compare LSRP with DVRP.
[Q2.12] Which algorithm is employed in OSPF to find the shortest path between nodes? Describe in detail its steps.
[Q2.13] Give a brief account of the reason why the EIGRP is referred to as a hybrid routing protocol.
[Q2.14] Differentiate between IBGP and EBGP.
[Q2.15] How to cope with the scaling problem in the BGP?

2.8 Problems

[P2.1] What is the maximum number of nodes in a RIP network?
[P2.2] Consider the bandwidth consumption of the RIP protocol. Assume a network consists of eight nodes and a node has three neighboring nodes.

- Estimate the number of messages exchanged per unit time.
- Estimate the size of the messages exchanged as a function of the size of the RIP network given a RIP message consists of a 4-byte header plus 20-byte per entry and up to 25 entries per message.
- Estimate the bandwidth consumption of a RIP network.

[P2.3] Assume that there are ten routers in the network and every router has four neighbors.

- Estimate the amount of memory required to store the information used by the distance-vector routing given the size of each entry that RIP requires is 10 bytes.
- Estimate the amount of memory required to store the information used by the link-state algorithm given each entry size is 10 bytes.

[P2.4] Find the shortest distance from node A to node J in the network in Figure 2.7 using Dijkstra's algorithm.

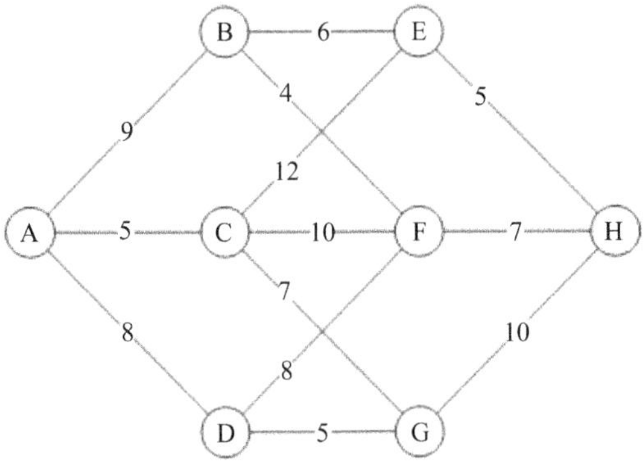

Figure 2.7 Figure for Problems 4 and 5

[P2.5] Find the shortest route from node A to other nodes in the network in Figure 2.7.

[P2.6] Using Dijkstra's algorithm, find which of the vertices A or B or C is the nearest to J in Figure 2.8.

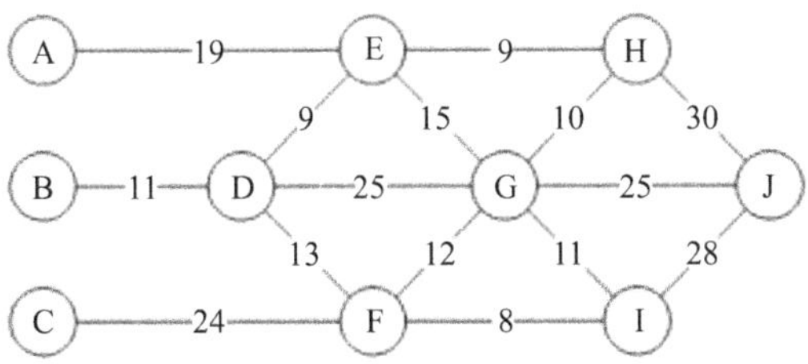

Figure 2.8 Figure for Problem 6

Chapter 3
Internet Protocol addressing

As discussed in Sections 1.5.1 and 1.5.2, Internet Protocol (IP) is a protocol at Internet layer of TCP/IP suite. In order to employ a variety of routing protocols to direct data packets over the internetwork, an address of every node in the network is of paramount importance. The IP addressing and routing are therefore the cornerstones of a computer network. Indeed, a packet includes not only the actual data to be conveyed but also a source IP address and a destination IP address along with some additional control information.

In this chapter, I will first present in Section 3.1 an overview of IP addresses with various types and addressing mechanisms. The first version of IP addresses, namely, IP version 4 (IPv4) will be discussed in Section 3.2 followed by various mechanisms for dividing the network with subnetting in Sections 3.3 and 3.4 as well as how to group the small networks via supernetting in Section 3.5. A brief introduction of the extended version of IP addressing called IP version 6 (IPv6) will be described in Section 3.6 along with discussions for further readings.

3.1 An overview of IP address

An IP address is a sequence of number which is assigned to each device in the internetwork. Basically, an IP address is used to identify the host, i.e., network interface and location, in the network. Globally managed by the Internet Assigned Numbers Authority, the IP addresses are distributed to regional Internet registries in different areas. These IP addresses are then assigned to every device connecting to the network by a network administrator. Depending on their applications in practice, such assignment of IP addresses can be either local or global, private or public, temporary or persistent addressing. Given the IP addresses of all devices in the network, routing protocols (cf. Chapter 2) can be implemented allowing intermediary devices to determine the best routes for the communications between hosts.

There are two versions of IP addresses including IPv4 and IPv6. The IPv4 was well known as the first version of the IP developed by the DARPA, while the latter is the latest version of the IP with a larger address space developed by the IETF.

An IP address basically consists of two main parts including network prefix and host number. In order to identify a device in a TCP/IP network, we need to know not only its IP address or host address but also the network address which is the prefix of an IP address determined through a mask with bitwise AND operation. The mask in computer networking is used as a bitmask to characterize either a network or a subnetwork (or subnet). The network/subnet address of a host is determined by

$$\text{Network/Subnet address} = \text{Host address } \textbf{AND (\textit{bitwise})} \text{ mask} \tag{3.1}$$

For simplicity in representation of the network/subnet address, a slash character, i.e., "/", followed by the number of bits for the network/subnet address can be used instead of writing the mask in full. This compact representation is also known as classless inter-domain routing (CIDR) notation.

Example 3.1. *A host has an IP address 158.92.1.1 with different masks as follows:*

- *Network mask 255.255.0.0: The network address is 158.92.0.0 with 16 first bits for network prefix and 16 remaining bits for host number. The IP address can be either written as 158.92.1.1 255.255.0.0 or simply 158.92.1.1/16 (cf. Figure 3.1).*

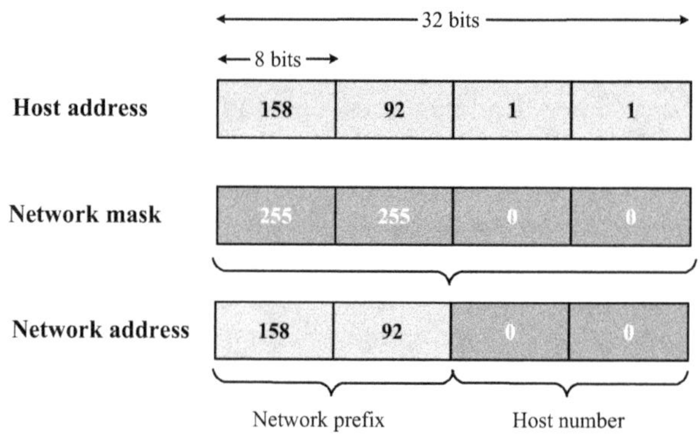

Figure 3.1 IP address with network mask in Example 3.1

- *Subnet mask 255.255.255.0: The subnet address is 158.92.1.0 with 16 first bits for network prefix, 8 bits for subnet number, and 8 last bits for host number. The IP address can be written as 158.92.1.1/24 (cf. Figure 3.2).*

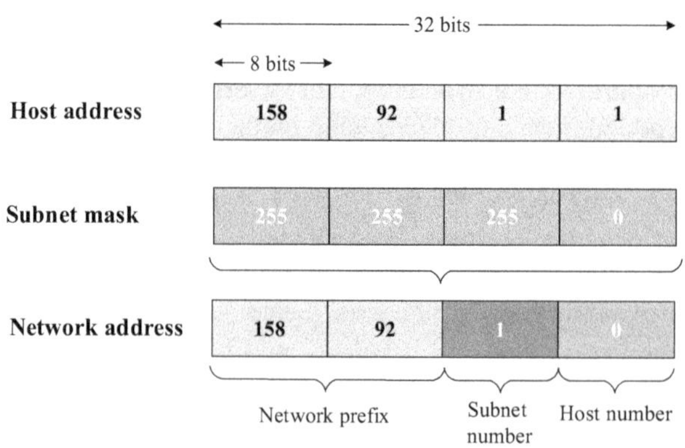

Figure 3.2 IP address with subnet mask in Example 3.1

3.1.1 Addressing types

Depending on their usage, there exist different address types, which can be categorized as follows:

- *Local addresses* versus *global addresses*: The local addresses are recognized locally within the network as private addresses, while the global addresses are used as public addresses which can be routed through the Internet.
- *Temporary addresses* versus *persistent addresses*: The temporary addresses are assigned to network devices for a short period of time with dynamic host configuration protocol, while the persistent addresses are allocated permanently to specified nodes in the network.

3.1.2 Addressing mechanisms

In order to manage the IP addressing of devices in a network, it is necessary to have different mechanisms to classify the IP addresses, to divide the network into different subnetworks and also to group the network addresses of different subnets. Specifically, the IP addresses can be managed using either one of the following mechanisms or a combination of these approaches:

- *Classful addressing*: The IP addresses are arranged in different classes, i.e., Class A, B, C, D, and E, with default network mask, namely, natural mask.
- *Subnetting*: An IP address hierarchy is created to divide a network into subnetworks sharing the same subnet mask.
- *Variable-length subnetting*: Different subnet masks are employed for subnets of different sizes.
- *Supernetting*: The IP addresses of different subnets in a network are aggregated prior to advertise to the Internet.

This chapter is focused on discussing the above-addressed mechanisms for IPv4 in the following sections. Their extension for IPv6 addressing can be deployed in a similar approach.

3.2 IP version 4 (IPv4)

An IPv4 address has a length of 32 bits which are represented in the form of four octets. Each octet has 8 bits,[1] and these four octets are separated by a dot and thus also known as dotted-quad format. For simplicity, an IP address is usually written in dot-decimal format consisting of four decimal numbers, each of which is in the range of 0–255, i.e., 00000000 to 11111111 in binary format, for example, 128.10.10.82 (cf. Figure 3.3).

The IPv4 addresses can be allocated either locally as private addresses or globally with classful network architecture.

[1]Note that the term "octet" is different from "byte" which is used as a storage unit of 8 bits.

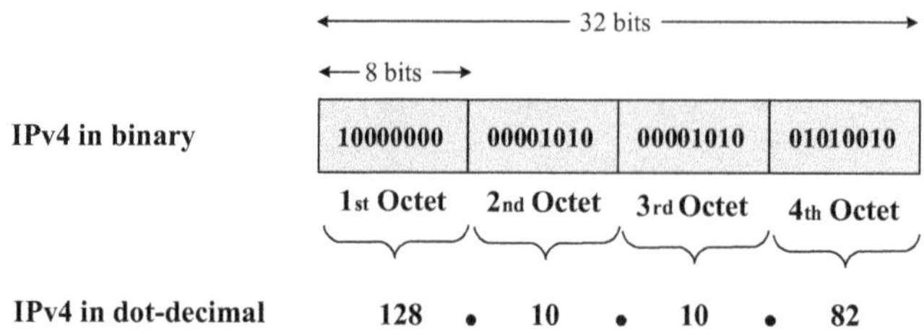

Figure 3.3 An IPv4 address

Table 3.1 Private IP address range

IP address range	Number of addresses	Mask	CIDR
10.0.0.0–10.255.255.255	2^{24}	255.0.0.0	/8
172.16.0.0–172.31.255.255	2^{20}	255.240.0.0	/12
192.168.0.0–192.168.255.255	2^{16}	255.255.0.0	/16

3.2.1 Private IP addresses

As an attempt to cope with the issue of address exhaustion, the IETF has reserved IPv4 address ranges for LANs within home and enterprise which are regarded as private networks. There are three private IP address ranges as listed in Table 3.1. These IP addresses are only valid inside the network and cannot be routed to the Internet. In order to connect to the Internet, an IP network address translator gateway is required to map between private and public address spaces.

3.2.2 Classful IP addresses

Initially proposed for the Internet since 1981, IP addresses have followed a classful network-addressing architecture. Given IP addresses of 32-bit length, the address spaces have been divided into five classes, namely, Classes A, B, C, D, and E (cf. Figure 3.4). Specifically, Class A, B, and C networks have different sizes, while Class D addresses are only for multicast and Class E addresses are reserved for future use. The classification of IP addresses is only based on the leading bits in the first octet. The bit patterns, address ranges, and natural masks of different IP classes are defined in Table 3.2 where N and H denote eight network bits and eight host bits, respectively.

Written in decimal format, the class of an IP address can be simply identified through its first number as follows:

- *Class A*: 0–127
- *Class B*: 128–191
- *Class C*: 192–223

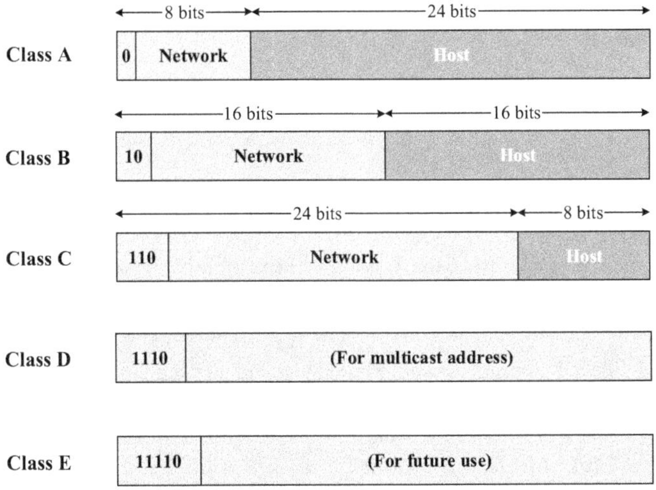

Figure 3.4 IP address classes

Table 3.2 IP address classes

Class	Leading bits	Bit pattern	Natural mask	CIDR	Number of networks	Number of hosts per network
A	0	*N.H.H.H*	255.0.0.0	/8	2^7-2	$2^{24}-2$
B	10	*N.N.H.H*	255.255.0.0	/16	2^{14}	$2^{16}-2$
C	110	*N.N.N.H*	255.255.255.0	/24	$2^{21}-2$	2^8-2
D	1110			Not defined		
E	1111			Not defined		

- *Class D*: 224–239
- *Class E*: 240–255

Note that two IP addresses are taken away from the number of addresses per network in Table 3.2 due to the usage of a network address with all-zeros host bits and a broadcast address with all-ones host bits. There are also two Class A network addresses 0.0.0.0/8 and 127.0.0.0/8 and two Class C network addresses 192.0.0.0/24 and 223.255.255.0/24 which are reserved and cannot be assigned to any network.

Let n_X, h_X, $N_{N,X}$, $N_{H,X}$, $X \in \{A, B, C\}$, denote the number of bits for Class X network, the number of host bits in Class X network, the number of Class X networks, and the number of hosts per Class X network, respectively. As shown in Table 3.2, $n_A = 8$ bits, $n_B = 16$ bits, and $n_C = 24$ bits. The length of an IPv4 address is 32 bits, so the number of host bits is $h_X = 32 - n_X$ [bits].

Since the leading bits are fixed and there are some reserved IP addresses, $N_{n,X}$ and $N_{h,X}$ can be computed as

$$N_{N,X} = \begin{cases} 2^{n_A-1} - 2, & \text{if } X = A, \\ 2^{n_B-2}, & \text{if } X = B, \\ 2^{n_C-3} - 2, & \text{if } X = C, \end{cases} \tag{3.2}$$

$$N_{H,X} = 2^{h_X} - 2 = 2^{32-n_X} - 2. \tag{3.3}$$

The total number of hosts in Class X network, denoted by $N_{T,X}$, is thus given by

$$N_{T,X} = N_{N,X} N_{H,X}. \tag{3.4}$$

Example 3.2. *Host IP addresses in different network classes:*

- *A host with an IP address 130.18.12.50/16 belongs to a Class B network having network address 130.18.0.0, broadcast address 130.18.255.255, and* $(2^{16} - 2)$ *IP addresses available for hosts in the range 130.18.0.1–130.18.255.254.*
- *A host with an IP address 200.168.120.5/24 belongs to a Class C network having network address 200.168.120.0, broadcast address 200.168.120.255, and 254 IP addresses available for hosts in the range 200.168.120.1–200.168.120. 254.*

3.3 Subnetting

In a small network with a limited number of colocated hosts, the usage of only a specific network address is possible to support all hosts sharing a common natural mask. However, within an enterprise or a campus where a network is divided into smaller networks or subnetworks, subnetting is crucial to allocate the IP addresses to hosts in different subnets.

For subnetting, some bits for host identifier are used to indicate subnet number (cf. Figure 3.5). The hosts in the same subnet share a common subnet address and

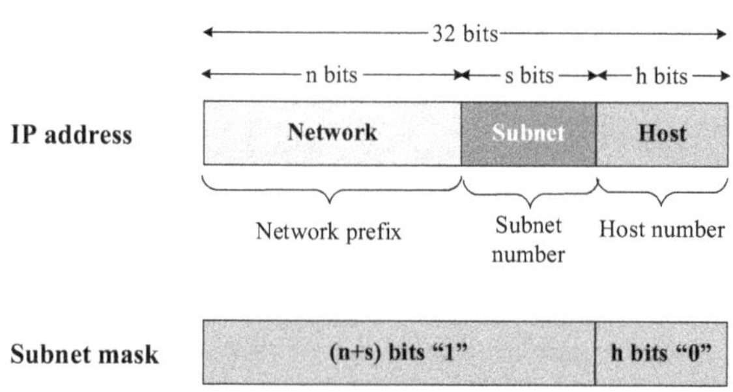

Figure 3.5 Subnetting

subnet mask. Let n, s, and h denote the number of bits for network, subnet, and host, respectively, i.e., $n + s + h = 32$ bits. The number of subnets per network, i.e., N_S and the number of hosts per subnet, i.e., N_{H_S}, are computed by

$$N_S = 2^s, \tag{3.5}$$

$$N_{H_S} = 2^h - 2 = 2^{32-n-s} - 2, \tag{3.6}$$

respectively. Notice that, in (3.6), two IP addresses are taken away for subnet address and broadcast address. The total number of hosts per network with subnetting, i.e., N_H, is thus given by

$$N_H = N_S N_{H_S} = 2^{s+h} - 2^{s+1}. \tag{3.7}$$

Remark 3.1. *It can be observed that the number of hosts with subnetting is less than those without subnetting due to the usage of additional IP addresses for subnet address and broadcast address in each subnet. Indeed, without subnetting, the total number of hosts per network is given as in* (3.3)*, i.e.,* $N_{H,no\ subnet} = 2^{s+h} - 2 > N_H$.

Example 3.3. *Given a network address 180.16.0.0/16, i.e.,* $n = 16$ *bits for network prefix. If we borrow 3 bits for subnetting, i.e.,* $s = 3$ *bits, we have a total of* $N_S = 2^3 = 8$ *subnets, each of which has* $N_{H_S} = 2^{13} - 2$ *hosts per subnet. All eight subnets have the same subnet mask 255.255.224.0 with CIDR /19. The subnet address, broadcast address, and host address range in each subnet can be assigned as in Table 3.3.*

Table 3.3 Subnetting in Example 3.3

Subnet	Subnet address	Host addresses	Broadcast address
1	180.16.0.0	180.16.0.1–180.16.31.254	180.16.31.255
2	180.16.32.0	180.16.32.1–180.16.63.254	180.16.63.255
3	180.16.64.0	180.16.64.1–180.16.95.254	180.16.95.255
4	180.16.96.0	180.16.96.1–180.16.127.254	180.16.127.255
5	180.16.128.0	180.16.128.1–180.16.159.254	180.16.159.255
6	180.16.160.0	180.16.160.1–180.16.191.254	180.16.191.255
7	180.16.192.0	180.16.192.1–180.16.223.254	180.16.223.255
8	180.16.224.0	180.16.224.1–180.16.255.254	180.16.255.255

Remark 3.2. *In Example 3.3, the third octet of the subnet addresses is multiples of 32, which is* 2^{8-s} *with* $s = 3$*. Note that this is only applied for* $n = 16$ *and* $s \leq 8$*. Extended for the scenario when* $8 < s \leq 16$*, the fourth octet of the subnet addresses would be multiples of* 2^{16-s}*, while the third octet is unchanged. A similar scenario can be observed when* $n = 24$ *and* $s \leq 8$*. In this case the fourth octet of the subnet addresses would be multiples of* 2^{8-s}.

A typical problem is to design a subnetting scheme for a network consisting of K subnets and n bits for network prefix. The kth subnet, $k = 1, 2, \ldots, K$, is required to have N_{H_k} hosts. The following steps can be employed:

Step 1 Identify the subnet with a maximum number of hosts, i.e.,

$$N_{H_S,\max} = \max\{N_{H_1}, N_{H_2}, \ldots, N_{H_K}\}.$$

Step 2 Find the minimum number of host bits that can support the maximum number of hosts using (3.6), i.e., solving

$$N_{H_S,\max} \leq 2^h - 2 \text{ or } h \geq \log_2\left(N_{H_S,\max} + 2\right) \tag{3.8}$$

The minimum number of host bits is thus $h_{\min} = \lceil \log_2\left(N_{H_S,\max} + 2\right)\rceil$, where $\lceil x \rceil$ denotes the least integer greater than or equal to x.

Step 3 Find the number of bits borrowed for subnetting, i.e., $s = 32 - n - h_{\min}$.

Step 4 Determine subnet mask and design subnets.

Example 3.4. *Given an IP address 198.16.10.0/24, design subnetting for laboratories with the required number of hosts in each lab as in Table 3.4.*

Table 3.4 Laboratories in Example 3.4

Lab	Users
A	60
B	30
C	25
D	20

In this example, Lab A has the maximum number of hosts with $N_{H_S,\max} = 60$. From (3.8), the minimum number of hosts bits is $h_{\min} = 6$ bits. The number of bits borrowed for subnetting is $s = 32 - 24 - 6 = 2$ bits and thus can support up to four subnets sharing the same subnet mask 255.255.255.192 with CIDR /26. Each subnet can have maximum $N_{H_S} = 62$ hosts satisfying the requirements of the number of users in all labs in Table 3.4. The subnet address, broadcast address, and host address range in each subnet can be assigned as in Table 3.5.

Table 3.5 Subnetting in Example 3.4

Lab	Subnet address	Host addresses	Broadcast address
A	198.16.10.0	198.16.10.1–198.16.10.62	198.16.10.63
B	198.16.10.64	198.16.10.65–198.16.10.126	198.16.10.127
C	198.16.10.128	198.16.10.129–198.16.10.190	198.16.10.191
D	198.16.10.192	198.16.10.193–198.16.10.254	198.16.10.255

3.4 Variable-length subnetting

The usage of fixed-length subnetting as in Section 3.3 can be applied to the network having subnets of the same or most likely similar number of hosts. For the networks having subnets of different sizes, e.g., Table 3.4, only one subnet mask taking into account the subnet with maximum number of hosts is not efficient. Such inefficiency can be easily noticed when subnets have considerably different sizes as in the following example.

Example 3.5. *A network has an IP address with a network prefix of* $n = 16$ *bits and subnet identifier of* $s = 6$ *bits (cf. Figure 3.6). The network can have* $N_S = 64$ *subnets, each of which can have* $N_{H_S} = 1022$ *hosts corresponding to* $h = 10$ *host bits. However, for the subnets with much less than 1,022 hosts, e.g., only 100 hosts, this is really a waste. In these small subnets, 7 bits are enough for the hosts, i.e.,* $h = 7$ *bits. In this case, the subnet identifier has a length of* $s = 9$ *bits, which can support up to 512 subnets.*

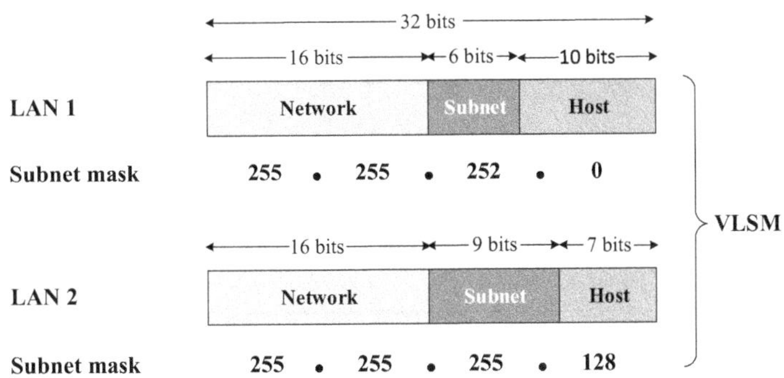

Figure 3.6 Variable-length subnet mask in Example 3.5

In order to deal with this practical issue, variable-length subnetting is crucial to establish an address hierarchy within a network having subnets of different sizes. A variable-length subnet mask (VSLM) should be used in which a longer mask is for subnets with fewer hosts and a shorter mask is for subnets with more hosts.

As can be observed in Example 3.5, VLSM also implies subnetting of subnets to reduce the address wastage. In other words, the VLSM is used to maximize the efficiency in addressing. The subnetting should be designed to guarantee that host addresses are not overlapped between subnets. Therefore, prior to design subnetting, it is suggested to sort the subnets in descending order of the number of hosts. The following steps can be employed:

Step 1 Sort subnets in descending order of the required number of hosts.

Step 2 Find the minimum number of host bits for each subnet that can support its required number of hosts using (3.6).

Step 3 Find the number of bits borrowed for subnetting in each subnet with respect to the number of host bits determined in Step 2.

Step 4 Determine subnet mask for each subnet, and design all subnets and host address range.

Example 3.6. *Given an IP address 199.10.10.0/24, design variable-length subnetting for an office consisting of four rooms with the required number of hosts in each room as in Table 3.6. Following the steps for the VLSM design, the subnetting for these four rooms can be sketched out as in Table 3.7 where for simplicity the network part in the host address range, i.e., 199.10.10, is eliminated. The subnet mask is also not shown, which can be deduced from the CIDR.*

Table 3.6 Rooms in Example 3.6

Room	Users
A	12
B	52
C	20
D	5

Table 3.7 Variable-length subnetting in Example 3.6

Room	Host bits	Subnet bits	CIDR	Subnet address	Host addresses	Broadcast address
B	6	2	/26	199.10.10.0	0.1–0.62	199.10.10.63
C	5	3	/27	199.10.10.64	0.65–0.94	199.10.10.95
A	4	4	/28	199.10.10.96	0.97–0.110	199.10.10.111
D	3	5	/29	199.10.10.112	0.113–0.118	199.10.10.119

3.5 Supernetting

While subnetting is employed to divide a network into subnetworks, supernetting is the process of combining subnetworks into a large network or supernetwork. Supernetting is viewed as network address aggregation. The requirement of supernetting is further illustrated in Example 3.7.

Example 3.7. *Assume that there are 12,000 computers in an enterprise to be networked. A Class B network can be used to support up to* $N_{H,B} = 2^{16} - 2 =$*65,534 hosts. However, this is a wastage of IP resources. Alternatively, 48 Class C network addresses can be employed for* $N_{H,C} = 48(2^8 - 2) =$*12,192 hosts. In this case, for routing, it is required to advertise 48 network addresses. Instead of advertising all of these addresses, address aggregation or route summarization can be carried out to form a supernet, and only the summarized address is advertised.*

In order to perform the address aggregation, the subnet ID should be aggregated with the following rules:

- The number of addresses in a CIDR block is always a power of two.
- There is no gap or hole in the address block sharing a common border.
- Two ends of the address block correspond to bit boundaries.

Example 3.8. *A network consists of four subnets with the following/24 IP addresses (cf. Figure 3.7):*

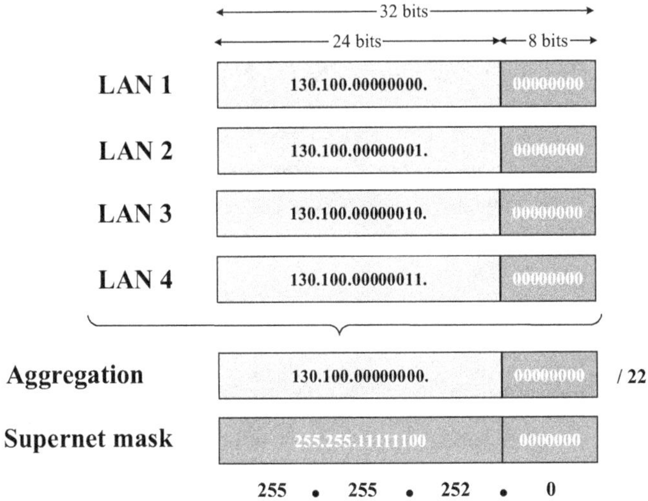

Figure 3.7 Supernetting in Example 3.8

- *130.100.0.0/24*
- *130.100.1.0/24*
- *130.100.2.0/24*
- *130.100.3.0/24*

Performing address aggregation, the above four IP addresses, can be grouped with CIDR /22, and only 130.100.0.0/22 is advertised to represent all subnets.

3.6 IP version 6

IPv6 is the latest version of the IP developed by the IETF in 1995. The IPv6 addresses are used as 128-bit identifiers for either interfaces or sets of interfaces. With such extensive address space, the IPv6 is aimed to replace the existing IPv4.

An IPv6 address consists of 128 bits, i.e., 16 octets, which are represented in eight groups of hexadecimal numbers. Each group has two octets, and these eight groups are separated by colons. For example, 0182:0A1C:001F:0000:0000:0014:0000:0003.

For convenience in representation, the IPv6 can be abbreviated in a shorter format by applying the following rules:

- Leading zeros in any group of 16 bits, i.e., two octets or four hexadecimal digits, can be omitted.
- Consecutive 16-bit zeros can be replaced by a double colon, i.e., "::," with a notice that only one double colon is used in the shortened form to allow the determination of the full address.

Example 3.9. *Given an IPv6 address written in full (cf. Figure 3.8) as*

$$0182 : 0A1C : 001F : 0000 : 0000 : 0014 : 0000 : 0003.$$

By applying the above rule, this address can be abbreviated by

$$182 : A1C : 1F :: 14 : 0 : 3.$$

The full address can be determined by filling the missing leading zeros and replacing the double colon with zeros satisfying there are totally eight groups of hexadecimal digits.

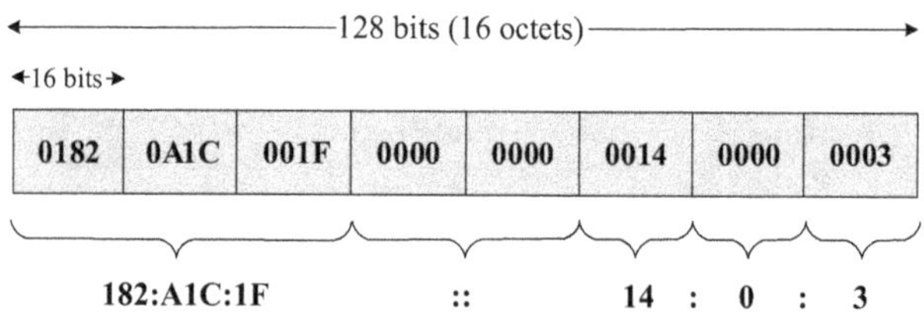

Figure 3.8 IPv6 address in Example 3.9

There are basically three IPv6 types including the following:

- *Unicast address*: An unicast address is used as an identifier for a single interface, and the packet is only sent to that interface.
- *Anycast address*: An anycast address is used as an identifier for a set of interfaces, and the packets will be sent to the nearest interface in this group.
- *Multicast address*: A multicast address is used as an identifier for a set of interfaces, and the packet is delivered to all interfaces.

An IPv6 packet typically has the format as in Table 3.8 where the header has a fixed size of 320 bits or 40 octets consisting of

- *Version* (4 bits): v6, i.e., "0110."
- *Traffic class* (8 bits): for differentiated services.
- *Flow label* (20 bits): for data types, e.g., speech, video, etc., with priorities.
- *Payload length* (16 bits): to identify the size of payload in bytes including possible extension headers. By default, payload length is 64 kB.
- *Next header* (8 bits): to identify the extension headers if present.

Table 3.8 IPv6 packet format

<table>
<tr><td colspan="2">Version (4 b)</td><td colspan="2">Traffic Class (8 b)</td><td colspan="2">Flow Label (20 b)</td></tr>
<tr><td colspan="3">Payload length (16 b)</td><td colspan="2">Next header (8 b)</td><td>Hop limit (8 b)</td></tr>
<tr><td colspan="6">Source address (SA) (128 b)</td></tr>
<tr><td colspan="6">Destination address (DA) (128 b)</td></tr>
<tr><td colspan="6">Possible extension header</td></tr>
<tr><td colspan="6">Payload</td></tr>
</table>

- *Hop limit* (8 bits): is set as a counter which is decremented by one at each intermediary device, e.g., router/gateway and the packet is dropped when this counter reaches zero.
- *Source address (SA)/destination address (DA)* (128 bits): is IPv6 address of the source/destination node.

There are also *extension headers* which have the size of multiple of 8 bits to carry additional information about the Internet layer; for instance, some have been defined including hop-by-hop header, end-to-end header, routing header, fragment header, authentication header, and privacy header.

3.7 Related works

For more information, the readers are suggested to refer to the following selected references:

IP Addressing Postel in 1981 [51], Mockapetris in 1987 [52], Braden in 1989 [53], Egevang and Francis in 1994 [54], Rekhter *et al.* in 1996 [55], Graziani and Johnson in 2007 [28], Dye *et al.* in 2011 [3], and Cotton *et al.* in 2013 [56].

IPv4 Perkins in 1996 [57] and in 2010 [58], and Touch in 2013 [59].

Subnetting Mogul in 1984 [60] and Tsuchiya in 1991 [61].

VLSM Pummill and Manning in 1995 [62].

CIDR Rekhter and Li in 1993 [63], Rekhter in 1995 [64], and Fuller and Li in 2006 [65].

IPv6 Hinden and Haberman in 2005 [66], Hinden and Deering in 2006 [67], Boucadair and Venaas in 2014 [68], and Deering and Hinden in 2017 [69].

3.8 Review questions

[Q3.1] Describe two versions of IP addresses and differentiate between them.

[Q3.2] Explain how to determine the network address and reasoning of using network mask.

[Q3.3] List all private IP address ranges and give an example of their usage.

[Q3.4] How many IPv4 address classes? Show their bit pattern/format with corresponding natural mask.
[Q3.5] Describe the steps for subnetting a network, given the number of required hosts in each subnet.
[Q3.6] Why do we need to consider variable-length subnetting and in which scenarios?
[Q3.7] Compare fixed-length subnetting versus variable-length subnetting with illustrations.
[Q3.8] Explain why supernetting is necessary in an enterprise via an example.
[Q3.9] What are the rules in address aggregation?
[Q3.10] What is the length of IPv6 addresses and how to represent them in abbreviated format?
[Q3.11] How many IPv6 types? Describe and differentiate between them.
[Q3.12] Describe the format of an IPv6 packet.

3.9 Problems

[P3.1] A small organization has a Class C IP address for 6 networks each with 20 hosts. What is an appropriate subnet mask?
[P3.2] A university has 200 LANs with 100 hosts in each LAN. Design an appropriate CIDR addressing scheme for supernetting.
[P3.3] Perform CIDR aggregation on the following /24 IP addresses: 128.56.24.0/24, 128.56.25.0/24, 128.56.26.0/24, 128.56.27.0/24.
[P3.4] What is the efficiency of IPv6 packets that carry 10 ms of 64 kbps voice?
[P3.5] Abbreviate the following IPv6 addresses:

- 0000:0000:0F56:6281:0AB0:6DB7:BB27:7323
- 0000:0000:0000:AD63:73BA:0000:87AB:0392
- 2918:00AD:0000:0000:0000:0000:0123:000A

[P3.6] Write the following IPv5 address in full:

- 0:10::78A:6D
- ::13B:4C:0:8
- ::1

[P3.7] Fill in Table 3.9 with Class C address and Table 3.10 with Class B address

Table 3.9 Subnetting of Class C address

Number of subnetting bits	Number of subnets/networks	Number of hosts/subnet	Total number of hosts	Subnet mask/ natural mask
0				
1				
2				
3				
4				
5				
6				

Table 3.10 Subnetting of Class B address

Number of subnetting bits	Number of subnets/networks	Number of hosts/subnet	Total number of hosts	Subnet mask/ natural mask
0				
2				
4				
6				
8				
10				
12				
14				

[P3.8] Find the network address, broadcast address, and host address range given a host address:

- 192.168.180.94/27
- 172.23.10.19 /21
- 147.252.238.20 255.255.240.0

[P3.9] Divide a network having an address 136.178.0.0/16 into 16 subnets. Show the binary and dotted-decimal forms of each subnet, as well as the subnet mask.

[P3.10] Given an IP address 196.168.1.0. Develop a subnetting scheme for the network in Figure 3.9.

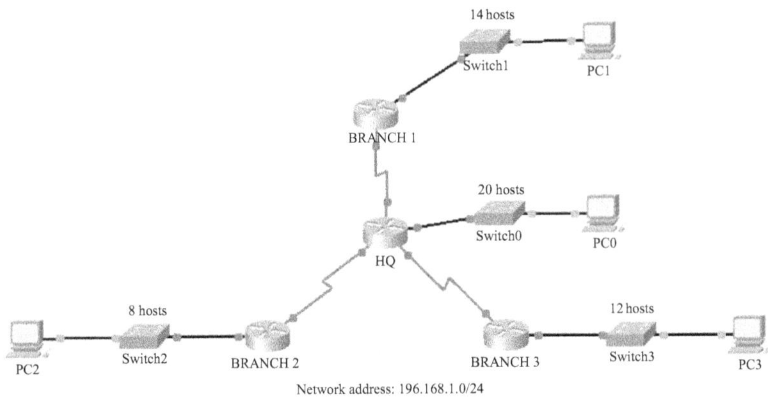

Figure 3.9 Figure for Problem 10

[P3.11] Given an IP address 129.130.0.0/16. Develop a variable-length subnetting scheme for the network in Figure 3.10.

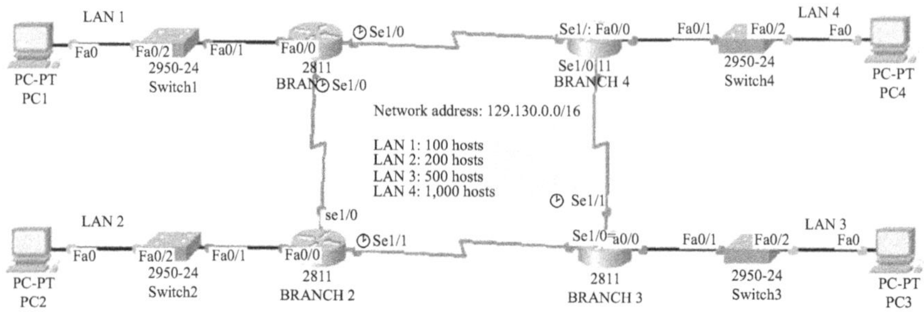

Figure 3.10 Figure for Problem 11

[P3.12] Given an IP address 150.100.0.0/16, the workgroups within an organization in Table 3.11 are required to be networked. Each workgroup is connected to a router having two Ethernet LANs and two serial interfaces. Design a variable-length subnetting scheme to meet the requirements of the organization.

Table 3.11 List of workgroups in Problem 12

Department	Number of workgroups	Number of hosts per workgroup
Administration	1	400
Finance	1	40
Sale	1	150
Manufacturing	15	60
Training	2	20

[P3.13] An organization has 1,200 hosts required to be networked with only Class C addresses, and the organization wants to advertise a single address. The address range available is 194.20.1.0–194.20.100.0.

- Discuss the criteria to select an appropriate block of addresses for this design.
- Design an addressing scheme to meet the network requirements.

[P3.14] A university has three sites, each of which has different departments with different requirements as shown in Table 3.12. The university has been allocated an IP address 132.128.0.0/16. Develop an IP addressing scheme to meet the design requirements with the most efficient use of address space.

Table 3.12 List of departments in Problem 14

Department	Site	Number of subnets	Number of hosts per subnet
Administration	A, B, C	2 per site	200
Finance	A	1	30
Students	A, B, C	20 per site	25
Academic	A, B, C	7 per site	150–200
Services	A	30	500
Headquarters	A	1	30

Chapter 4
Network analysis, architecture, and design

The first three chapters have reviewed basic concepts of internetworking and network fundamentals with different network models, routing protocols, and IP-addressing mechanisms. Given these background knowledge and existing technologies, it is absolutely necessary to understand how to model and design a network as well as validate its performance in different settings or configurations. The characteristics of systems and services also need to investigate with proper use of terminologies in computer-networking context.

In this chapter, I will first introduce in Section 4.1 a brief overview of various approaches and processes for network analysis, architecture, and design. Details of the network analysis, architecture, and design processes will be sequentially presented in Sections 4.2–4.4. As a typical network-design model, the hierarchical design approach will be discussed in detail in Section 4.5 along with its layers in Section 4.6. Various network-design approaches and their evolution will be summarized in Section 4.7. Finally, Section 4.8 will provide further discussion on network management and security.

4.1 An overview of network analysis, architecture, and design

Bandwidth has been well regarded as a key-performance metric in every network design. A network designer at all times aims at achieving a high bandwidth. The bandwidth is indeed of great significance, but it is not the only one component of network resources. There are other metrics that need to be considered in a practical network design such as latency, reliability, maintainability, and availability. Due to the heterogeneity of the practical networks, there exist a number of constraints and limitations, for instance, hardware configuration, network topology, transmission media, capital expenditures (CAPEX), and operating expenses (OPEX).

Prior to performing network architecture modeling and design in a real system, an initial analysis is crucial for drafting a preliminary network model. This analysis process basically consists of requirement analysis, flow analysis, and/or risk analysis. Based on the analyses, the relationships between functions within the network can be set and modeled via a descriptive network architecture. Such network modeling is helpful in determining an appropriate technology, equipment, and connectivity for network-design process as well as performance evaluation.

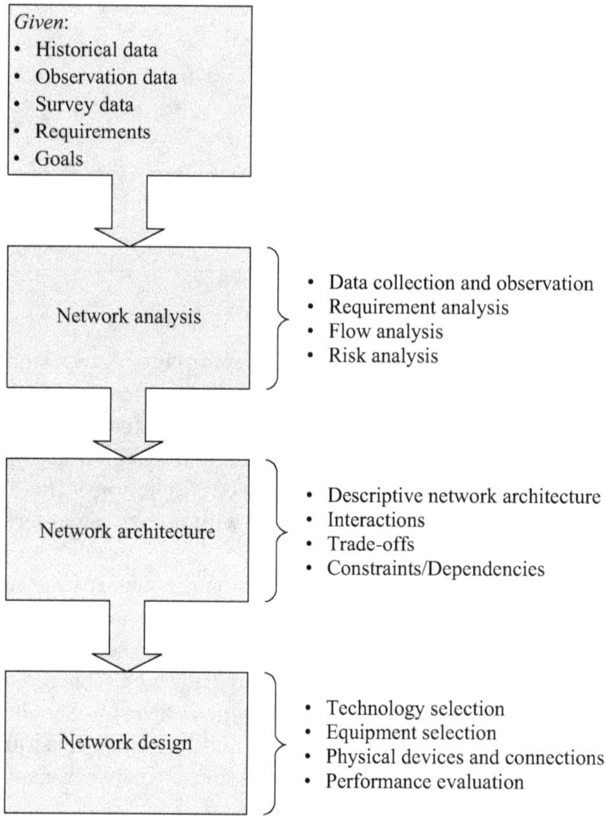

Figure 4.1 Network analysis, architecture, and design process

The information flows in network analysis, architecture, and design should go through the following steps (cf. Figure 4.1):

- Defining problem(s) to be addressed;
- Establishing and managing customer expectations;
- Monitoring the existing network(s), system(s), and environment;
- Analyzing the collected data;
- Developing a set of options to solve the developed problem(s);
- Optimizing options based on various trade-offs;
- Selecting one or more options;
- Planning the implementation;
- Designing network; and
- Evaluating network performance.

In the following sections, I will briefly introduce the key processes in accordance with those described above.

4.2 Network analysis process

In order to perform any network analysis, a network designer needs to have a background knowledge on networking and prerequisite skills. Specifically, it is required to know network components from different network devices, such as switches, hubs, and routers, to requirements and performance metrics for validation. Therefore, it is imperative to study characteristics of network components and their behaviors under various situations of network settings.

In the network-analysis process, the data fed in should consist of as much information as possible. These information can be collected via paperwork by asking the customer to fill in a survey form or through the observation and validation of the network designer. This may consist of the following:

- State of the current network (if existing),
- Historical problems of the network (if any),
- Network goals from the customer, and
- Requirements from users, applications, devices, and network in general.

After collecting and analyzing all the above facts, as the output of the network analysis process, the following information can be drawn:

- Descriptions for the requirements of the network in either a sorted list or a tabular format,
- Demands of the traffic flows in different networks/subnetworks,
- Mappings of applications and devices to the network under design or development.

4.3 Network architecture process

Network architecture is a process which uses information from network analysis process to develop a high-level and end-to-end structure for the network. Various network functions are built during the network-architecture process, which consist of network addressing, routing protocols, network management, performance evaluation, and security control. The network architecture process hence plays a decisive role as a critical requirement in every network design.

The network architecture process aims at developing a reference architecture for the real network design. Various simulation software and tools, such as Packet Tracer, Wireshark, NS2/NS3, etc., can be used to model the network architecture, validate the network performance, and also to optimize the network settings subject to various constraints. A well-developed reference network architecture should provide the following:

- A description of the complete network considering all of its architectural network components,
- A compilation of the relationship between network functions, and
- A description of their interactions, trade-offs, dependencies, and constraints employed to arrive at the developed architecture.

4.4 Network design process

Network design is a process to provide physical details for the developed network architecture. The network design process basically consists of the following tasks:

- Evaluate and choose appropriate technologies for each area/part of the network under design.
- Develop strategies to connect these technologies across the network.
- Learn how to set design goals, such as minimizing CAPEX/OPEX or maximizing network performance, as well as how to achieve these goals.
- Apply the trade-offs, dependencies, and constraints developed as part of the network-architecture process in solving the optimization problems.
- Develop evaluation criteria for the proposed design according to various performance metrics.

Two typical trade-offs in the network design are generally discussed for every network design, which are as follows:

- *Cost versus performance*: It is expected to have a high-performance network with high bandwidth, low latency, and high reliability. However, such enhanced performance normally requires a higher cost in both CAPEX and OPEX. This is indeed a challenging problem that every network designer has to deal with to balance the cost and network performance when considering various network constraints as well as requirements and affordability of customers.
- *Simplicity versus functionality*: Taking into account the practical network deployment, the simplicity, and adaptability of a network design should be substantial in facilitating the network-design process, while providing diverse functionalities as per requests. An equilibrium between the network simplicity and network functionality should be determined or optimized subject to the proposed constraints and/or dependencies of the network components and functions.

For ensuring the network connectivity, hierarchy and interconnectivity models are normally deployed across the network components (cf. Figure 4.2). The network hierarchy design depends on the degree of concentration of networks or traffic flows at interconnection points as well as the number of tiers or layers of interconnection points within the network. The network hierarchies provide a separation between tiers of points for a structured network when the network grows in size with a large number of devices, applications, and services. The hierarchical design is also helpful in determining the size of the network, addressing mechanism, scaling of network technologies, and service levels within the network. In addition to the hierarchy, interconnectivity is deployed to improve the network reliability with redundancy playing the role as backup links.

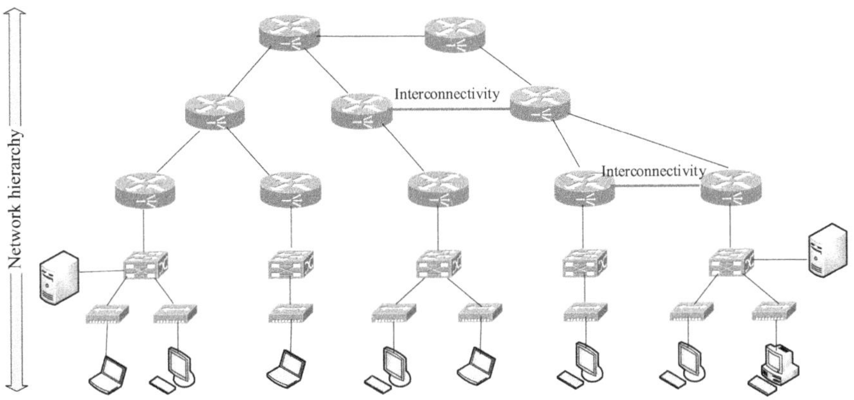

Figure 4.2 Network hierarchical design with interconnectivity

4.5 Network hierarchical design

As described in the previous section, hierarchical design has been employed in most of the networks. Some benefits of a hierarchical design model can be listed as follows:

- *Scalability*: The hierarchy allows the network to grow but still be able to control and manage. All functionalities are localized, and potentials problems can be easily identified to resolve. PSTN is a typical example of a very large-scale hierarchical network.
- *Implementation*: With a hierarchical design, a network designer can assign clear functionality to each layer, which helps facilitate network implementation.
- *Troubleshooting*: Since the functions of each layer are well defined, the problems in the network can be easily isolated for fault detection and correction. The network can be even temporarily segmented to reduce the scope and spreading of the problems.
- *Predictability*: With the functional layers, the behavior of a hierarchical network is predictable. In a sense, this is helpful for capacity planning to deal with network growth in future. Such design approach also facilitates network modeling with anticipated performance.
- *Compatibility*: The well-organized underlying infrastructure with hierarchical design enables the adaptation of diverse applications and protocols.
- *Manageability*: The above-listed benefits of the hierarchical design contribute to a great manageability of the network. As defined in the ITU standard for network management, five domains of FCAPS can be maintained, including **F**ault Management, **C**onfiguration Management, **A**ccounting Management, **P**erformance Management, and **S**ecurity Management.

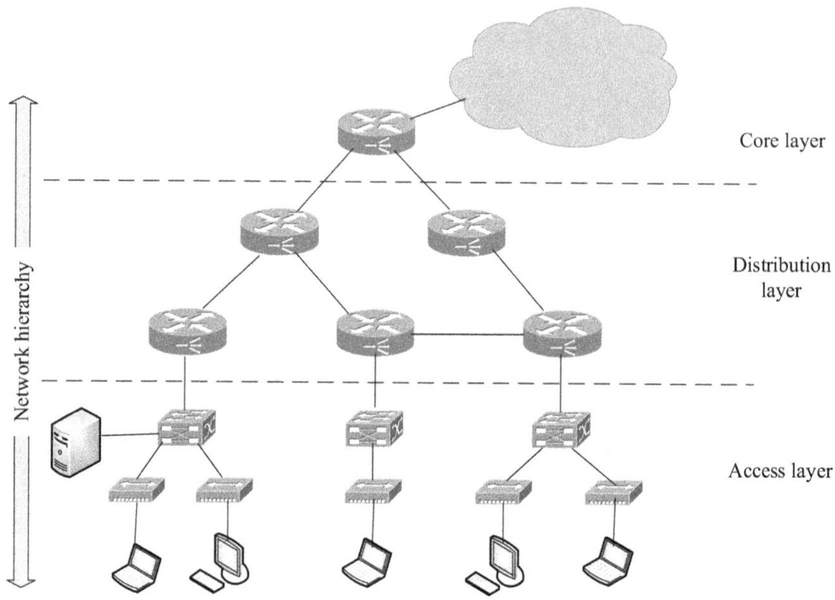

Figure 4.3 Network hierarchical layers

4.6 Network hierarchical layers

Basically, a network hierarchical design consists of core layer, distribution layer, and access layer (cf. Figure 4.3). In the following sections, these three layers will be described in detail.

4.6.1 Core layer

Core layer is at the top of the hierarchical design model. The core layer provides fast wide-area connections between geographically remote sites, for instance, between CANs within an organization or a large area connecting WANs (cf. Sections 1.1 and 1.3). In the core layer, there are rarely any hosts, and the links, i.e., core links, are usually point-to-point links.

The services provided at the core layer are typically leased from telecom service providers, e.g., Frame Relay (cf. Section 1.3.3.6). The major function at this layer is to transport data efficiently between remote sites by providing redundant paths and load sharing with rapid convergence and high bandwidth efficiency.

4.6.2 Distribution layer

Distribution layer is referred to as the layer-distributing network services to multiple LANs within a campus or an enterprise network environment (cf. Section 1.2). The campus backbone is deployed at this layer which is typically based on Fast Ethernet.

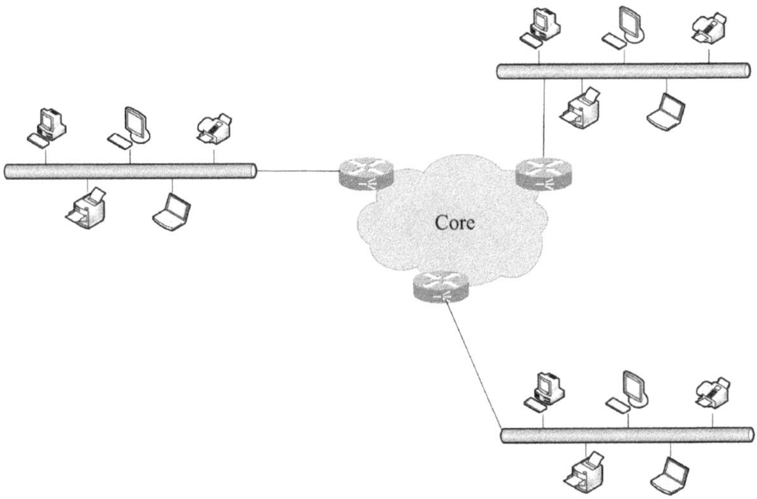

Figure 4.4 One-layer design

The distribution layer is generally implemented on large sites which require to interconnect buildings. As an intermediary to the top core layer, this layer mainly supports policy-based connectivity that allows to control access to services, path metrics, and also advertisements throughout the network.

4.6.3 Access layer

Access layer generally indicates a LAN or a group of LANs employing either Ethernet or Token Ring to enable user access to network services. The access layer consists of servers and workstations of all kinds.

The hosts, i.e., user workstations, within a workgroup of common interest can be grouped and connected to the backbone of the network at distribution layer. The access layer not only supports logical segmentation but also allows broadcast traffic between the workgroups.

A practical network may have different number of layers and may not have all the above three layers. Examples of some hierarchical designs are shown in Figures 4.4 and 4.5 for one-layer and two-layer designs, respectively.

Although the hierarchical design is shown to have a number of advantages (cf. Section 4.5), the flows are forced to go through hierarchy, which may affect the performance. When the connection between the access layer and distribution layer is broken or any intermediary nodes are faulty or malfunctioning, the communications between workgroups in different LANs or WANs are also lost. Hence, it is required to provide redundancy with interconnectivity between layers. Furthermore, content delivery network can be considered to bypass the network hierarchy for enhanced performance (cf. Figure 4.6).

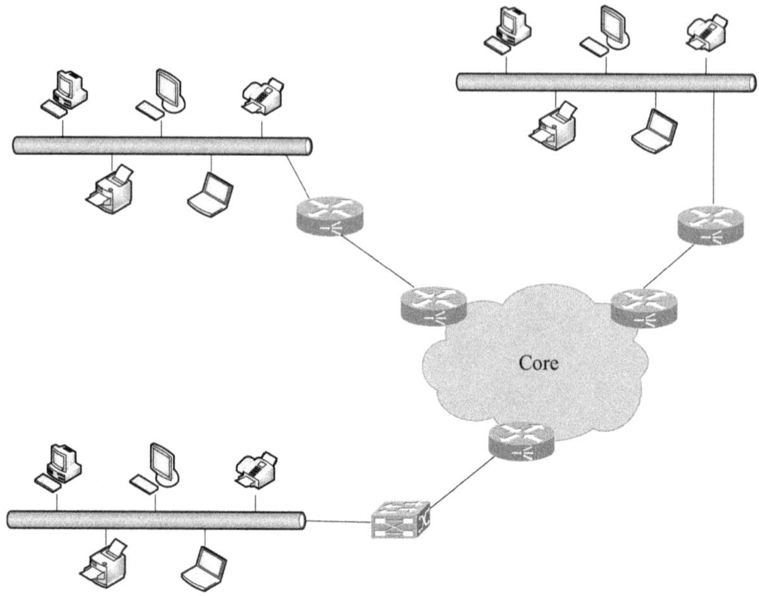

Figure 4.5 Two-layer design

4.7 Network design approaches and evolution

In summary, a systematic approach for network design consists of the following main steps:

- *Step 1*: Define problem(s)
- *Step 2*: Analyze customer's requirements
- *Step 3*: Analyze network environment
- *Step 4*: Analyze current systems/networks (if any)
- *Step 5*: Propose solutions for the developed problem(s)
- *Step 6*: Perform theoretical modeling and analysis (optional)
- *Step 7*: Model the network with simulation software/tools
- *Step 8*: Determine the optimal solution
- *Step 9*: Plan the implementation
- *Step 10*: Design the practical network or test bed
- *Step 11*: Evaluate the network performance.

In the above steps for network design and analysis, network services and service-based networking should be taken into account along with various performance metrics, such as throughput, latency, and reliability. Parameters in network configuration, such as bandwidth, flow allocation, networking, and physical design, should also be evaluated to realize their impacts on the network performance, and hence characterizing services with concluding remarks.

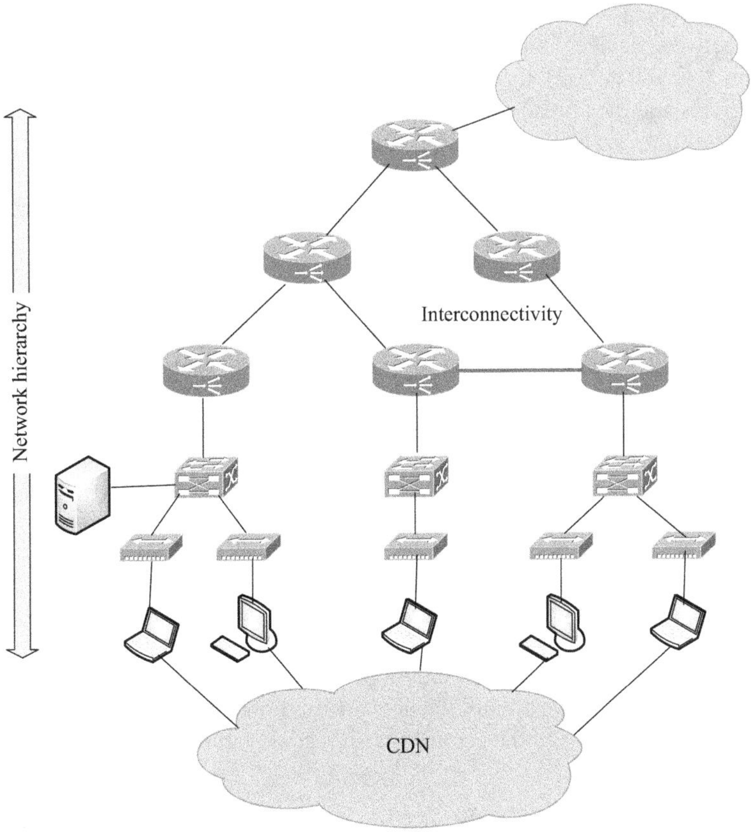

Figure 4.6 Network hierarchy over content-delivery network (CDN)

Network design is in fact an "art" in which a network designer is required to have a good knowledge on technologies, services, and protocols. Given these background along with working experience, appropriate network technologies can be selected to meet the practical network-design problem. There are, however, too many options with a number of available technologies, interconnection strategies, routing protocols, and topologies. The design problem can be viewed as an optimization problem where different performance metrics are required to be either individually or jointly optimized subject to various constraints in practical networks as well as QoS requirements from the customers.

4.8 Network management and security

The network design steps in Section 4.7 can only be realized with a good network management. Such a well-managed network however exists together with attackers and eavesdroppers who are trying to overhear the data or even harm the communications between devices across the networks. The security is therefore required to be maintained in any network together with the network management to defend against cyber threats and data breaches. The network security is a broad field consisting of a number of policies and mechanisms in the literature from various perspectives at different layers in the OSI reference model (cf. Section 1.4). The study of the network security is only briefly introduced in this book. For a thorough understanding, the readers are suggested to refer to the relevant works provided in Section 4.9.

A network designer should build a set of diagrams which can be used to prepare the environment for the network design. These diagrams may include network design goals and criteria for technology evaluation which will be discussed in Sections 4.8.1 and 4.8.2, respectively. Such schematic and illustrative representations are really helpful for the network implementation with equipment settings, cable deployment, etc. and also facilitate the network management afterwards. In particular, when the design is based on an existing network, a network transition plan should be carefully developed taking into account other factors, such as compatibility, adaptability, and extendability.

4.8.1 Network design goals

In a typical network, design goals should be established subject to requirements of the customers and specification of the network environment built or collected in the above initial steps. Some of the common design goals can be listed as follows:

- Network performance should be optimized over various metrics, such as maximize network throughput, minimize latency, maximize reliability, minimize energy consumption, minimize outage, maximize coverage, etc.
- Network deployment and/or operation should be at the lowest cost, i.e., minimize CAPEX and/or OPEX.
- Management and maintenance should be as simplest as possible in a flexible manner, i.e., minimize complexity in terms of computation, algorithm, mechanism, policies, etc.
- Network security should be borne in mind at the highest priority, i.e., minimize threats/attacks, minimize secrecy outage, optimize authentication, etc.
- Network should be adaptable to not only current user needs but also possible extension in future, i.e., maximize network availability in the long term with backup and extension plans.

Depending on the critical requirements in different networks, the above design goals should be prioritized as in a wish list where we set primary goal(s), secondary goal(s), etc. In order to satisfy these goals, the network designer is also required to consider various constraints in the network. These constraints could be limited resources or facilities, and cost in terms of both CAPEX and OPEX.

4.8.2 Technology evaluation criteria

The next step in the network-management process is the technology management to meet the design goals. This technology management is based on the criteria used to evaluate the technology selected for the network design. Given a number of existing technologies nowadays, the technology evaluation is a challenging process where its starting point is the list of design goals. A combination of various technologies is normally considered in the practical network design. In particular, their adaptability is crucial in the network management to make sure that they can support dynamic behavior in the network, such as reconfiguration of user groups, addresses, routing, and location.

The technologies are generally characterized through their capabilities with functions provided in the network, such as network throughput, data flow, broadcast, reliability, backup, security, etc. Specifically, both wired and wireless media in the network with either connection-oriented or connectionless communication protocols need to be managed properly guaranteeing that the capacity and service plans are in a good progress.

As per required in any network design, the functions and features of technology should be adaptable to users and applications with strict performance requirements. The lack of communication infrastructure due to limited resources and costs should also be managed in a flexible way with appropriate technologies and communication protocols. For instance, some questions that could be raised are as follows: (i) How to manage the multiple access in a crowded network with the usage of wireless techniques and protocols? and (ii) How to optimize the data flows in the network which is normally covered with asymmetric flows?

Considering the wireless users, their mobility together with their resources is required to be investigated and managed in the network plan. The requirements of multiple data rates for different services are also vital in these days in the selection of the communication medium type with respect to different operating distances. Moreover, the scalability of the network depends on the network connectivity in which hierarchy and interconnectivity models are normally deployed for easy handling of the data flows. Various medium types and technologies employed in a LAN can be referred to in Section 1.2.3 in Chapter 1.

4.8.3 Network-management design

Network management is crucial in any networks from small to large sizes. The network management is not simply polling simple network management protocol (SNMP) information across the network, but it also requires monitoring, storage, and evaluation of different metrics, such as the number of bytes or IP packets in/out of an interface and the number of errors at an interface.

This process should be logically designed to manage the whole network, and thus it is generally integrated into the network design. As described in the previous Sections 4.8.1 and 4.8.2, the network design goals and criteria for technology evaluation should be taken into account to define and characterize the management for a network design. Then, based on this, we can plan for monitoring and configuring the network, as well as for troubleshooting in the case of network faults. These network management protocols require to be examined regularly to update the instrumentation requirements.

In summary, the network management involves the following tasks:

- *Event notification via monitoring*: The event could be a problem or a failure of any components in the network, e.g., hosts, switches, hubs, routers, cables, NICs, etc. A threshold is set based on the characteristics of the individual element, which can be measured either between elements or across the entire network. The characteristics may include capacity, delay, availability, reliability, utilization, etc. The monitoring allows us to notify the status of the events in the network for real-time analysis.
- *Performance evaluation via monitoring*: The monitoring process is not only for examining events but also for evaluating the performance based on the collected data. In order to analyze the events in the real time, it is required to have short polling intervals which can only be supported in the network with high-speed CPU, memory, and sufficient storage capacity. The performance metrics are validated with reference to the preset baselines for every component within the network. In the performance evaluation, link and element characteristics should be specific to the type of the element being managed. For the sake of visualization, the processed data, either real-time or time-averaging data, can be displayed, and the old data is also required to archive for backup (if necessary).
- *Network configuration*: For network configuration, a characteristic set should be generated for all elements in the network. Specifically, for each element, a table of configuration parameters is generated along with techniques or methods which can be employed to adjust these parameters. It is therefore required to understand the advantages and disadvantages of these methods.
- *Network troubleshooting*: It is crucial to understand the effects of problems or faults in the network and how to fix them. The troubleshooting task is a complicated process requiring a thorough understanding of the characteristics and parameters of all elements in the network. Through troubleshooting, problems can be identified, noticed, isolated, and resolved.
- *Planning*: The above monitoring, evaluation, configuration, and troubleshooting tasks are all helpful for planning of the network management. For instance, troubleshooting task helps us plan ahead of potential issues in the network, while monitoring and evaluation support the reconfiguration of the network in the case of faults.

The network management is generally designed with manageable resources by employing appropriate network-management protocols and instrumentation. In the context of computer networks, the instrumentation is the set of facilities that allows us to access characteristics and configuration parameters of elements in the network. The network-management protocols are used to couple this instrumentation with the monitoring, processing, visualization, and storage. Specifically, SNMP is widely used for the network management. It can provide facilities for not only collecting but also configuring parameters from the network elements. The thresholds can be set in the form of traps for automatic notices. In the SNMP, there is a management information base which is a collection of parameters accessible by the SNMP for organizing elements in the network. The SNMP also supports remote monitoring allowing the MIP to provide information about the elements as well as the entire network.

Similar to network design, network management also requires an architecture. A network designer should consider the data flows for the information management. The monitoring can be realized via either in-band or out-of-band monitoring with the employment of centralized and/or distributed monitoring nodes. Due to these additional tasks, capacity and delay requirements for the network-management need to be taken into account. For instance, a monitoring node is deployed in every subnet of a LAN. In this case, the traffic for network management data should be designed so as not to overload the network.

4.9 Related works

For more information, the readers are suggested to refer to the following selected references:

Network Analysis, Architecture, and Design Minoli in 1993 [70], Partridge in 1994 [71], Spohn *et al.* in 2002 [1], Pioro and Medhi in 2004 [72], Stallings in 2006 [2], and McCabe in 2007 [73].

Network Hierarchy Raza and Turner in 1999 [74], Ravasz and Barabasi in 2003 [75], and Cisco Networking Academy in 2014 [76].

Network Management McCabe in 2007 [73], Oppenheimer in 2010 [77], and White *et al.* in 2010 [78].

Network Security Denning in 1982 [79], Stallings in 2002 [80], Kent and Seo in 2005 [81], Jiang *et al.* in 2005 [82], Shiu *et al.* in 2011 [83], Liao *et al.* in 2013 [84], Pescatore in 2014 [85], Deka *et al.* in 2015 [86], Le *et al.* in 2015, 2016 and 2017 [87–90], Vien *et al.* in 2016 and 2017 [91–93], and Karim and Vien in 2017 [94].

4.10 Review questions

[Q4.1] Describe network-analysis process. What are the required inputs and expected outputs of this process?

[Q4.2] What is the aim of network architecture process?

[Q4.3] How to validate a network architecture?

[Q4.4] What are key tasks in network design process?

[Q4.5] Explain two typical trade-offs in network design. Provide example(s) in telecommunication systems for each of them.

[Q4.6] Describe three main layers in a network hierarchy.

[Q4.7] Give an example of employing interconnection within the network hierarchy and demonstrate how these interconnectivity links can improve the network reliability.

[Q4.8] With a flowchart, describe steps in a network design.

[Q4.9] Discuss some common network design goals with examples.

[Q4.10] What are the criteria in the technology evaluation?

[Q4.11] Describe tasks in network management. From your point of view, what is the most challenging task in the network-management process?

4.11 Problems

[P4.1] Select any simulation software and do the research on how to develop a reference model using that software. Discuss its advantages and disadvantages.

[P4.2] Network hierarchical design is stated to allow us dealing with network growth in the future with its predictability. Elaborate this further via an example to show its extendability.

[P4.3] Consider a university network having two campuses, each of which has ten LANs. Develop a network hierarchical model for this network with the employment of interconnectivity links.

[P4.4] Do the research on techniques in content delivery network (CDN) and show how they can be integrated into the traditional network hierarchy.

[P4.5] In the network design, there is a step requiring us to determine the optimal solution of the network model, i.e., Step 8. Discuss potential approaches and/or available techniques to solve this optimal problem.

Chapter 5
Network-requirement analysis

Network-requirement analysis is a must-have process in every network design. Although this step seems to be simple, it plays an important role in the future network modeling and implementation. First, it is vital to understand the existing systems and network services. The requirements from the customer also need to be collected, based on which requirement analysis model and framework can be built.

In this chapter, I will first introduce various network services and requirements in Section 5.1. Service characteristics and levels will be described in Section 5.2 along with discussion on requirements of service performance. Section 5.3 will provide detailed steps of requirement analysis process where user, application, host, and network requirements will be sequentially presented in Sections 5.4–5.7. A typical requirement analysis model will be developed in Section 5.8. This chapter further provides analyses of reliability, maintainability, and availability (RMA) of the network and its components in Section 5.9.

5.1 Network services and requirements

Network-design process is not simply modeling and implementation. It requires a careful preparation taking into account several factors in a whole. A network designer should set a clear design goal according to practical circumstances. It could be either maximizing network performance or minimizing the cost, i.e., CAPEX or OPEX, or balancing both of them. The network architecture/topology and required functionalities also need to be considered in advance of the modeling and deployment. Identifying the services and performance levels in accordance with the requirements and design goals is crucial prior to deciding any network technologies.

With respect to network services and service-based networking, levels of various performance metrics, such as latency, bandwidth, reliability, availability, and security, should be taken into account. Starting from the first generation in the 1960s with only basic connectivity to the recently developed fourth generation, the service-based networking is expected to provide self-configuration and administration.

A system is defined as a set of components that are designed to work together as a whole. These components allow connectivity, communications, and services to users in the network. A system in general consists of four generic components (*cf.* Figure 5.1) which are as follows:

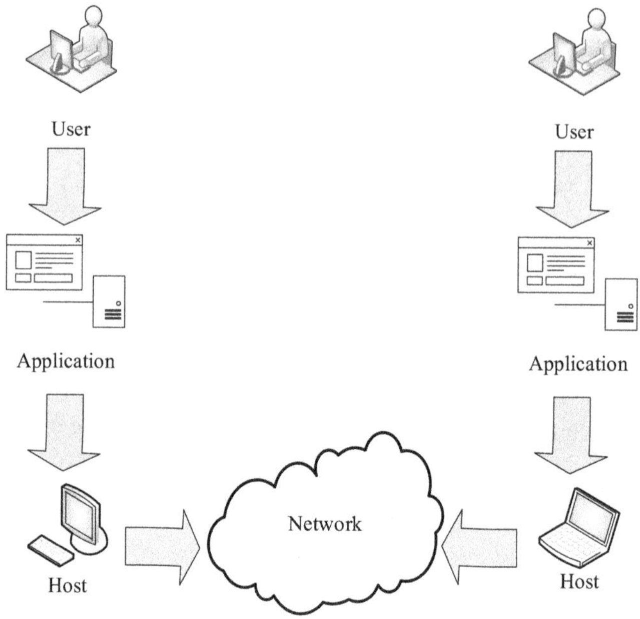

Figure 5.1 System components

- *Users*: Developers and customers both are users in the system.
- *Applications*: Applications can be either generic to the entire network or specific to a particular user per request.
- *Hosts*: Hosts in a system are either computing devices, servers, or specialized equipment which are connected together.
- *Networks*: Networks consist of all component interfaces within the system to provide various services.

In IP networks, services and their descriptions are developed by the Internet Engineering Task Force (IETF). The IETF standard reports and specifications also define levels of performance, e.g., capacity, delay, availability, and reliability, and functions supported in the system, e.g., management, scheduling, security, accounting, and billing.

5.2 Service characteristics

The performance levels and functions offered by either hosts, applications, or networks also reflect service characteristics that can be offered to meet the requests. The service characteristics, which will be described in detail in the following section, are of paramount significance in configuring network elements, e.g., hubs, switches, routers, OSs, etc. Indeed, they are helpful in not only network configuration but also monitoring and validation of various services in the system.

In order to facilitate the description of service levels in the network, service characteristics are grouped together. A service level can be then configured, measured, and verified instead of a number of service characteristics. The service levels are accordingly beneficial in network accounting and billing. These service levels can be described by the following factors:

- *Committed Information Rates (CIRs)*: In a network where virtual circuit is employed for a specified service, CIR is defined as a guaranteed data rate, and the bandwidth (cf. Definition 1.2 in Chapter 1) should not fall below this committed figure.
- *Classes of Service (CoS)*: CoS is literally a priority value that can be employed to differentiate traffics of different kinds.
- *Types of Service (ToS)*: ToS is employed to designate the priority of datagram allowing to determine how the datagram should be routed for different services.
- *Quality of Service (QoS)*: QoS is specified to guarantee a certain level of performance to a data flow in the network targeting either low-latency, high-throughput, or high-reliability services.

Service requests in general are classified as either best effort service or specified service with deterministic functions. In addition, service levels need to be defined so as to meet these requests. There are a variety of typical service performance requirements with respect to different metrics as follows:

- *Reliability*: is a measure of the accuracy of information delivery in the system. This can be represented by either bit error rate (BER) (cf. Definition 5.1) or packet error rate (PER) (cf. Definition 5.2). The reliability can also be reflected through the completion time to notify the accuracy of the information. If the completion time of a data transmission takes longer than that is specified for every ToS, then the data is regarded as lost or corrupted.
- *Capacity*: is a measure of the amount of information that can be reliably transferred over the communication channels. In computing, the bandwidth is regarded as the theoretical capacity of one or more elements in the system (cf. Definition 1.1). The capacity is also referred to as the maximum bit rate which is measured in bits per second (bps) (cf. Theorems 5.1 and 5.2 for the well-known Nyquist's and Shannon's theorem to calculate the channel capacity of noiseless and noisy channels, respectively). In the real system, throughput is normally measured as the achievable capacity of the system or its component which is lower or much lower than the theoretical capacity.
- *Latency*: is a measure of time differences in transmission of information between nodes across the system incorporating application-processing time and task-completion time. Note that the latency is different from network delay which specifies only the duration for data to travel across the network. The network delay generally consists of processing delay of the packet headers at the end node, queuing delay of the packets in the waiting queues, transmission delay

for pushing packets onto the communication link, and propagation delay for the signals to travel from the source node to the destination node.

Definition 5.1. *BER is defined as the number of bit errors divided by the total number of transmission bits during a time interval. These bit errors can be caused be noise, interference, and distortion in digital transmission. The BER is normally represented in percentage with no unit.*

Definition 5.2. *PER is defined as the number of incorrectly received packets over the total number of packets transmitted over the media. The PER is expressed in percentage, and the packet is incorrectly received when there is at least a bit error.*

From the above definition of the BER and PER, their relationship can be deduced as follows:

Lemma 5.1. *Consider a data transmission with data packets having length of N bits. Let P_e and P_N denote the BER and PER, respectively. The PER can be determined through the BER as*

$$P_N = 1 - (1 - P_e)^N. \tag{5.1}$$

Proof. As defined in Definition 5.2, a packet is incorrectly received when there is at least a bit error. This means that a packet is only correct when all N bits in that packet are correct, i.e.,

$$1 - P_N = (1 - P_e)^N. \tag{5.2}$$

The lemma is proved. □

Corollary 5.1. *For large packet size with low BER, i.e., $N \to \infty$, $P_e \to 0$, (5.1) can be approximated as*

$$P_N \approx NP_e. \tag{5.3}$$

Proof. When $N \to \infty$ and $P_e \to 0$, it can be approximated that

$$(1 - P_e)^N \approx 1 - NP_e. \tag{5.4}$$

Substituting (5.4) into (5.1), we have (5.3). The corollary is proved. □

An illustration of PER is shown in Figure 5.2 where BER $P_e = 0.001$.

Theorem 5.1 (Nyquist's Theorem). *Consider a bandwidth-limited signal consisting of V discrete signal states transmitted with a bandwidth of B [Hz] over a noiseless channel. The channel capacity, denoted by C, can be computed by Nyquist's theorem as follows:*

$$C = 2B\log_2 V. \tag{5.5}$$

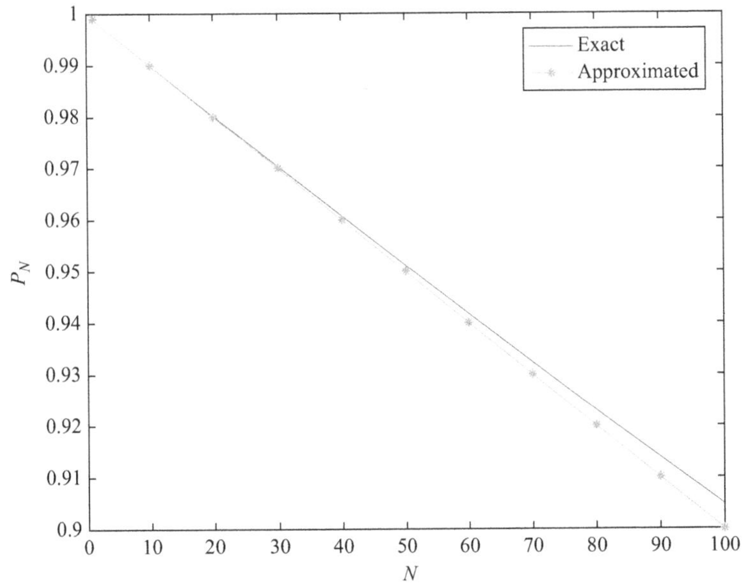

Figure 5.2 Packet error probability (PER) versus the number of bits

A special case of binary data without modulation, i.e., $V = 2$, the channel capacity can be simply given by

$$C = 2B. \tag{5.6}$$

Example 5.1. *A noiseless 10 kHz channel can transmit binary data at a maximum rate of 20 kbps.*

Definition 5.3. *The signal-to-noise ratio (SNR) γ is defined as the ratio of signal power to noise power. The SNR is a dimensionless quantity, which can also be expressed in logarithmic decibel scale as*

$$\gamma_{\text{dB}} = 10 \log_{10} \gamma. \tag{5.7}$$

Theorem 5.2 (Shannon's Theorem). *Over a noisy channel having bandwidth of B [Hz] and SNR of γ (cf. Definition 5.3), the channel capacity C, in bps, is determined by Shannon's theorem as follows:*

$$C = B \log_2 (1 + \gamma). \tag{5.8}$$

Example 5.2. *A noisy 10 kHz channel link having SNR of 10 dB can transmit binary data at a rate up to 34.59 kbps, while a link with 0 dB SNR, i.e., signal power is equal to noise power, results in a lower capacity of 10 kbps (cf. Figure 5.3).*

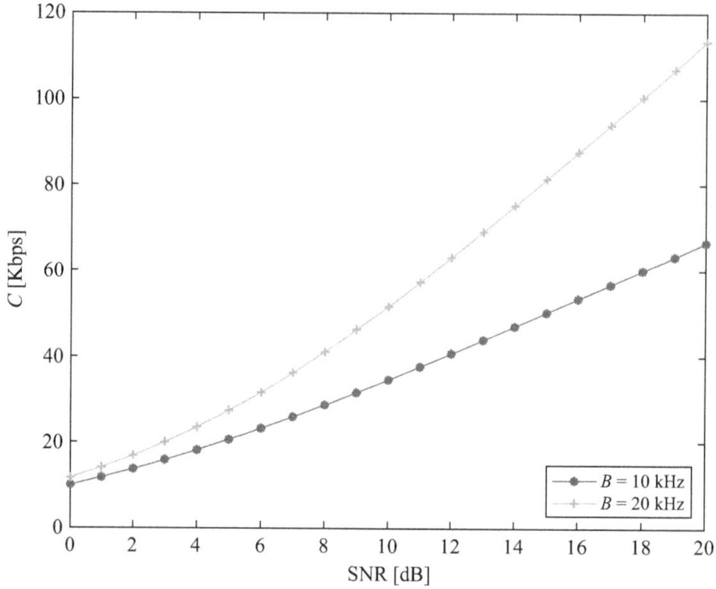

Figure 5.3 Channel capacity versus signal-to-noise ratio (SNR)

5.3 Requirement-analysis process

Prior to performing the analysis, it is critical to understand network environment. Specifically, system requirements and characteristics need to be first identified and gathered. Then, services and performance thresholds can be determined.

The knowledge of network environment helps the network designer to have a better understanding as well as to anticipate the behavior and characteristics of the network under development. Furthermore, payoffs should be taken into account in the requirement analysis.

In fact, there exist a variety of objectives and options for selecting the network technologies. Either stand-alone or combined approaches can be considered, where the former requires us to solve an optimization problem for an optimal solution, while the latter needs to find the adaptation of various strategies in the network. The compatibility of network components is also necessary to be investigated with respect to various user and application requirements. The network designer therefore should have a clear understanding of the location and mechanism to offer services in the network.

Requirement-analysis process basically comprises user, application, host, and network-requirement analyses following the structure of a generic system between two end nodes (cf. Figure 5.4). In the following sections, each of these tasks will be described in detail.

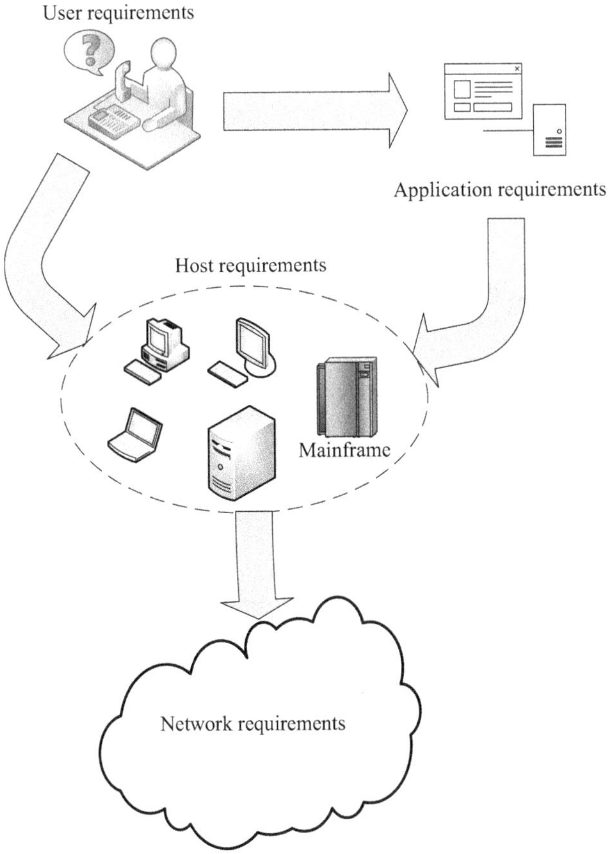

Figure 5.4 Requirement-analysis process between two end nodes

5.4 User-requirement analysis

Targeting at user requirements, a system should be adapted to various user environment to offer high service quality. The system is commonly desired to provide high data bandwidth with reliable information transfer and storage. Some general user requirements can be listed as follows:

- Timeliness of data,
- Interactivity between users,
- Quality of supplied services,
- Adaptability in different networks,
- Security of data communications and storage from attacker, and
- Affordability of networking.

Apart from user expectation for high QoS, a network designer also needs to have knowledge on

- number of users in the network,
- location and distribution/density of users,
- expected growth of the network, and
- redundancy for enhanced reliability or data backup.

5.5 Application-requirement analysis

As a second component in the generic system, application-requirement analysis is related to analyzing the performance of services that can be supported within the network. The performance metrics may include the following:

- *Latency*: is used to validate the interactivity and timeliness.
- *Reliability*: is a measure of the quality, adaptability, and security of communications and services.
- *Capacity*: reflects the affordability of the system, the number and density of users, and also predicts the network growth with possible redundancy.

Given these three specific application requirements, there are three corresponding application types as follows:

- *Real-time applications*: The real-time applications are the applications that require to have a strict timing relationship between source node and destination node considering the duration of both data transfer and processing. The delay requirement is specified for each application to enable the interactivity in real time. The speech and non-buffered videos are typical examples of applications requiring real-time services. On the other hand, non-real-time applications are referred to as the applications with a loose delay requirement, such as web browser and email applications.
- *Mission-critical applications*: In mission-critical applications, the reliability measure is taken into account. With deterministic and/or guaranteed reliability, the mission-critical applications aim to provide accurate delivery of data. In fact, the loss of reliability means the information is undetectable or unrecoverable, which may cause the loss of revenues and even loss of life due to corrupted sensitive data. In these applications, best effort delivery is employed given the constraints/requirements of the reliability.
- *Controlled-rate application*: In most applications, a high bandwidth is normally expected for the best QoS. This requirement is critical in controlled-rate applications where the different applications and services may require different rate. For instance, voice, video, teleconference/teleservice, and file transfer have their own specified requirements of data rate. In these controlled-rate applications, similar to the mission-critical application, the data delivery follows the best effort approach subject to the constraint on the required bandwidth.

In order to analyze the application requirements, the applications should be first categorized in different application groups as defined above. The application group allows us to easily determine the general performance characteristics. Some examples of application groups are as follows:

- *Visualization applications*, e.g., weather monitoring/forecast and medicine management, require high performance in terms of data rate and latency. These applications belong to real-time and/or controlled-rate applications.
- *Telemetry control applications*, e.g., aircraft control, require high-reliability and low-latency data communications. These applications are thus regarded as mission-critical and/or real-time applications.
- *Cloud computing applications*, e.g., parallel computing with cluster/distributed cloud-based servers, require interactivity with low-latency data communication and computation. These applications are normally classified as real-time applications.
- *Web applications*, e.g., Outlook Web based mail services and web-based inventory check, are delay sensitive by some means but do not require high performance. These applications are thus tolerant of bandwidth and/or reliability.
- *Data transport applications*, e.g., fire transfer with FTP, require high reliability but delay- and bandwidth-tolerant. These applications are listed as mission-critical applications depending on the importance of data to be delivered.
- *Teleservice applications*, e.g., teleconference and webinars, provide multicast services to a number of users. They are expected to provide high capacity and low latency, while still maintaining a reliable communication. Taking the capacity and latency into account, these applications are grouped into real-time and controlled-rate applications.
- *Operation, administration, and maintenance applications* in general require high reliability, while the capacity and latency are optional. These applications are grouped as mission-critical applications.

The applications after being categorized with specified requirements will be connected to the next component of the system, i.e., host requirement analysis, via an interface providing detailed information about group, type, location, and performance characteristics of all applications in the system.

5.6 Host-requirement analysis

Host-requirement analysis is the third step in a generic system (cf. Figure 5.4). Generally, a system consists of three types of hosts as follows:

- *Generic computing devices*: e.g., PCs, laptops, workstations, etc.
- *Servers*: e.g., web server, application server, database server, mail server, etc.
- *Specialized equipment*: e.g., sensors, supercomputers, parallel computing systems, etc.

Each kind of hosts is characterized by either some or all of the following performance measures:

- Storage performance
- Processor performance
- Memory performance
- Capacity performance
- OS performance
- Device driver performance.

In addition to the performance criteria, the location information of each equipment is critical in the network design and planning.

5.7 Network-requirement analysis

Network-requirement analysis, though the last step in a generic system, plays an important role in the system. This process considers the system as a whole linking all requirements from the previous steps. In addition, the existing network is required to be validated for its adaptability and interoperability. Specifically, the network size, services, location, and performance characteristics all need to be taken into account in the network-requirement analysis. A network interface of the system therefore consists of the following components:

- Network size and optional scaling;
- Location and/or relocation information;
- Network and support services;
- Adaptability and interoperability; and
- Network performance in terms of latency, reliability, and capacity.

Some basic functional requirements to be planned at the network level can be listed as follows:

- *Management requirements*: How to configure network given existing network settings and user requirements as constraints? Then, how to monitor network events and evaluate network performance? Lastly, how to trace and correct faults, a.k.a. troubleshooting, in the running network?
- *Security requirements*: It is important to define security policies together with other requirement analysis at the starting point to make them manipulable in the future network design. This is also helpful for risk analysis.
- *Financial requirements*: Available funds need to be verified considering both the CAPEX and OPEX.
- *Enterprise requirements*: Required services in the system, e.g., data type, voice, video, printer, phone, fax, etc., need to be discussed in advance.

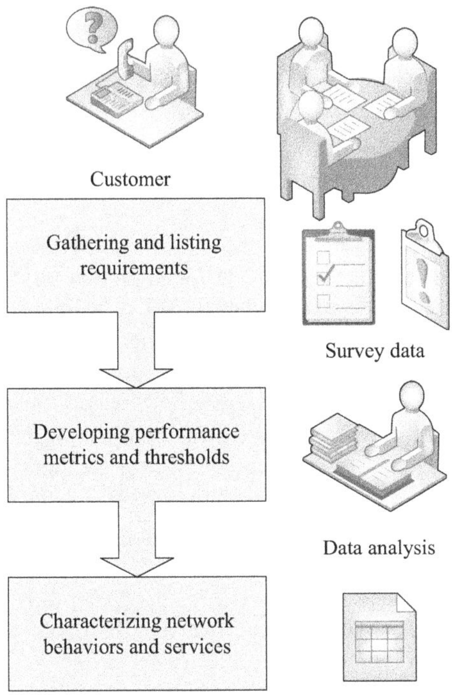

Figure 5.5 Requirement-analysis model

5.8 Requirement-analysis model

From the above analyses, guidelines for these requirements should be outlined for all steps in a generic system. Basically, a requirement-analysis process model consists of the following three steps (cf. Figure 5.5):

- *Step 1 (gathering and listing requirements)*: This step requires us to determine initial conditions of the network under design; for instance,
  - *Design type*: new, modification, validation, etc.?
  - *Design scope*: network size, coverage area, number of sites, etc.?
  - *Design goals*: high bandwidth, low latency, high security, etc.?
  - *Other issues*: related to politics, administration, finance, environment, etc.?

  In order to collect all the above information, a network designer should work with users in a constructive manner. These could be done by either asking the customer fill in a survey form listing a number of relevant questions or organizing meetings for Q&A.
- *Step 2 (developing performance metrics and thresholds)*: As discussed in the previous sections, there are three specific performance metrics to be considered, which include the following:

- *Reliability*: is represented by error/loss rate, e.g., BER or PER. The availability, stability, and maintainability also need to be taken into account, for which each needs to be set with a threshold. These will be described in detail in Section 5.9.
- *Capacity*: is a measure of the maximum achievable data rate or peak data rate. The minimum data rate with respect to data packet size and duration should be defined as a threshold.
- *Latency*: is normally validated through end-to-end or round-trip delay where application-processing time and task-completion time are negligible compared to the duration of the data travel. The requirement of timeliness and delay variants caused by human response and interaction can also be considered.

- *Step 3 (characterizing network behavior and services)*: Service levels should be configured by using network-management protocols, such as Simple Network Management Protocol at the application layer (cf. Sections 1.4.1.1 and 1.5). The behavior also needs to be characterized based on usage patterns and/or history data, for example, frequency and duration of application sessions and traffics.

5.9 Reliability, maintainability, and availability analyses

In order to analyze performance requirements of application, service, and network, besides practical measurements, statistical metrics should be considered to provide an anticipated performance given available numerical data. Specifically, RMA are three typical performance measures which are defined as follows:

- *Reliability*: is a statistical indicator for the frequency of network failures or the unscheduled outages of service. The reliability, in this sense, is a function of the failure rate which can also be represented by mean time between failures (MTBF) in a system. Here, the failure rate is defined as the number of failures during the operation in a unit of time.
- *Maintainability*: is a statistical measurement of the time for restoring system from faulty status to be fully operational. The maintainability is thus represented by mean time to repair (MTTR).
- *Availability*: is defined as the ratio of time that a device is in operation over its total time including both available and unavailable duration of the device. The availability shows the relationship between the frequency of mission-critical failures and the time to restore service. It depends on both MTBF and MTTR (cf. Figure 5.6).

Definition 5.4. *The reliability function, a.k.a. survival function, is defined as the probability that the lifetime of a device exceed a given specified time. Let T denote the device lifetime, $R(t)$ denote the reliability of that device in a duration t, and* $\Pr[A]$ *denote the probability of an event A. By definition, $R(t)$ can be computed by*

$$R(t) \triangleq \Pr[T > t]. \tag{5.9}$$

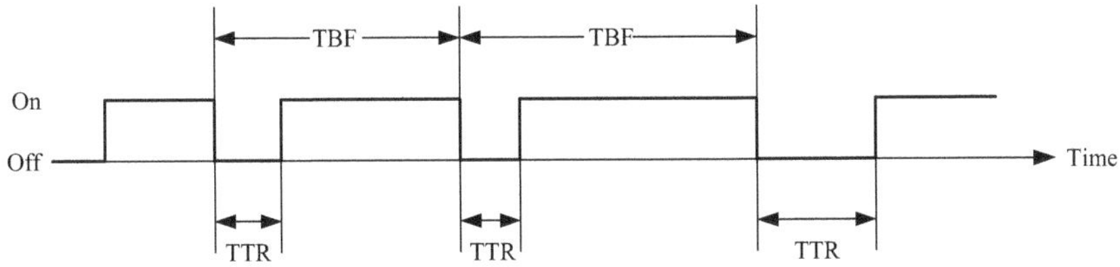

Figure 5.6 Time between failure (TBF) and time to repair (TTR) of a system

Lemma 5.2. *When the lifetime of a device follows an exponential distribution with a failure rate λ, its reliability in a duration t can be computed by*

$$R(t) = e^{-\lambda t}. \tag{5.10}$$

Proof. According to (5.9), $R(t)$ can be given by

$$\begin{aligned} R(t) &= 1 - \Pr[T \le t] \\ &\overset{(a)}{=} 1 - F_T(t) \\ &\overset{(b)}{=} e^{-\lambda t}, \end{aligned}$$

where $F_T(t)$ denotes the cumulative distribution function (cdf) of T, (a) follows from the definition of the cdf, i.e., $F_T(t) \triangleq \Pr[T \le t]$, and (b) follows due to the cdf of an exponential random variable given by $F_T(t) = 1 - e^{-\lambda t}$ (details of the cdf and exponential distribution will be discussed in Chapters 11 and 13). □

Definition 5.5. *MTBF of a device is defined as the average operating time of the device between failures. Statistically, the MTBF can be expressed as its average lifetime, i.e.,*

$$MTBF = E[T], \tag{5.11}$$

where $E[\cdot]$ denotes the expectation operator and $E[T]$ is the expected value, a.k.a. average value or mean, of T.

Lemma 5.3. *When the lifetime of a device follows an exponential distribution with a failure rate λ, its MTBF is given by*

$$MTBF = \frac{1}{\lambda}. \tag{5.12}$$

Proof. The proof follows from the expected value of an exponential random variable, i.e., $E[T] = 1/\lambda$. □

Remark 5.1. *Equation* (5.12) *in Lemma 5.3 means the MTBF of the device is the reciprocal of its failure rate. This is indeed true from Definition 5.5 of the MTBF.*

Consider a system consisting of N devices. The devices can be connected in either series or parallel or series–parallel combination. Let $R_S(t)$ and $R_n(t)$, $n = 1, 2, \ldots, N$, denote the reliability of the whole system and the nth device, respectively, in a duration t. The lifetime of the devices is assumed to be independently but not identically distributed. The system reliability in each case can be determined as follows:

Lemma 5.4. *The reliability of a system having N devices connected in series (cf. Figure 5.7) is given by*

$$R_S(t) = R_1(t)R_2(t)\cdots R_N(t) = \prod_{n=1}^{N} R_n(t). \tag{5.13}$$

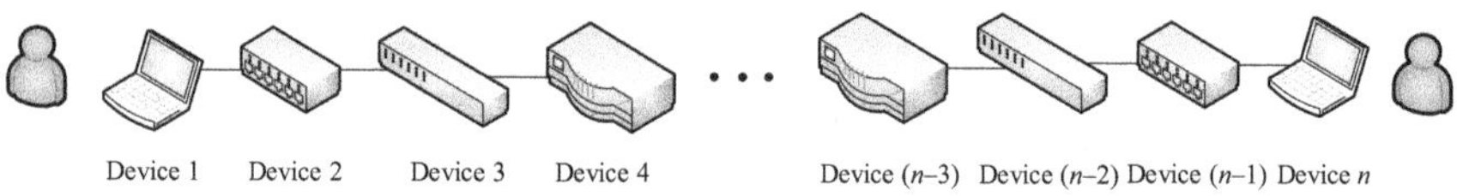

Figure 5.7 Devices connected in series

Proof. With serial connection as in a chain, a system only works when all connected devices work properly. The system reliability therefore depends on all of its devices. Let T_S and T_n, $n = 1, 2, \ldots, N$, denote the lifetime of the system and the nth device, respectively. It can be noticed in a serial connection that the lifetime of the system is the minimum lifetime of all devices, i.e.,

$$T_S = \min\{T_1, T_2, \ldots, T_N\} = \min_{n=1,2,\ldots,N}\{T_n\} \tag{5.14}$$

From the definition of reliability (cf. (5.9) in Definition 5.4), the system reliability is given by

$$\begin{aligned}
R_S &\triangleq \Pr[T_S > t] \\
&\overset{(a)}{=} \Pr[\min\{T_1, T_2, \ldots, T_N\} > t] = \Pr\left[\min_{n=1,2,\ldots,N}\{T_n\} > t\right] \\
&= \Pr[\{T_1 > t\} \cap \{T_2 > t\} \cap \cdots \cap \{T_N > t\}] = \Pr\left[\bigcap_{n=1}^{N}\{T_n > t\}\right] \\
&\overset{(b)}{=} \Pr[T_1 > t]\Pr[T_2 > t]\cdots\Pr[T_N > t] = \prod_{n=1}^{N}\Pr[T_n > t], \\
&= R_1(t)R_2(t)\cdots R_N(t) = \prod_{n=1}^{N} R_n(t),
\end{aligned}$$

where (a) follows from (5.14) and (b) follows due to the fact that the lifetime of devices is independently distributed. The lemma is proved. □

Corollary 5.2. *In a serially connected system, when the lifetime of devices follows an exponential distribution, each of which has a failure rate λ_n, $n = 1, 2, \ldots, N$, the system reliability is given by*

$$R_S(t) = e^{-\sum_{n=1}^{N} \lambda_n t}. \tag{5.15}$$

Proof. The proof follows from (5.10) in Lemma 5.2 and (5.13) in Lemma 5.4. □

Lemma 5.5. *The reliability of a system having N devices connected in parallel (cf. Figure 5.8) is given by*

$$\begin{aligned} R_S(t) &= 1 - (1 - R_1(t))(1 - R_2(t)) \cdots (1 - R_N(t)) \\ &= 1 - \prod_{n=1}^{N} (1 - R_n(t)). \end{aligned} \tag{5.16}$$

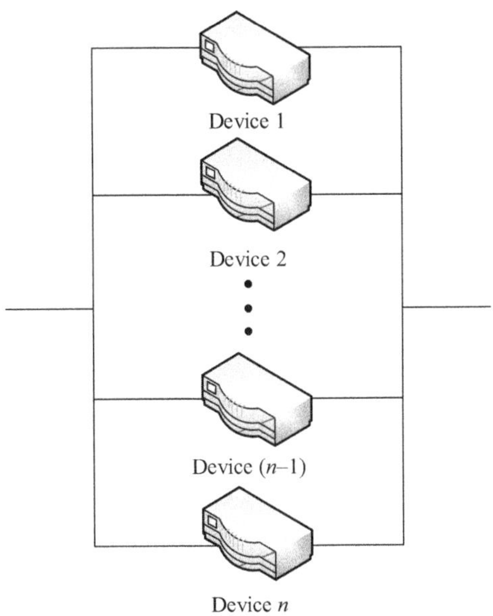

Figure 5.8 Devices connected in parallel

Proof. When N devices are connected in parallel over N branches, the system still can work when one out of N devices works fine. This means that the lifetime of the system is the maximum lifetime of all devices within the system, i.e.,

$$T_S = \max\{T_1, T_2, \ldots, T_N\} = \max_{n=1,2,\ldots,N} \{T_n\} \tag{5.17}$$

Following the same approach as in the proof of Lemma 5.4, the system reliability can be determined by

$$
\begin{aligned}
R_S &\triangleq \Pr[T_S > t] \\
&\overset{(a)}{=} \Pr[\max\{T_1, T_2, \dots, T_N\} > t] = \Pr\left[\max_{n=1,2,\dots,N}\{T_n\} > t\right] \\
&= \Pr[\{T_1 > t\} \cup \{T_2 > t\} \cup \dots \cup \{T_N > t\}] = \Pr\left[\bigcup_{n=1}^{N}\{T_n > t\}\right] \\
&\overset{(b)}{=} 1 - \Pr[\{T_1 \le t\} \cap \{T_2 \le t\} \cap \dots \cap \{T_N \le t\}] = 1 - \Pr\left[\bigcap_{n=1}^{N}\{T_n \le t\}\right] \\
&\overset{(c)}{=} 1 - (1 - \Pr[T_1 > t])(1 - \Pr[T_2 > t]) \cdots (1 - \Pr[T_N > t]) \\
&= 1 - \prod_{n=1}^{N}(1 - \Pr[T_n > t]), \\
&= 1 - (1 - R_1(t))(1 - R_2(t)) \cdots (1 - R_N(t)) = 1 - \prod_{n=1}^{N}(1 - R_n(t)),
\end{aligned}
$$

where (a) follows from (5.17), (b) follows from $\Pr[A \cup B] = 1 - \Pr[\bar{A} \cap \bar{B}]$ generalized for N events, and (c) follows from the complementary property $\Pr[\bar{X}] = 1 - \Pr[X]$ and property of independently distributed random variables. Here $\bar{X}$ denotes the complement of an event X. The lemma is proved. □

Corollary 5.3. *In a parallel-connected system, when the lifetime of the devices follows an exponential distribution, each of which has a failure rate λ_n, $n = 1, 2, \dots, N$, the system reliability is given by*

$$R_S(t) = 1 - \prod_{n=1}^{N}(1 - e^{\lambda_n t}). \tag{5.18}$$

Proof. The proof follows from (5.10) in Lemma 5.2 and (5.16) in Lemma 5.5. □

Remark 5.2. *Connecting devices in parallel is also to provide backup(s) in practical systems. If the same device is deployed across all N branches, i.e., $\lambda_n = \lambda_0$ $\forall n = 1, 2, \dots, N$, the system reliability is given by*

$$R_S(t) = 1 - (1 - e^{-\lambda_0 t})^N. \tag{5.19}$$

Remark 5.3. *The two connection cases in Lemmas 5.4 and 5.5 are also well known in electronic circuits. In a practical system, the connection of devices is much more complicated. They can be connected in series-parallel hybrid or combination. The reliability of the system can be found by first dividing it into simple serial and parallel branches. Then, compute the reliability of each branch and group them together for the whole system.*

Definition 5.6. *The availability of a device is defined by the ratio of the average working time of the device to the total time of the device in the system. Let A, T_{op}, and T_{total} denote the availability, operation time, and total time, respectively. A is computed by*

$$A \triangleq \frac{E[T_{op}]}{E[T_{total}]} \tag{5.20}$$

Note that the average operation time of the device is in fact the MTBF, and the total time consists of the operation time, i.e., MTBF, and the time when the device is unavailable and needs to be repaired, i.e., MTTR. The availability A can be given by

$$A = \frac{MTBF}{MTBF + MTTR}. \tag{5.21}$$

Lemma 5.6. *When both the lifetime and repair time of a device follow exponential distributions with a failure rate λ and a repair rate δ, respectively, its availability is given by*

$$A = \frac{\delta}{\delta + \lambda}. \tag{5.22}$$

Proof. As shown in Lemma 5.3, the MTBF is the reciprocal of its failure rate λ. Similarly, when the repair time of the device follows an exponential distribution with a rate δ, the MTTR can be computed by

$$MTTR = \frac{1}{\delta}. \tag{5.23}$$

Substituting (5.12) and (5.23) into 5.21, the lemma is proved. □

Remark 5.4. *Unavailability is sometimes considered as opposed to availability metric. The unavailability is defined as the ratio of time that the device is unavailable, i.e., being repaired, over the total time in the system. The unavailability, denoted by U, is simply given by*

$$\begin{aligned} U \triangleq 1 - A &= \frac{MTTR}{MTBF + MTTR} \\ &= \frac{\lambda}{\delta + \lambda}. \end{aligned} \tag{5.24}$$

5.10 Related works

For more information, the readers are suggested to refer to the following selected references:

System Components Bertsekas and Gallager in 1992 [95], Stallings in 2003 [25], and White in 2010 [96].

Service Characteristics Braden *et al.* in 1994 [97], Huston in 2000 [98], Evans and Filsfils in 2007 [99], Marchese in 2007 [100], Braun *et al.* in 2008 [101], Papadimitriou in 2010 [102], and Baker *et al.* in 2010 [103].

Reliability Rubino in 1999 [104] and Proakis and Salehi in 2001 and 2007 [105,106].

Capacity Nyquist in 1928 [107], Shannon in 1948 and 1949 [108,109], Marks in 1991 [110], and Cover and Thomas in 2006 [111].

Latency Ramaswamy *et al.* in 2004 [112], Stallings in 2006 [2], and Kurose and Ross in 2012 [113].

Requirement Analysis Jain in 1991 [114], Kotonya and Sommerville in 1998 [115], Hay in 2003 [116], McCabe in 2007 [73], and Laplante in 2013 [117].

RMA Analyses Smith in 2011 [118], Ayers in 2012 [119], Elsayed in 2012 [120], and Lienig and Bruemmer in 2017 [121].

5.11 Review questions

[Q5.1] List generic components of a system. Differentiate between hosts and users.

[Q5.2] How to describe service levels and based on what factors?

[Q5.3] Describe typical performance metrics used to validate services in a system.

[Q5.4] Differentiate between latency and network delay in a communication system.

[Q5.5] What are bit error rate (BER) and packet error rate (PER)?

[Q5.6] Derive the relationship between PER and BER.

[Q5.7] Describe Nyquist's theorem for noiseless channel. Show the impact of the number of discrete signal states on the channel capacity.

[Q5.8] Describe Shannon's theorem for noisy channel. Discuss the impact of channel bandwidth and signal-to-noise ratio (SNR) on the channel capacity.

[Q5.9] Explain the requirement-analysis process via a flowchart.

[Q5.10] Differentiate between real-time applications, mission-critical applications, and controlled-rate applications. Provide an example for each kind of application.

[Q5.11] In the host-requirement analysis, discuss how to characterize the performance of each kind of hosts. Choose a specific equipment and list all of its performance measures.

[Q5.12] What are functional requirements at the network level?

[Q5.13] Describe steps in the requirement-analysis model. From your opinion, which step is principal and time-consuming?

[Q5.14] What are reliability, maintainability, and availability analyses?

[Q5.15] Illustrate the difference between mean time between failures (MTBF) and mean time to repair (MTTR).

[Q5.16] What is the relationship between the device lifetime and its failure rate with the assumption that the lifetime follows an exponential distribution?

[Q5.17] Derive the reliability of a system consisting of N devices in two following cases:

- N devices are connected in series.
- N devices are connected in parallel.

[Q5.18] How to determine the availability of a device? Show its relationship with the MTBF and MTTR.

5.12 Problems

[P5.1] Consider a noiseless 20 kHz channel.

- Find the number of discrete signal states required to achieve a channel capacity of 64 kbps.
- Given a fixed channel bandwidth, how many states are required to achieve a double channel capacity in general?

[P5.2] In Theorem 5.2, the signal-to-noise ratio (SNR) γ is given by

$$\gamma = \frac{S}{N_0 B},$$

where S [W] is signal power, N_0 [W/Hz] is noise power spectral density, and B [Hz] is channel bandwidth.

- Rewrite (5.8).
- Derive the channel capacity for a very low SNR regime when B goes to infinity.

[P5.3] Over a noisy channel having bandwidth $B = 10$ MHz and $S/N_0 = 10^5$

- Find the channel capacity.
- Find the maximum channel capacity that can be achieved over broadband given fixed S/N_0.

[P5.4] The mean time between failures (MTBF) for a newly installed component is given as 2,500 h.

- What is the reliability over a week, assuming 8 h/day and 5 days/week working?
- How would the reliability change for the component if a backup is used?
- Calculate the overall reliability of three identical components connected in series without backup.
- Calculate the overall reliability of the above three-component system assuming that each has a backup individually.
- Assume the mean time to repair (MTTR) of a component is 0.5 h.
  - What is the long-term availability of this component?
  - What is the availability for the given time interval?

[P5.5] A system of components is connected as in Figure 5.9. The MTBF and MTTR of each type of component are given in Table 5.1.

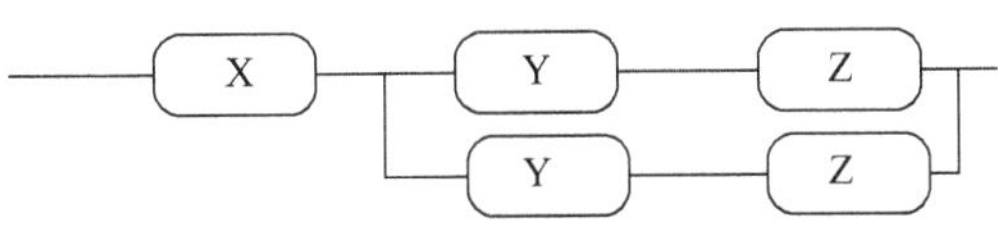

Figure 5.9 Figure for Problem 5

Table 5.1 Table for Problem 5

Component	MTBF (h)	MTTR (h)
X	5,000	6
Y	80,000	12
Z	150,000	48

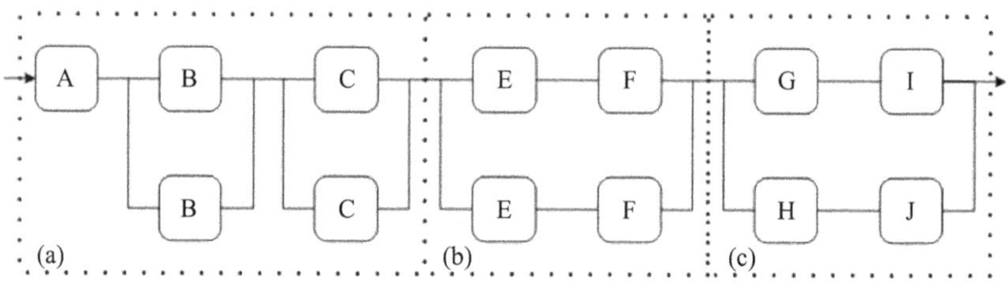

Figure 5.10 Figure for Problem 7

- Find the long-term availability and the availability of each component over time.
- Find the reliability of each component and the overall system reliability over time.
- Draw a block diagram if Y and Z are individually backup. Find the new overall system reliability and compare with the reliability of the original setting.
- Find the best backup strategy to improve the system reliability given three above components. Draw a block diagram to depict your strategy. Calculate the new system reliability.

[P5.6] Prove the best backup strategy for N components serially connected in a system is when each component is individually backed up.

[P5.7] Given a system consisting of three subsystems connected in series as in Figure 5.10 where the components in each subsystem are grouped by dotted lines. Let $R_X(t)$ denote the reliability of component X at a specific time t.

- Find the expression for each of the following reliabilities:
  - Reliability of the first subsystem $R_1(t)$ in terms of $R_A(t)$, $R_B(t)$, and $R_C(t)$.
  - Reliability of the second subsystem $R_2(t)$ in terms of $R_E(t)$ and $R_F(t)$.
  - Reliability of the third subsystem $R_3(t)$ in terms of $R_G(t)$, $R_H(t)$, $R_I(t)$, and $R_J(t)$.
  - Reliability of the overall system $R_0(t)$ in terms of $R_1(t)$, $R_2(t)$, and $R_3(t)$.

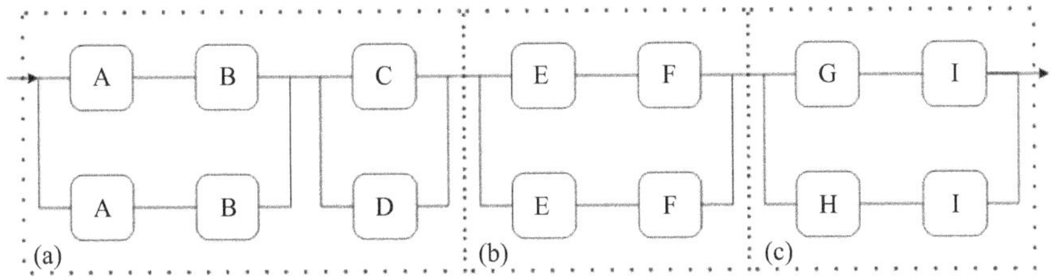

Figure 5.11 Figure for Problem 8

- Redraw the diagram using only the components given above showing the layout to give the highest reliability. Find the formula of $R_1(t)$, $R_2(t)$, $R_3(t)$, and $R_0(t)$ for the new configuration.

[P5.8] A system is set as in Figure 5.11 where the components in each subsystem are grouped by dotted lines. Denote the reliability of component X at a specific time t by $R_X(t)$.

- Find the reliability of each subsystem and overall system.
- Redraw the diagram using only the components given above showing the layout to give the highest reliability. Find the formula of reliability for the new configuration.

Chapter 6
Network flow analysis

Given established requirements with specifications of user, application, host, and network via the requirement analysis step as discussed in Chapter 5, this chapter will analyze these requirements further to provide an end-to-end perspective based on the characteristics of elements within the network. This process is known as flow analysis. The flow in the context of computer or mobile networks is understood as "data" flow, while it may have different meanings adaptable to different models and environment. With the aim of supporting a number of services with different requirements and specifications, the flow analysis is a vital process in every network planning and design.

In this chapter, I will first introduce the concepts of flow, different flow types, and an overview of flow analysis process in Section 6.1. Basic components of a flow will be briefly shown in Section 6.2. Section 6.3 will outline various flow models in detail. Flow boundaries, distribution, and specification will be sequentially described in Sections 6.4–6.6. Finally, this chapter will be closed by sketching guidelines and procedures for network flow analysis.

6.1 Flows

A flow is defined as a set of application and protocol information having common attributes, such as source and destination addresses, information type, routing protocol, etc. During a single session of a running application, there always exists a flow between two end nodes, which is associated with the end-to-end communication between a source (or sender) and a destination (or receiver). The flows therefore can be examined through the communications on the basis of links within the network, which can be then adapted for the entire network.

The network design is generally aimed for either capacity planning or service planning. The flows are studied in the context of these two network planning schemes, which are categorized as follows:

- *Capacity planning*: In this planning scheme, the flows are designed for best effort services. There is no guarantee of data delivery or QoS requirement with unspecified delivery bit rate and time.
- *Service planning*: Different from capacity planning, service planning targets at services with specified QoS, such as data rate, latency, reliability, etc.

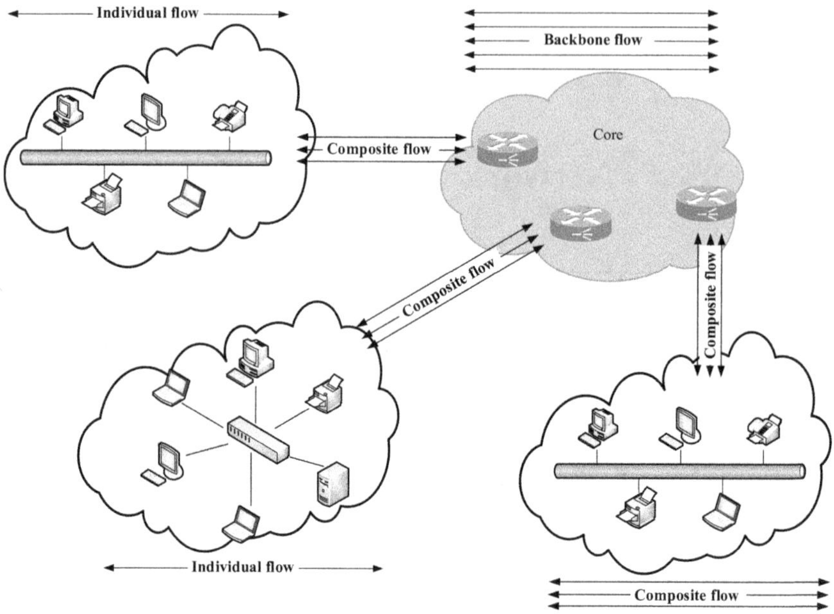

Figure 6.1 Flow types

There are basically following three flow types (cf. Figure 6.1):

- *Individual flow*: is the flow for a single session of an application. The individual flow is devised for specified requirements.
- *Composite flow*: is the combination of individual flows that share the same path and links within the same network. The composite flow is normally used in capacity planning for best effort services.
- *Backbone flow*: is formed by composite flows to link different networks following network hierarchy.

These above flow types have implications on capacity planning in the network. They are useful for developing routing protocols and addressing schemes, especially in a large network with a number of flows.

Identifying data flow types is important to help us select appropriate technology and protocols. Such flow analysis process not only provides an end-to-end perspective on link requirements but also establishes flow boundaries in the network hierarchy. The flow analysis indeed allows us to gain some insights into the network hierarchy degree and redundancy (if necessary). Useful information for determining interconnection strategy should also be attained through the flow analysis. Switching, routing, or hybrid techniques are expected to be developed with respect to different flow specifications for either capacity or service plan.

6.2 Data sources and sinks

Discussing about flows, they always need to have two ends. One end is called source (or transmitter or sender), while another end is called sink (or receiver or destination). In the context of data communications in networks, the sources and sinks imply data sources and data sinks, respectively, which can be defined as follows:

- *Data source*: is a node that produces data to convey to another node or multiple nodes within the network. The data source can be either a network device or a group of devices coordinated with each other in the data transmission. Some examples of data sources are computing servers, parallel systems, and computing clusters.
- *Data sink*: is a node that receives data from another node or multiple nodes. These nodes play the role as the sources of this sink node. The data sink simply acts as a data repository. Some examples of data sinks are display devices (e.g., monitor and projector) and disks/tapes for data storage.

In practical networks, most devices can do both tasks of sending and receiving data. Therefore, they act as either sources or sinks. Some can even do both the data transmission and reception simultaneously, which is known as full-duplex system, while others can support only half-duplex systems where the nodes can transmit and receive data from each other but not simultaneously. A typical example of half-duplex device is a walkie-talkie with a push-to-talk button allowing a person to either talk or listen to another person, but not both at the same time. For full-duplex communications, the telephone that we are using almost everyday can be taken as an example. The telephone device with the employment of a pair of simplex links for two communication paths allows people at both ends speak and hear at the same time.

As an initial step of the flow analysis, the components of the flows, i.e., the data sources and sinks, need to be indicated to find out where flows are generated and where they are going to. Then, we can create flow models for the network design and analysis.

6.3 Flow models

Flows are generally variant in different network infrastructures depending on both logical and physical designs. In order to simplify the flow analysis, the flows should be first mapped into well-known flow models; for instance, peer-to-peer (P2P) model, client–server model, cooperative computing model, and distributed computing model.

Flow models are characterized by the following:

- *Directionality*: The flows in practice are generally asymmetric. Each flow may have its own requirements in each direction.
- *Hierarchy*: The degree of flows can be described through network hierarchy. Users, hosts, addresses, and networks (subnetworks) are commonly grouped via network hierarchy. Such hierarchical grouping helps in assigning appropriate

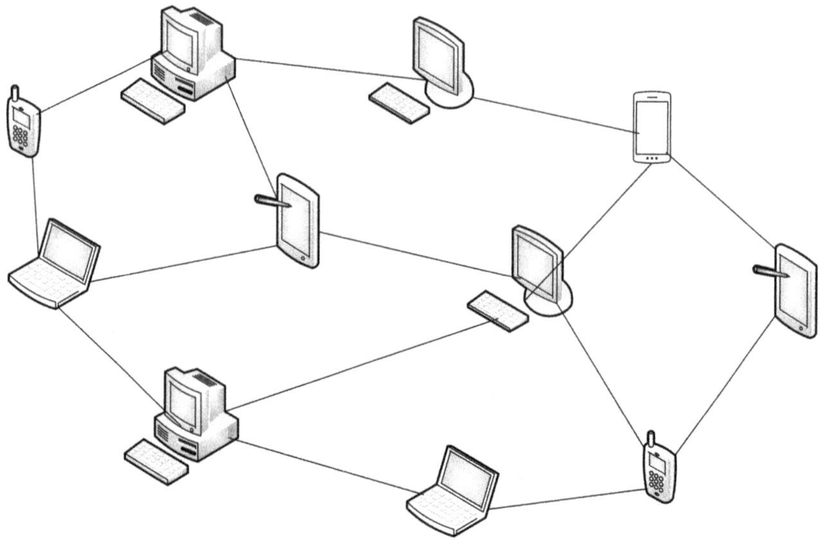

Figure 6.2 Peer-to-peer model

flows for each of these groups. Furthermore, it facilitates the identification of critical flows which are the most important ones in the network, such as the backbone flows to link different networks.

These two characteristics of the flows can be realized in different models which will be discussed in the following subsections.

6.3.1 Peer-to-peer model

P2P model is applied when users or hosts and applications within the network are likely similar. All users have the same access requirements of an application.

In the P2P model, all nodes play the same role, and thus, there is no obvious directionality and hierarchy (cf. Figure 6.2). A typical example of the P2P flow can be seen in tele-service applications (e.g., teleconferencing) where people with different devices connect remotely with each other to share audio, video, and data via a telecommunications system. The communication media could be over Internet or LAN/WAN, and the supported devices could be computers, normal handsets, and/or smart phones. The P2P model can also be deployed over the wireless in the form of device-to-device (D2D) communication systems which have been developed as an underlay to cellular networks. Various applications of the P2P as well as D2D systems can be referred to the relevant works provided in Section 6.8.

Since there is no guarantee of data delivery over the P2P model, this model is normally used in the capacity planning for best effort services.

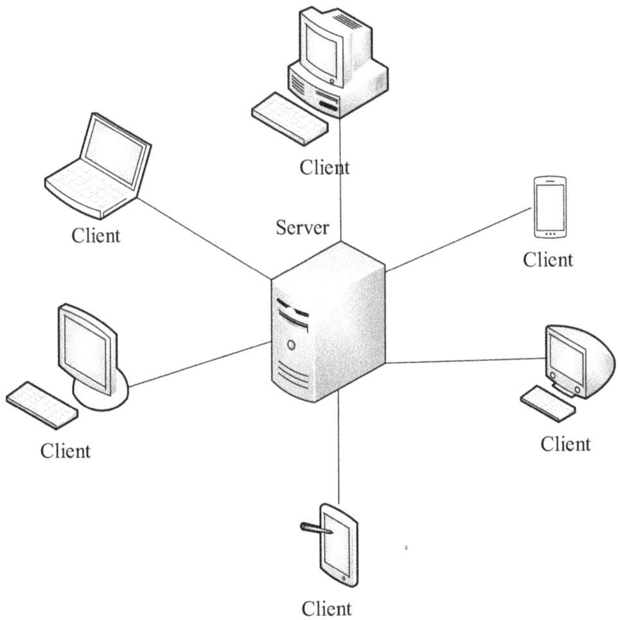

Figure 6.3 Client–server model

6.3.2 Client–server model

Client–server model is the most applicable flow model which has both flow characteristics, i.e., directionality and hierarchy. In this model, the clients are people who request the services, while the servers are service providers which could be web server, application server, database server, mail server, etc. (cf. Figure 6.3). Examples of client–server model can be found in the World Wide Web (WWW) on the Internet and network printing within the LAN.

The communications in the client–server model are carried out via two ways between the clients and the servers following the network hierarchy. One way is for the request from the client, while the other is for the response from the server. Generally, there are more responses consuming higher network resources than the requests, which mean that the two-way data flow is asymmetric, mainly in the direction to the clients. In a very large network where the response flows are much larger than the request ones, such asymmetric data flow can be approximated as unidirectional flow.

6.3.3 Cooperative computing model

Cooperative computing model is an hierarchical flow model where applications collaboratively work and share data to accomplished a required task or service. There is a manager with higher level application to control multiple client–server applications (cf. Figure 6.4). The cooperative computing model is thus more hierarchical than the client–server flow (cf. Section 6.3.2). Examples of cooperative computing model can

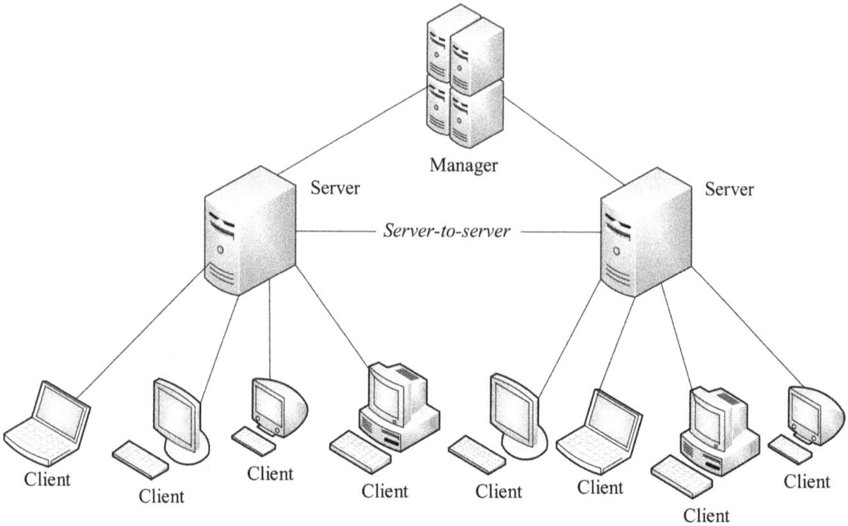

Figure 6.4 Cooperative computing model

be found in monitoring systems (e.g., weather/climate, water, etc.) via either wired or wireless communications between sensors or computing nodes.

In the cooperative computing flow model, apart from server-to-client flows as in client–server model, there are server-to-server flows to connect the servers for collaborative work and server-to-manager flows to connect the servers with the manager at the top level.

6.3.4 Distributed computing model

Distributed computing model is a specialized flow model where the flows are mainly either between computing nodes or between a manager and its computing nodes (cf. Figure 6.5).

In distributed computing flow model, depending on the amount of information transferred between computing nodes, we can categorize them as follows:

- *Loosely coupled nodes*: when these nodes have little or no information to be transferred between them.
- *Closely coupled nodes*: when the information is frequently transferred across these nodes.

The computing nodes are required to accomplish various computing tasks which are classified as follows:

- *Coarse granularity*: Multiple tasks are assigned to the computing nodes in the manner that each of them is dedicated to a single computing node.
- *Fine granularity*: A task is divided into several subtasks to be concurrently manipulated at different computing nodes.

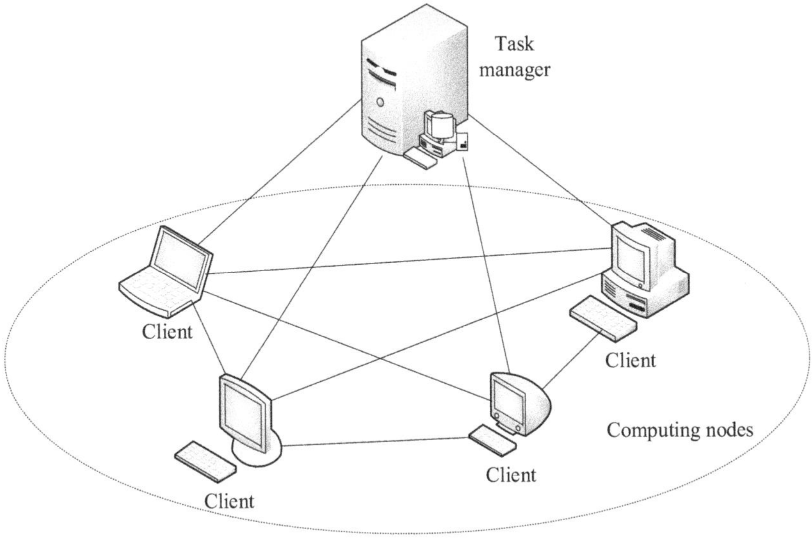

Figure 6.5 Distributed computing model

Considering the categories of both computing nodes and tasks, following are the two specified models related to distributed computing:

- *Computing cluster*: is a group of loosely coupled computing nodes to accomplish a coarse-granularity task. In this case, the flows are asymmetric, which are mainly from the nodes to task manager. Hence, the node-to-manager or node-to-server flows are critical flows. Since the task for each node is independent of that for another node, the flows in the computing cluster are not synchronized.
- *Parallel computing system*: is a set of closely coupled computing nodes to carry out a fine-granularity task. Specifically, each task is divided into several subtasks for computing nodes to concurrently work on along with continuously sharing information between them. The communications between nodes is critical in the parallel computing system; hence, node-to-node flows are critical flows.

In the above two computing models, flow characteristics vary depending on the degree of granularity and coupling. Therefore, in order to optimize flow performance, multicasting should be taken into account subject to the constraints on the task requirements and capabilities of computing nodes.

6.4 Flow boundaries

In a large system consisting of a number of components connecting over multiple networks (subnetworks), it is vital to define boundaries between different group of components for simplifying network design and analysis.

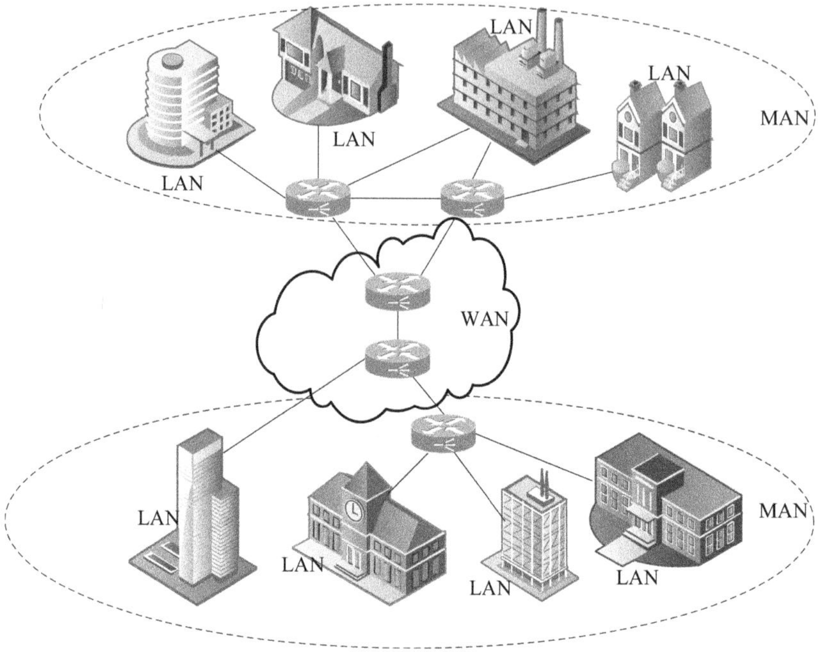

Figure 6.6 Flow boundaries

The flow boundaries are defined as the separations between large portions within a system. They are helpful in indicating not only network flow hierarchies but also directionality and consolidation of the flows. Flow boundaries are thus required to be determined as part of the flow analysis.

A simple way to set up the boundaries is based on geographic areas, such as between campuses, buildings, and floors. The network topology is also a factor to define the flow boundaries, such as LAN/WAN, LAN/MAN, and MAN/MAN separations (cf. Figure 6.6).

Another separation approach is to use logical boundaries to separate between backbones, between network access points in the Internet, between WANs, or between areas with different service requirements. For simplicity, service providers and network engineers normally consider the network boundaries between the WANs in the logical model.

6.5 Flow distributions

After determining flow boundaries between different networks, the next step in the flow analysis is to assign flow distributions, through which the location of backbone flows can be determined. As defined in Section 6.1, the backbone flows link different networks over network hierarchy. A network engineer should first identify the regions

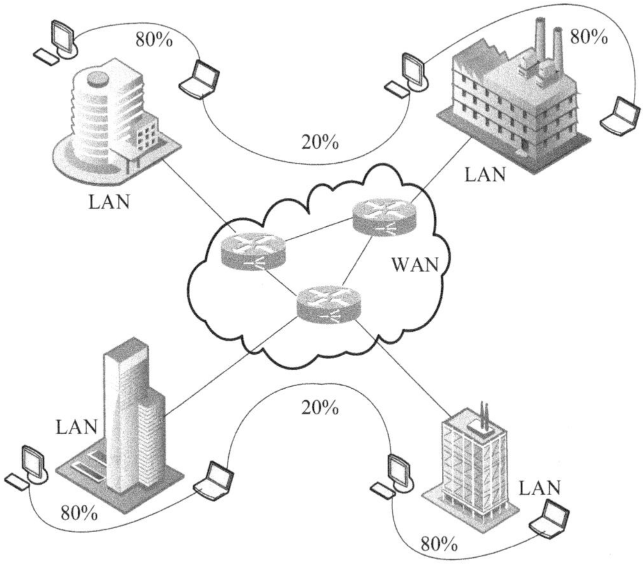

Figure 6.7 Flow distribution

for the flow distributions, e.g., LANs, MANs, and WANs, or more specifically with campus area networks and home area networks.

The flow distributions with identified backbones are helpful in showing when the flows stay in one region or transit from one region to another region. A typical example of flow distribution is within a LAN and between LANs in a WAN. A simple 80/20 rule can be applied where 80% of the traffic flow is within the LAN and 20% is across the WAN (cf. Figure 6.7). This means that the WAN capacity is approximately 25% of the LAN capacity. For example, 2.048 Mbps T1/E1 WAN circuit connects 10 Mbps LANs.

6.6 Flow specification

Along with the identified flows, boundaries, and their distributions, we need to build a flow specification, a.k.a. flowspec, taking into account the requirements of each application. Based on this collective requirement list, we can determine how to group these requirements and assign them to the corresponding flows. The flowspec can be then constructed.

In general, the flowspec has three forms as follows:

- *One-part flowspec*: is used for capacity planning to provide best effort services. The flows are thus named best effort flows. There are no specified flows and no guarantee of data delivery.

- *Two-part flowspec*: contains not only best effort flows but also specified or predictable flows. This flowspec can be applied for both capacity and service planning.
- *Multipart flowspec*: consists of all the above in addition to provide more details and requirements on individual components of the flows.

Given the list of all flows and their characteristics, a flowspec algorithm can be developed to bring all these characteristics together into a specification in the form of combined flows. All characteristics for each flow, including capacity/throughput, delay/latency, and RMA (if any available), should be combined to describe the expected overall performance.

The development of flowspec algorithm should meet the following conditions:

- For best effort flows: only capacity requirements are involved in the calculation for these flows.
- For specified flows: all performance characteristics (if available), i.e., capacity/throughput, delay/latency, and RMA, are computed for these flows. For each of these characteristics, it is aimed at solving an optimization problem to maximize the overall performance of the specified flows.
- When delay and/or reliability requirements are indicated, these can be considered as constraints in the optimization problem and should be treated individually in the calculation for the flowspec.
- Capacities produced by one-part, two-part, and multipart flowspecs are regarded as baseline performance only, while the performance modifiers rely on various factors in practical network deployment.

The flow characteristics in flowspec algorithm can be computed for different flowspec types as follows:

- *One-part flowspec*: Considering capacity planning with L best effort (BE) flows, let $C_{BE,i}$, $i = 1, 2, \ldots, L$, denote the capacity of the ith flow. The total capacity C_{BE} is given by

$$C_{BE} = \sum_{i=1}^{L} C_{BE,i}. \tag{6.1}$$

- *Two-part flowspec*: Considering both capacity and service planning with L BE flows and K specified (SP) flows, let $C_{BE,i}$, $i = 1, 2, \ldots, L$, and $C_{SP,j}$, $j = 1, 2, \ldots, K$, denote the capacity of the ith BE flow and the jth SP flow, respectively. The total capacity for BE services, i.e., C_{BE}, is given as in (6.2), while the total capacity in specified environment, i.e., C_{SP}, is computed by

$$C_{SP} = \sum_{j=1}^{K} C_{SP,j}. \tag{6.2}$$

In addition, the reliability and delay of the SP flows denoted by R_{SP} and D_{SP}, respectively, should be considered in two-part flowspec.

- *Multipart flowspec*: Along with performance characteristics devised as in one-part and two-part flowspecs, the specified services of the jth flow, $j = 1, 2, \ldots, K$, should be guaranteed at the minimum of $R_{SP,j}$, $D_{SP,j}$, and $C_{SP,j}$.

6.7 Flow analysis model

The above flow analyses with identified models, distributions, characteristics, and flowspec algorithm are all combined into developing guidelines for the flow analysis which basically consists of the following steps (cf. Figure 6.8):

- *Step 1—Applying flow models*: In this initial step, application map is first built. Then, flow model is developed, in which data sources and sinks are identified along with their locations, flow directionality, and hierarchy.
- *Step 2—Establishing flow boundaries*: Based on the identified flow model, data sources, and sinks, this step sets up flow boundaries and identifies which flows cross these boundaries. In particular, critical flows are determined among those crossing the boundaries.

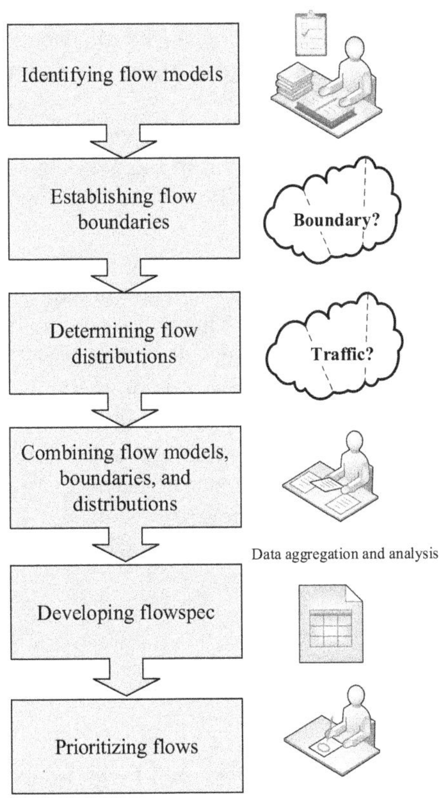

Figure 6.8 Flow analysis model

- *Step 3—Applying flow distributions*: Determining flow distributions is critical in network design and optimization. The flow distributions show the expected traffic volumes of a flow when it crosses the boundaries. This step is also helpful in determining the necessity of generating a backbone flow across the network. There are some specific guidelines with respect to flow models as follows:
  - *P2P model*: Due to not having obvious flow directionality and hierarchy, there is no particular flow distribution in this model.
  - *Client–server model*: Flow distribution 50/50 can be applied for the case when clients cross a boundary to reach a server, otherwise flow distribution 80/20 is used as in the LAN/WAN separation.
  - *Cooperative computing model*: Flow distribution 50/50 can be applied for the case when the flows from different servers must cross a boundary to arrive at a server, otherwise applying 80/20 rule.
  - *Distributed computing model with computing cluster*: Flow distribution 50/50 is applied since the computing nodes are remote from the task manager, and all of them have to cross a boundary.
  - *Distributed computing model with parallel computing system*: Flow distribution 20/80 is applied since all the traffic flows between computing nodes are across flow boundaries.

  Note that the above guidelines can be changed to adapt with practical network design given different requirements of applications, host and user distribution, network topology/model, etc.
- *Step 4—Combining flow models, boundaries, and distributions*: This step simply combines the information collected in the previous three steps to indicate the likelihood of making use of backbone flows and their location in network hierarchy.
- *Step 5—Developing flowspec*: Summarizing all the above steps, this step helps in providing capacity and service plans along with requesting, configuring, and verifying services. These also facilitate the selection of appropriate technology to meet service requirements. In order to develop the flowspec, as outlined in Section 6.6, we need to collect a list of applications and their requirements, application map, data sources and sinks, flow models, boundaries, distributions.
- *Step 6—Prioritizing flows*: The flowspec developed in Step 5 enables us to prioritize flows according to the number of users, distance, location, and the performance characteristics of each flow.

6.8 Related works

For more information, the readers are suggested to refer to the following selected references:

Information Flows Ford in 1956 [29], Ford and Fulkerson in 1956 [30], Omidyar and Pujolle in 1996 [122], Brownlee *et al.* in 1999 [123], and Floyd and Fall in 1999 [124], Gamal and Kim in 2012 [125], and Frana in 2018 [126].

P2P Model Shen *et al.* in 2009 [127], Fitzek and Charaf in 2009 [128], Gurler and Tekalp in 2013 [129], Kumar and Lee in 2014 [130], Yin *et al.* in 2017 [131], Wu *et al.* in 2017 [132], and Jiang *et al.* in 2018 [133].

D2D Systems Asadi *et al.* in 2014 [134], Trestian *et al.* in 2015 [135], Wu *et al.* in 2016 [136], and Ali *et al.* in 2016, 2017, and 2018 [137–139].

Client–Server Model Sinha in 1992 [140], Ramesh and Perros in 2000 [141], Kanitkar and Delis in 2002 [142], AbdulFatah and Majumdar in 2002 [143], and Zhu *et al.* in 2016 [144].

Cooperative Computing Taylor in 1990 [145], Richardson *et al.* in 1998 [146], Wu *et al.* in 2013 [147], You and Huang in 2018 [148], and Sheng *et al.* in 2018 [149].

Distributed Computing Casavant and Kuhl in 1988 [150], Morse in 1994 [151], Bianchi in 2000 [152], Younis and Fahmy in 2004 [153], Nishida and Nguyen in 2013 [154], Kandaswamy *et al.* in 2018 [155], and Pascale *et al.* in 2018 [156].

Flow Boundaries Estrin *et al.* in 1989 [157] and Quoitin *et al.* in 2003 [158].

Flow Distributions McCabe in 2007 [73] and Oppenheimer in 2010 [77].

Flow Specification Partridge in 1992 [159] and Marques *et al.* in 2009 [160].

Flow Analysis and Control Cruz in 1991 [161], Parekh and Gallager in 1993 [162], Jane *et al.* in 1993 [163], Sen and Wang in 2004 [164], Popovski and Yomo in 2007 [165], Katti *et al.* in 2008 [166], Sohn and Kim in 2008 [167], and Liu and Chen in 2015 [168].

Flow Virtualization Dally in 1992 [169], Jagadeesan *et al.* in 2014 [170], Lara *et al.* in 2014 [171], Kreutz *et al.* in 2015 [172], Hassan *et al.* in 2017 [173], Li *et al.* in 2018 [174], and Jia *et al.* in 2018 [175].

6.9 Review questions

[Q6.1] What is a flow between two end nodes?

[Q6.2] Describe three common flow types within a network.

[Q6.3] Give an example of data sources and data sinks for each of the following kinds of end nodes:

- The end nodes are data sources only.
- The end nodes are data sinks only.
- The end nodes are either data sources or data sinks.

[Q6.4] How to characterize flow models?

[Q6.5] What is peer-to-peer (P2P) model?

[Q6.6] Describe the communications in client–server model.

[Q6.7] What kinds of flows in cooperative computing model?

[Q6.8] What are loosely coupled nodes and closely coupled nodes in distributed computing model?

[Q6.9] Describe two kinds of computing tasks at nodes in distributed computing model.

[Q6.10] Differentiate between cooperative computing and distributed computing models.

[Q6.11] Explain the usage of three flowspec forms.

[Q6.12] What conditions are required in the development of flowspec algorithm?

[Q6.13] Explain steps in a flow analysis model and show how to determine the appropriate flow distribution in different networks.

6.10 Problems

[P6.1] Discuss how to design flows for capacity planning and service planning schemes in an enterprise subject to various service requirements in different areas/buildings/offices.

[P6.2] Provide an example of a communication network where peer-to-peer model is incorporated in client–server model.

[P6.3] Assume the servers are located in the cloud. This computing model is also known as cloud computing. Do the research on this model and show how to adapt cooperative computing and distributed computing models for the cloud-based servers.

[P6.4] Apart from the popular 80/20 rule, investigate other flow distribution rules, e.g., 50/50, 20/80, etc., and their applications in practical networks.

[P6.5] In the flow analysis model, discuss criteria for prioritizing flows based on flowspec. Show their corresponding impacts on the network flow performance in the whole network.

Chapter 7

Network performance evaluation

In the previous chapters, we have gone through the whole network-design process starting from the initial requirement analysis to flow analysis with detailed guidelines for configuring network elements and entire network, as well as mechanisms for developing network and flow specification. In order to validate the effectiveness of a network design, its performance needs to be evaluated. In general, there are three approaches to evaluate the network performance, including benchmarking, simulation, and analytical modeling.

In this chapter, I will sequentially discuss these three approaches along with their merits and demerits as well as their applications. Section 7.1 will first present the benchmarking technique, followed by the introduction of simulation and analytical modeling approaches in Sections 7.2 and 7.3, respectively. Also, in this chapter, a typical system and its component will be provided in Section 7.4 as illustrations of different performance-evaluation techniques.

7.1 Benchmarking: merits and demerits

Benchmarking is an evaluation technique which has been well employed in various network models to measure network performance while a standard set of task or workload is running. Taking into account different approaches and solutions over a number of tests and trials, the benchmarking process in particular allows a fair comparison between them.

The benchmarking technique is helpful in building a new system and also when modifying the existing system. Some of its merits can be intuitively observed as follows:

- *Reliability*: With a number of trial runs given various standard sets, the results obtained with benchmarking are shown to be more accurate than other approaches.
- *Applicability*: After a rigorous validation, the benchmarking results can be directly applied in practice.
- *Understandability*: As a practical approach, the benchmarking is easy to grasp.
- *Marketability*: Due to its practicability, the benchmarking approach can be easily marketed to be sold.

However, apart from these above merits, the benchmarking technique has some essential demerits as follows:

- *Cost*: The benchmarking requires a high cost in building the system, instrumentation, and testing before the product is able to come into use.
- *Time*: Such high complicated process with benchmarking technique is definitely time-consuming.
- *Personnel*: The success of the product after a number of tests and trial runs requires a lot of efforts, and human factor plays a key role in the benchmarking process.
- *Extendability*: Some changes in the system or network environment may have considerable impacts on the results. The benchmarking results are hard to be extended for different cases; hence, different benchmarking processes are needed.

7.2 Simulation: merits and demerits

Simulation is a well-known approach for modeling a system as well as evaluating network performance. We can build a simulation model of the system under investigation by using various existing simulators, such as Riverbed Modeler, discrete-event network simulator with NS3 for Internet systems, or OMNeT++ based on C++ programming language.

The simulation approach can provide a simple modeling of the network for assessing its performance and generally does not require high mathematical skills. The simulation has shown to provide a number of advantages as follows:

- *Adaptability*: The simulation program can be run and rerun with several alternative parameter sets to obtain various results for validation.
- *Flexibility*: The simulation is flexible in the sense that the simulation models allow us to collect results at different stages during the simulation process to keep track of all generated data.
- *Extendability*: With the above merits of flexibility and adaptability, the simulation models are generally easy to extend for different networks.

Although the simulation has shown a number of merits, it has some demerits that need to be considered in the performance-evaluation process, such as the following:

- *Personnel*: In order to construct a new simulator or simulation software, it normally requires a lot of human efforts, time, and work commitment, especially with inexperienced engineer or a programmer with no specialized knowledge in that simulator.
- *Time*: The simulation is shown to be much faster than benchmarking. However, a large number of simulation runs are required to elaborate and validate the correctness of the results, which is a time-consuming process depending on the length of the simulation program. Although the simulation programs are easy to run and rerun for a number of times, very minor changes of the system parameters may require new simulation runs.

- *Managing*: Some simulation programs are prohibitively bulky with a number of functions, controls, variables, etc., Understanding all of these may be easy, but managing them is a challenging task that every programmer is facing. A manageable approach should be considered at the beginning of programming or writing with any simulator.

7.3 Analytical modeling: merits and demerits

Analytical modeling, a.k.a. mathematical modeling, has been well established to provide an abstract model of a network or system instead of performing simulation with various simulators or experiment with hardware implementation. The mathematical model allows us to provide exact or approximate solution of a problem, such as evaluating channel capacity, network or system throughput, latency, reliability, security, etc. Some of these, e.g., reliability, latency, and capacity, have been discussed in Section 5.2.

In order to analyze network performance via mathematical modeling, cutting-edge mathematical techniques and tools can be employed, which are shown to be very computationally efficient. Different mathematical methods can be applied, such as differential/integral calculus, algebraic methods, probability theory, stochastic process, queuing theory, etc. In particular, these techniques can be easily employed for several cases when the model has been validated. Due to its convenience, the analytical modeling has emerged as a promising approach.

Specifically, the mathematical modeling has some merits as follows:

- *Computability*: A good mathematical modeling should provide a fast computation of various performance measures. Different mathematical tools or software, e.g., MATLAB or MATHEMATICA, can also be used for the computation.
- *Adaptability*: The derived expressions through the mathematical analysis can be easily employed to achieve the results for different cases.
- *Optimality*: Based on the derived expression of the performance measures, we can maximize the network performance. Various optimization problems can be developed to find the optimal network parameters subject to constraints in accordance with requirements which could be from different perspectives.

The analytical modeling, however, has some disadvantages due to its inherent restrictions and requirements as follows:

- *Mathematical skills*: High-level mathematical skills are required to derive the expressions of various performance metrics which could be in either approximated or closed forms.
- *Artificiality*: Several assumptions are considered in analytical modeling to simplify the analysis. These assumptions are sometimes too artificial and unrealistic to be able to apply for practical cases.

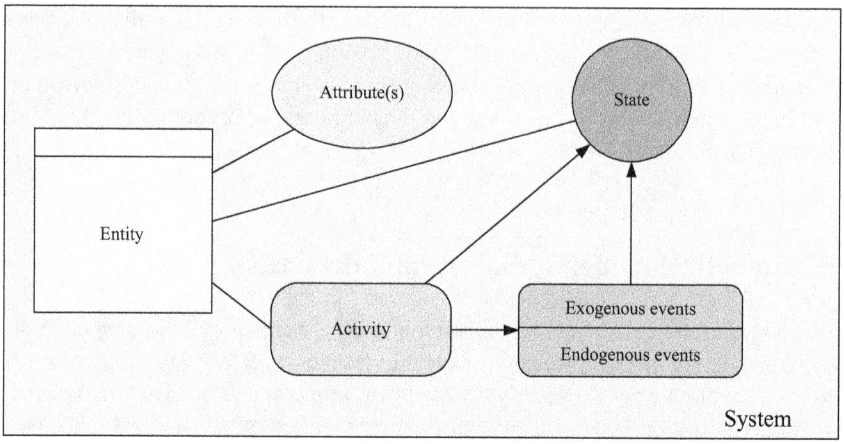

Figure 7.1 System components in a simulation model

- *Generalization*: The analytical modeling can generalize the problem with a number of formulas. However, it does not provide detailed information with specification as in simulation approach.
- *Accuracy*: Due to various assumptions, the results obtained by the analytical modeling are not as accurate as in benchmarking.
- *Practicality*: Exploiting the mathematical formulas, the derived expressions can provide a quick validation of the performance, but some of them can only be derived in approximated forms. With the evolved complex analytical models, it is promising that these problems can be solved in a general closed form to be able to compare with those in practice.

7.4 Systems and system environment

The above performance-evaluation techniques can be applied to either each component of the system or the entire system. In order to understand the applications of these techniques, this section will introduce some basic concepts of systems and their components.

In the context of network performance evaluation, a system is defined as a group of objects that are working together in a collaborative manner to accomplish several tasks. Any changes in system environment may affect the behavior of the system. It is therefore vital to understand the functionality and operation of the system and its environment.

In general, a system has the following components (cf. Figure 7.1):

- *Entity*: is an object within a system.
- *Attribute*: is a property of an entity.

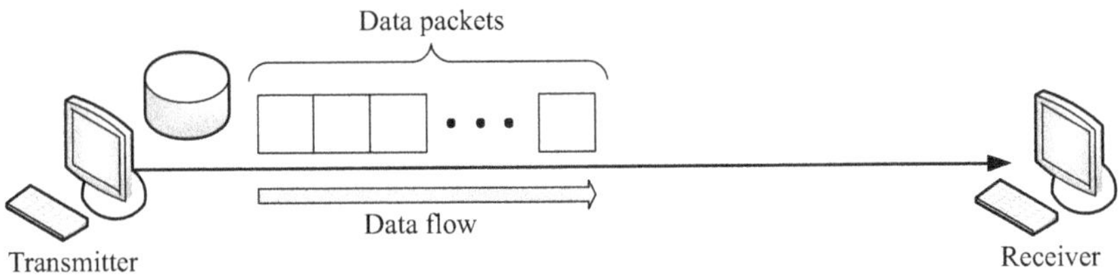

Figure 7.2 Data flow between two end devices

- *Activity*: is an action over a specified period of time.
- *State*: is a collection of related variables which is necessary to describe the system. The state in general is a function of time. In some systems, for simplicity in the analysis, the state is assumed to be quasi-static, which is invariant for a fixed period of time.
- *Event*: is an instantaneous occurrence of an entity. The event may result in a change of the state.

The activities and events causing the changes of the system could have either internal or external causes. These two kinds of activities and events are defined with the following terms:

- *Exogenous activities and events*: refer to the activities and events in the environment that affect the system.
- *Endogenous activities and events*: refer to the activities and events occurring within a system.

Considering the change of state over time, a system can be classified into the following two types:

- *Discrete system*: is a system in which the state variables change in a discrete set of points over time.
- *Continuous system*: is a system in which the state variables change continuously over time.

Although these two kinds of systems are clearly defined, they may be used interchangeably depending on the state variables under consideration. For example, in a communication system, the number of packets arrived is discrete, but the data flow is continuous (cf. Figure 7.2).

Example 7.1. *A typical example of the above concepts can be found in a bank (cf. Figure 7.3). In the bank, every customer is an entity, and account balance is an attribute of the considered entity. The customer is assumed to go to the bank to make deposits. In this case, making deposits is an activity of the entity. The arrivals of the customer and the completion of the services are events in the system. To describe the processing of the system, the number of the customers waiting in the queue and the*

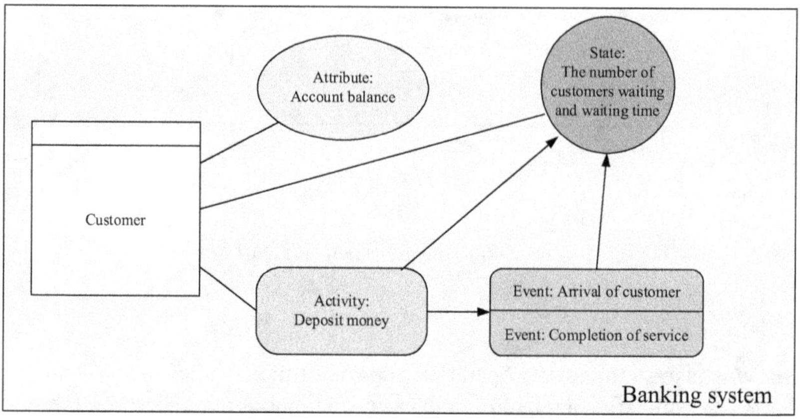

Figure 7.3 Components in a banking system

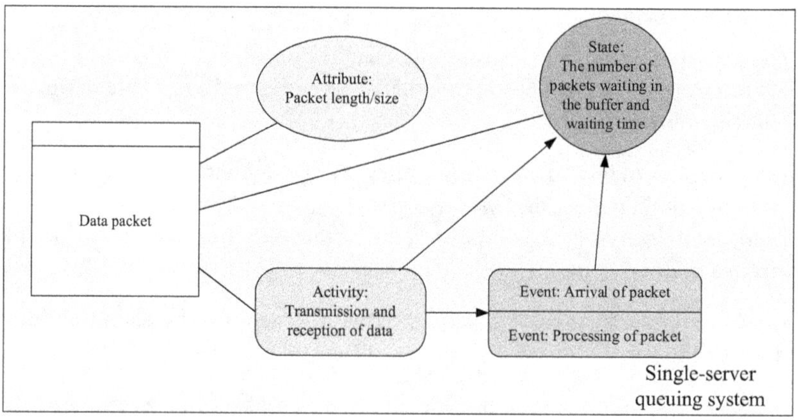

Figure 7.4 Components in a single-server queuing system

waiting time are state variables. The arrival of the customer is an exogenous event, while the completion of the service for the customer is an endogenous event.

Example 7.2. *Analogically, consider a single-server queuing system for data communications (cf. Figure 7.4). The entities in this system are the packets, and their length is an attribute. The transmission and reception of the packets are activities of these entities. In this case, the event can be the arrival of the packets at the server, and the state variable is the number of packets waiting in the queue to be sent. Details of the queuing system will be discussed in Chapter 15.*

7.5 Related works

For more information, the readers are suggested to refer to the following selected references:

Benchmarking Gray in 1992 [176], Madhavan *et al.* in 2009 [177], Harcho-Balter in 2013 [178], and Nambiar and Poess in 2016 [179].
Analytical Modeling Wolfram Mathematica [180], MATLAB [181], Abell and Braselton in 2017 [182], and Wolfram in 2017 [183].
Simulation and System Components Varga in 1999 [184], Banks *et al.* in 2010 [185], Wehrle *et al.* in 2010 [186], Davis in 2010 [187], Rimoldi in 2016 [188], Choi and Liu in 2016 [189], Attaway in 2017 [190], NS3 [191], OMNeT++ [192], MATLAB [181], and Simulink [193].

7.6 Review questions

[Q7.1] What is benchmarking? What are its advantages and disadvantages?
[Q7.2] What are the advantages and disadvantages of simulation in general?
[Q7.3] How to perform analytical modeling and what skills are indispensable?
[Q7.4] Discuss the merits and demerits of the analytical modeling.
[Q7.5] List components in a simulation model and explain each of them in the context of a computer network.
[Q7.6] What are exogenous and endogenous events and activities?
[Q7.7] Differentiate between discrete and continuous systems. Provide an example for each kind of these systems.
[Q7.8] Describe the components within a banking system and a single-server queuing system. Illustrate their relationship in a diagram for each system.

7.7 Problems

[P7.1] Select a network-simulation software. Discuss about its advantages and disadvantages in modeling a real system.
[P7.2] Compare the complexity of analytical modeling and simulation approaches. Any simulation software or tool can be selected.
[P7.3] Discuss the accuracy of the results derived from analytical modeling compared with those obtained from simulation software.
[P7.4] Provide an example to show the artificiality and impracticality of the analytical modeling.
[P7.5] Build a table comparing merits and demerits of three approaches to evaluate network performance.
[P7.6] Develop a simulation model for a multi-server queuing system. Illustrate its components and their relationship in a diagram.

Chapter 8
Network simulation models

In the previous chapter, we have learnt different techniques for network-performance evaluation. Every technique has its own merits and demerits. Depending on specific requirements and capabilities, we can select one or a combination of multiple techniques. Specifically, simulation is the most popular approach due to its simplicity, flexibility, and adaptability with a number of tools and software. This chapter will discuss in detail simulation models for emulating and evaluating a system.

In this chapter, I will first introduce simulation techniques used for emulating and evaluating a system in Section 8.1. The selection of appropriate simulation tools along with their advantages and disadvantages will be also discussed. Simulation models of a typical system will be presented in Section 8.2. Detailed steps and phases of a simulation study will be outlined in Section 8.3. Simulation examples will be then provided in Section 8.4 to show the effectiveness of network simulation models.

8.1 Simulation for emulating and evaluating a system

Simulation is well known as the imitation of the operation of a system in real world. A simulation model is developed to study the behavior and characteristics of the system. Prior to being deployed, the simulation model needs to be validated. After proving validity and accuracy, the simulation model can be employed as an analysis tool. It can be used to study the system at the design stage as if in a practical system, allowing us to evaluate various performance metrics of the system in different circumstances. Some applications of simulation can be found in manufacturing systems, construction engineering, military, transportation, traffic, health care, logistics, business and project management, etc.

As discussed in Chapter 7, there exist different approaches for evaluating network performance. The question is when simulation should be an appropriate tool. In order to answer this question, it is crucial to know the benefits as well as potential application of simulation for different purposes.

Specifically, a simulation model should

- enable experimentation with a system or a subsystem,
- simulate changes and evaluate the effects of these changes on system behavior,
- allow us to gain knowledge to improve the system,

- give a better understanding of system variables and their interaction,
- be used to verify the findings with analytical approaches or methodologies,
- be employed and tested prior to implementing a new system,
- allow us to investigate different capabilities of a system,
- be developed for different learning purposes,
- support visualization control and/or produce animation shows, and
- be exploited to treat some interactions within a complex system where a real implementation would be challenging.

In fact, the above scenarios when employing simulation would be benefited due to a number of its advantages. The simulation could

- allow new policies, procedures, algorithms, etc. that can be tested and validated without interrupting ongoing process and operation of a real system;
- help us to test new hardware/systems to verify their functionalities and compatibility and/or to validate their performance prior to buying them;
- test various hypotheses on certain phenomena that can occur during the operation;
- allow us to either speed up or slow down the phenomena by compressing or expanding the simulation time;
- provide insight about the interaction of variables within the system along with their impacts on the system performance;
- perform bottleneck analyses of the system; and
- help us to understand the operation of an entire system by answering various "What If" questions.

However, the simulation is sometimes not appropriate in certain circumstances when

- the solution of the problem can be intuitively observed by common sense,
- the problem can be easily solved via analytical method,
- it is easier to perform benchmarking technique,
- simulation software/tools are too expensive to invest in,
- resources and time are limited or even not available due to human and/or system,
- data/parameters required for the simulation are not able to be obtained,
- system behavior is too complex to perform simulation, and
- expected output of the simulation is unreasonable and infeasible.

The above limitation is due to the following disadvantages of the simulation, including the following:

- Special training and experiences may be required for specific software.
- Some simulation results can be difficult to interpret for their meanings and reasoning.
- Learning new simulation software is time-consuming.
- Simulation software is too expensive, while analytical method is not too hard to implement.

8.2 Simulation models of a system

As discussed in Section 7.4, a system is a group of objects working together to accomplish several tasks. In simulation, the system needs to have a model which is simply defined as a representation of the system. Such modeling is helpful for studying the system.

A model is designed for one or several goals in the simulation. Therefore, we may have different simulation models of the same system for different objectives; for instance, when evaluating different performance metrics over different timescale of the same system. While the validation of a real system could be very complicated in practice, its simulation model would facilitate this process by simplifying the system. However, the simulation model needs to maintain sufficient details to represent the real system allowing us to draw valid performance evaluation results.

The simulation model is regarded as a special type of mathematical models which use symbolic notation and mathematical equations to represent a system. Indeed, the simulation model is usually exploited to validate various performance metrics of the system which are expressed in the form of mathematical equations.

In order to verify and validate a simulation model, a number of tests or simulation runs are required. Considering the change of the model over time, there are two types of simulation models, including the following:

- *Static simulation models*: These models are used to represent a system at a particular point of time.
- *Dynamic simulation models*: These models are used to represent systems as they change over time.

For simplifying the modeling and analysis, the system can be assumed to be invariant within a specific period of time. In this case, it is regarded as a static simulation model.

Within a system, there may have one or several variables affecting the performance of the system. The simulation models should represent these variables and thus can be classified into two other kinds as follows:

- *Deterministic simulation models*: In these models, there is no random variable, and all system parameters are invariant over time.
- *Stochastic simulation models*: These models contain one or multiple random variables as inputs. In the simulation, these variables are randomly generated in every run. Such random changes can imitate the behavior of a real system over time, allowing us to provide statistical estimate of the true characteristics of the system.

Considering the change of state over time of the random variables within a system, there are also two kinds of the simulation models, including the following:

- *Discrete simulation models*: are used to represent the discrete systems in which the state variables change in a discrete set of points over time.

- *Continuous simulation models*: are used to represent the systems in which the state variables change continuously over time.

A typical model that is frequently considered in network design is "discrete, dynamic, and stochastic simulation model." In this model, there is one or multiple random variables which change in a discrete set of points over time.

8.3 Steps/phases in a simulation

A simulation study basically consists of the following steps:

- *Step 1*: Formulate the problem that needs to be solved.
- *Step 2*: Set objectives and overall project plan.
- *Step 3*: Form model concept to meet the proposed plan and objectives.
- *Step 4*: Collect data required for simulation model.
- *Step 5*: Translate the model with simulation software/toolbox.
- *Step 6*: Verify the simulation model. If the model is verified, then go to Step 7. Otherwise, go back to Step 5 to check the model-translation process.
- *Step 7*: Validate the simulation model. If the model is shown to be able to provide the desired results, then go to Step 8. Otherwise, there may be something wrong in the model conceptualization in Step 3 or the data collection in Step 4.
- *Step 8*: Design experiment to evaluate various performance metrics.
- *Step 9*: Run experiment and analyze the system performance.
- *Step 10*: Verify if the number of simulation runs is enough to represent the real statistics of the considered performance metrics. If the number of runs is enough, then go to Step 11. Otherwise, go back to Step 9 to run more simulations. In case that more changes in the design are required, go back to Step 8 to amend the experimental design.
- *Step 11*: Write report and analyze the results for the entire system given the collected experimental data.
- *Step 12*: Implement the design.

For simplicity, the above 12 steps can be grouped into four phases (cf. Figure 8.1) as follows:

- *Phase 1—Formulation and planning*: consists of Steps 1 and 2.
- *Phase 2—Model construction*: consists of Steps 3–7.
- *Phase 3—Model running and data collection*: consists of Steps 8–10.
- *Phase 4—Data analysis, documentation, and implementation*: consists of Steps 11 and 12.

8.4 An example of simulation model

In order to understand further the steps in a simulation study, let us consider an example of a discrete simulation model to represent a queuing system.

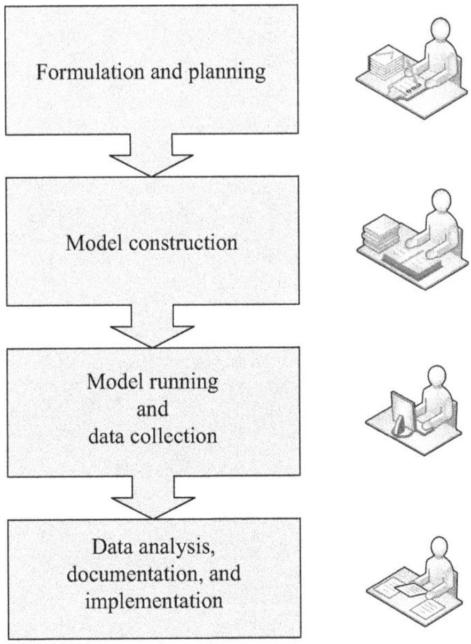

Figure 8.1 Simulation phases

A queuing system is described by calling population, nature of arrivals, service mechanism, system capacity, and queuing discipline (details of queuing system will be discussed in Chapter 15). Specifically, a single-channel queue is considered, having an infinite calling population, random arrivals, random service, unlimited system capacity, and first-come first-serve queuing discipline.

In this simulation, the queuing system consists of a server, data packets being served and data packets waiting in the queue to be served. The state of the system is the number of data packets in the system, and the status of the server is either busy or idle. An event is a set of circumstances that causes an instantaneous change in the state of the system, e.g., arrival and departure events (cf. Example 7.2 in Chapter 7). There is a simulation clock to record the duration of the simulation as well as the number of simulation runs.

The simulation is carried out as in Figure 8.2 with the following phases:

- *Phase 1*: Determine the characteristics of each of the inputs to the simulation model. Then, formulate the problem with objectives and constraints (if any) for the simulation.
- *Phase 2*: Construct a simulation model consisting of p inputs x_i, $i = 1, 2, \ldots, p$, and the corresponding response y_i.
- *Phase 3*: Run simulation. For each simulation run, generate a value for each of the p inputs and evaluate the function to calculate the response y.
- *Phase 4*: Aggregate all simulation data for writing report prior to implementing for a real system.

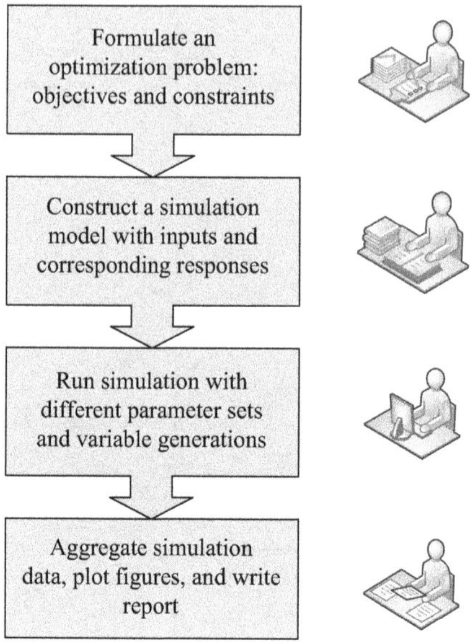

Figure 8.2 Simulation phases in a queuing system

8.5 Related works

For more information, the readers are suggested to refer to the following selected references:

Simulation Models and Process Jain in 1991 [114], Walrand *et al.* in 1999 [194], Robinson in 2004 [195], Banks *et al.* in 2010 [185], Wehrle *et al.* in 2010 [186], and Murray-Smith in 2012 [196].

Deterministic Simulation Models Kumar and Davidson in 1978 [197], Poole and Raftery in 2000 [198], and Kleijnen in 2007 [199].

Stochastic Simulation Models Tijms in 1986 [200], Ripley in 1987 [201], Kulkarni in 1995 [202], and Pinsky and Karlin in 2011 [203].

Discrete Simulation Models Varga in 1999 [184], Fishman in 2001 [204], Banks *et al.* in 2010 [185], and Choi and Kang in 2013 [205].

Continuous Simulation Models Zeigler *et al.* in 2000 [206], Cellier and Kofman in 2006 [207], Birta and Arbez in 2007 [208], and Bandyopadhyay and Bhattacharya in 2014 [209].

8.6 Review questions

[Q8.1] Discuss when simulation should be selected for modeling a system and evaluating its performance.

[Q8.2] What are limitations and disadvantages of simulation?

[Q8.3] Differentiate between static and dynamic simulation models.
[Q8.4] In what systems the simulation models are considered as stochastic?
[Q8.5] Based on what criteria a simulation model can be categorized as discrete or continuous?
[Q8.6] List all steps and phases in a simulation process.
[Q8.7] Provide an example of a simulation model experiencing all of the steps/phases.

8.7 Problems

[P8.1] Provide an example for each type of the following simulation models:
- Static simulation model
- Dynamic simulation model
- Deterministic simulation model
- Stochastic simulation model
- Discrete simulation model
- Continuous simulation model.

[P8.2] Produce a table summarizing all the above simulation models highlighting the differences between them.
[P8.3] Find a research article on networking which uses simulation to verify the mathematical analysis. Do the research on the considered simulation approach to show its necessity and effectiveness in network-performance evaluation.
[P8.4] Provide an example of a complex system that simulation should be performed prior to implementing in a real system.
[P8.5] Discuss on how to perform bottleneck analyses of the system with simulation.
[P8.6] Investigate why the simulation outputs are sometimes regarded as unfeasible and/or unreliable in practical networks.
[P8.7] In a simulation model, how to verify if the number of simulation runs is enough to represent the real statistics of the performance metrics under investigation, i.e., Step 10 in the simulation process? Provide an example of a simulation model where the number of runs is not large enough to validate its performance.

Chapter 9
Statistical models in network simulation

In practical networks, most of the parameters are random and variant over time. Statistical models are crucial in network simulation to represent the real network. Specifically, queuing systems have been developed as a statistical model of the waiting lines in telecommunication networks.

In this chapter, I will first introduce statistical models used for modeling queuing systems with different distribution types in Section 9.1. Section 9.2 will present Poisson point process (PPP) and its variants including homogeneous PPP and inhomogeneous PPP which have been well developed to model the number of events occurred in the statistical models. This chapter will be closed by outlining detailed steps for developing input models in Section 9.3, which include data collection, distribution identification, parameter estimation, and testing.

9.1 Statistical models in queuing systems

Queuing models have been constructed for a number of applications in different areas, such as telecommunications, computing, industrial engineering, data management, etc. In the context of telecommunication networks, the queuing models will hereafter consider the queuing of data packets.

In a queuing system, the interarrival time of data packets and service time at servers are generally random. In order to model the queuing system in a probabilistic way, some statistical models have been developed for the interarrival and service time distribution as follows:

- *Exponential distribution*: If the duration is completely random, it is generally assumed to follow an exponential distribution which is used to represent the time between events in a PPP.
- *Normal distribution*: If the duration is fairly constant but with some random variability which can be either longer or shorter, it can follow a normal distribution.
- *Truncated normal distribution*: If the duration also follows normal distribution but it is restricted within a specific interval, it can be modeled by truncated normal distribution with bounded random variable.

- *Gamma and Weibull distribution*: This distribution is a general form of exponential distribution when both the shape and scale are taken into account.

Details of the above distributions and their properties will be discussed in the following chapters. In addition, the number of packets arrived in the system during a specific interval follows a Poisson distribution, a.k.a. PPP, which will be discussed in Section 9.2.

For evaluating the reliability and maintainability, the time to failure is another random variable that needs to be considered. In order to represent this, some statistical models can be used, such as follows:

- *Exponential distribution*: in case that the failures are random.
- *Normal distribution*: in case that the failures are due to wear after continuous use.
- *Gamma and Erlang distribution*: is generalized from exponential distribution to characterize standby redundancy for failures where each component follows an exponential distribution.
- *Weibull distribution*: to model failures caused by a large number of defects in a system.

For applications in different areas, in order to model a queuing system, various distributions can be employed. For example, in inventory and supply chain, Gamma distribution is used to model the lead time. The number of units demanded per order can follow either Poisson distribution for a general demand or negative binomial distribution, a.k.a. Pascal distribution, for a larger number of demands or geometric distribution as a special case of the negative binomial distribution when at least one demand should occur. There are some other useful distributions that can be found in different applications, such as Bernoulli, binomial, Erlang, uniform, triangular, beta, etc.

The above random phenomena can be either discrete or continuous over time, and thus, the distributions can be classified as

- *Discrete distributions*: are used to describe random phenomena having only countable values.
- *Continuous distributions*: are used to describe random phenomena in which the variable can take an infinite and uncountable value.

Apart from the well-known distributions, there are empirical distributions in which the parameters are observed values from experiment with data sampling. These distributions are helpful when it is impossible to figure out the exact or approximated statistical distribution of the random phenomena. Since this approach is based on the practical experiment, its inherent advantage is that there is no assumption when observing and analyzing the data. However, a number of samples are required for an entire range of possibilities, which is sometimes unfeasible to cover all cases.

9.2 Poisson point process

PPP has been well studied in probability and statistics field to model the number of points randomly located in Euclidean space which can be in one, two, three, or higher dimensions. In queuing theory, the PPP is defined in one dimension, i.e., on a line, as a stochastic process to represent a counting process in a queuing model, while in higher dimensions, it can be used in wireless networks to model the locations of scattered objects, e.g., transmitters and receivers, in space.

The PPP is a special case of a more general process known as Markov process or birth–death process. The Markov process is a stochastic or random process in which the future probabilities are determined by the current state of the process but not on how that state was reached. The Markov process is characterized as "memoryless." As will be shown later, the PPP also has the memoryless property of the Markov process.

In order to understand the PPP, Poisson random variable and its distribution will be discussed in this section. The Poisson distribution can be defined as in Definition 9.1.

Definition 9.1. *The Poisson distribution is a discrete distribution of the number of events that occur randomly in a given interval of time or space.*

Let us denote the number of events occurring in a given interval and its average value by X and λ, respectively. Here, λ is also known as the event rate or rate parameter of Poisson distribution. We have the following definition of the Poisson random variable.

Definition 9.2. *X is a Poisson random variable if the probability of observing x, $x \in \mathbb{N}$, events in a specified interval is given by*

$$\Pr[X = x] = \frac{e^{-\lambda}\lambda^x}{x!}, \tag{9.1}$$

where $x!$ denotes the factorial of x computed by

$$x! \triangleq \prod_{k=1}^{x} k. \tag{9.2}$$

Notice that $0! = 1$, and thus, $\Pr[X = 0]$ still exists which is given by $\Pr[X = 0] = e^{-\lambda}$.

The Poisson random variable has the same average value, a.k.a. mean or expected value, and variance which are given by

$$E[X] = V[X] = \lambda, \tag{9.3}$$

where $E[\cdot]$ denotes the expectation operator and $V[\cdot]$ denotes the variance operator.

Let $N(t)$ denote a counting function that represents the number of events occurred in $[0, t]$. If the event rate, i.e., λ, is a constant, we have homogeneous PPP or stationary PPP (SPPP). Otherwise, we have inhomogeneous PPP or nonstationary PPP (NPPP). In the following subsections, the SPPP, NPPP, and their operations will be sequentially presented in Sections 9.2.1–9.2.3.

9.2.1 Stationary Poisson point process

Definition 9.3. *A counting process $\{N(t), t \geq 0\}$ is a homogeneous PPP or SPPP with mean rate λ if*

- *Arrivals occur one at a time.*
- *$N(t)$ has stationary and independent increments.*
- *$N(t)$ follows Poisson distribution having the following probability function*

$$\Pr[N(t) = n] = \frac{e^{-\lambda t}(\lambda t)^n}{n!}, n \in \mathbb{N} \tag{9.4}$$

Similar to Poisson random variable, $N(t)$ has equal mean and variance which are given by

$$E[N(t)] = V[N(t)] = \lambda t. \tag{9.5}$$

Due to the stationary increment, the number of events in the interval $(a, b]$, $a \leq b$, denoted by $N(a, b]$ follows Poisson distribution with

$$\Pr[N(a, b] = n] = \frac{e^{-\lambda(b-a)}(\lambda(b-a))^n}{n!}, n \in \mathbb{N}. \tag{9.6}$$

The average number of events in $(a, b]$ is given by

$$E[N(a, b]] = \lambda(b - a). \tag{9.7}$$

The number of events is also independent in any disjoint intervals, i.e., $(a_1, b_1]$, $(a_2, b_2]$, $\ldots$, $(a_K, b_K]$, and thus

$$\begin{aligned} \Pr[\{N(a_1, b_1] &= n_1, N(a_2, b_2] = n_2, \ldots, N(a_K, b_K] = n_K\}] \\ &= \prod_{i=1}^{K} \frac{e^{-\lambda(b_i - a_i)}(\lambda(b_i - a_i))_i^n}{n_i!}. \end{aligned} \tag{9.8}$$

Example 9.1. *Packet arrivals at the server occur randomly following a PPP $N(t)$ with a rate of $\lambda = 10$ packets per minute. $N(t)$ has the following probability function:*

$$\Pr[N(t) = n] = \frac{e^{-10t}(10t)^n}{n!}, n \in \mathbb{N}.$$

The probability of observing ten packets in a minute is

$$\Pr[N(1) = 10] = \frac{e^{-10}10^{10}}{10!} \approx 0.125.$$

Extended for the number of events in the interval $(a + c, b + c]$, we have

$$\Pr[N(a + c, b + c] = n] = \Pr[N(a, b] = n]. \tag{9.9}$$

From (9.9), it can be observed that the distribution of $N(a + c, b + c]$ is independent of c, while it only depends on the interval length, i.e., $b - a$. This accordingly verifies the memoryless property of the SPPP.

In order to model the waiting lines in queuing systems with fixed arrival rate, the SPPP is widely used since it approximately fits well in the actual behavior of these

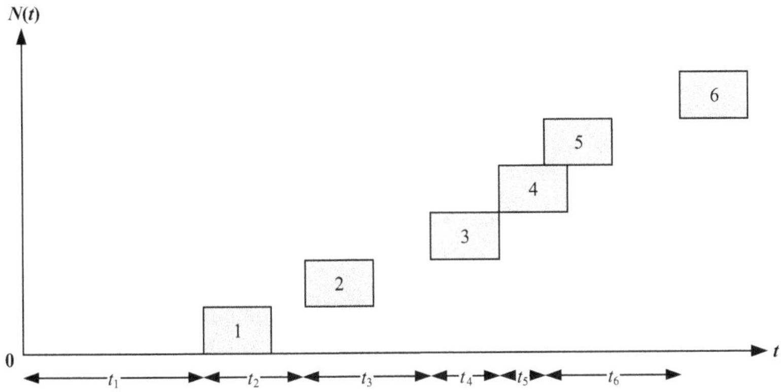

Figure 9.1 Interarrival time in Poisson point process $N(t)$

systems in practice. Specifically, in communication networks, the arrival of data packets is random over time with a relatively short duration, which can be described by the interarrival time modeled by an exponential random variable and the number of arrivals in a given time interval represented by a Poisson random variable (cf. Figure 9.1). Due to the stationary, memoryless, and independent property of the SPPP, the packet arrivals in a time interval have the same rate and have no influence on the arrivals in other nonoverlapping time intervals. It can be noticed that this scenario likely occurs when the data packets are generated from a large number of independent sources and the characteristics of the interarrival distribution are invariant.

Theorem 9.1. *In the SPPP, the time difference between events, a.k.a. interarrival time, is exponentially distributed and independent with a mean of* $1/\lambda$.

Proof. Let us consider the interarrival time $t_1, t_2, \ldots,$ of a Poisson process, where t_i is the elapsed time between the ith arrival and the $(i+1)$th arrival.

The first arrival occurs after time t if and only if there is no arrival in the interval $[0, t]$, i.e.,

$$\Pr[t_1 > t] = \Pr[N(t) = 0] = e^{-\lambda t}. \tag{9.10}$$

In general, the $(i+1)$th arrival occurs after time $t_i + t$ if and only if there is no arrival in the interval $(t_i, t_i + t]$. From (9.6), we have

$$\Pr[t_{i+1} - t_i > t] = \Pr[N(t_i, t_i + t] = 0] = e^{-\lambda t}, \tag{9.11}$$

and thus

$$\Pr[t_{i+1} - t_i \leq t] = 1 - e^{-\lambda t}. \tag{9.12}$$

Equation (9.12) is indeed the cumulative distribution function (cdf) of the exponentially random variable with a mean of $1/\lambda$. This means the interarrival time is exponential distributed and independent. The lemma is proved. □

This invariant average interval implies that the SPPP is stationary and memoryless.

Adapting to queuing systems with a service facility, the next customer is served as soon as the one in service leaves the system. The time between service completions is indeed the service time. Therefore, if the time between completions is exponentially distributed, then the service time itself is exponentially distributed. This means that the service is also an SPPP.

9.2.2 Nonstationary Poisson point process

Definition 9.4. *A counting process $\{N(t), t \geq 0\}$ is an inhomogeneous or NPPP characterized by a rate $\lambda(t)$ which varies over time. $N(t)$ has the following probability function:*

$$\Pr[N(t) = n] = \frac{e^{-\Lambda(t)}(\Lambda(t))^n}{n!}, n \in \mathbb{N}. \tag{9.13}$$

Here, $\Lambda(t)$ denotes the average number of arrivals over time t, which can be determined by

$$\Lambda(t) = E[N(t)] = \int_0^t \lambda(s)ds.$$

Due to the stationary increment, the number of events in the interval $(a, b], a \leq b$, denoted by $N(a, b]$ follows Poisson distribution with

$$\Pr[N(a, b] = n] = \frac{e^{-\Lambda(a,b)}(\Lambda(a, b))^n}{n!}, n \in \mathbb{N}, \tag{9.14}$$

where $\Lambda(a, b)$ denotes the average number of events in $(a, b]$ given by

$$\Lambda(a, b) = E[N(a, b]] = \int_a^b \lambda(t)dt.$$

The number of events is also independent in any disjoint intervals, i.e., $(a_1, b_1], (a_2, b_2], \ldots, (a_K, b_K]$, and thus

$$\begin{aligned}\Pr[\{N(a_1, b_1] &= n_1, N(a_2, b_2] = n_2, \ldots, N(a_K, b_K] = n_K\}] \\ &= \prod_{i=1}^{K} \frac{e^{-\Lambda(a_i,b_i)}(\Lambda(a_i, b_i))_i^n}{n_i!}.\end{aligned} \tag{9.15}$$

Example 9.2. *Suppose e-mail arrivals to a server follow a PPP having rates of 2 per minute in the morning from 8* AM *to 12* PM, *and then 0.5 per minute in the afternoon from 12 to 4* PM.

Set $t = 0$ corresponds to 8 AM, *an NPPP $N(t)$ over time $[0, t]$ has the following rate function:*

$$\lambda(t) = \begin{cases} 2 & \text{if } 0 \leq t < 4, \\ 0.5 & \text{if } 4 \leq t < 8. \end{cases}$$

The probability distribution of the number of arrivals over time $[0, t]$ *is expressed by*

$$\Pr[N(t) = n] = \frac{e^{-\Lambda(t)}(\Lambda(t))^n}{n!}, n \in \mathbb{N},$$

where the average number of arrivals by time t, i.e., $\Lambda(t)$*, is given by*

$$\Lambda(t) = \begin{cases} \int_0^t 2ds = 2t & \text{if } 0 \le t < 4, \\ \int_0^4 2ds + \int_4^t 0.5ds = \frac{t}{2} + 6 & \text{if } 4 \le t < 8. \end{cases}$$

The average number of arrivals between 11 AM *and 2* PM *is computed by*

$$\Lambda(3, 6) = \int_3^4 2ds + \int_4^6 0.5ds = 3.$$

The probability distribution of the number of arrivals between 11 AM *and 2* PM *is thus given by*

$$\Pr[N(3, 6) = n] = \frac{e^{-\Lambda(a,b)}(\Lambda(a, b))^n}{n!} = \frac{e^{-3}3^n}{n!}.$$

9.2.3 *Operations of Poisson point process*

We can perform mathematical operations on the PPP to generate different PPPs. Some of them are as follows:

- *Superposition/Pooling*: The union of a countable set of PPPs $N_1(t)$, $N_2(t)$, ..., forms a pooled process which is also a PPP due to the fact that a distribution resulting from the sum of Poisson distributions retains the Poisson distribution. Let $N(t)$ denote the union set, i.e.,

  $$N(t) = \bigcup_i N_i(t). \tag{9.16}$$

  $N(t)$ is a PPP with a mean of

  $$\Lambda(t) = \sum_i \Lambda_i(t), \tag{9.17}$$

  where $\Lambda_i(t)$ is the average number of arrivals over time t given by (9.4) for the case of NPPP. The superposition property of the PPP means that if the arrivals to a multi-server facility are PPP with each server having exponential service times, the departures also constitute a Poisson stream.
- *Splitting/Thinning*: A PPP with a mean $\Lambda(t)$ can be split into multiple subprocesses, each having a probability p_i, $i = 1, 2, \ldots$. The ith subprocess is also a PPP with a mean rate of

  $$\Lambda_i(t) = p_i\Lambda(t). \tag{9.18}$$

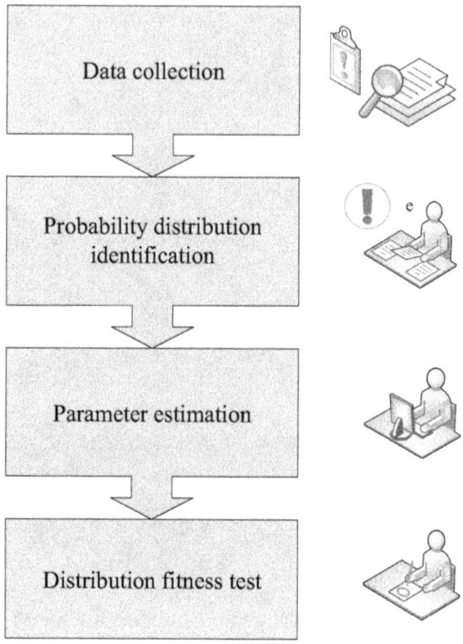

Figure 9.2 Input modeling in network simulation

9.3 Input modeling

In any simulation model, the initial step is always to determine what kind of input models should be suitable for the simulation. The development of an input model is of great importance for building an appropriate simulation model.

Basically, there are four steps for developing an input model (cf. Figure 9.2), which are as follows:

- *Step 1 (data collection)*: to collect data from the real system.
- *Step 2 (probability distribution identification)*: to identify a probability distribution to represent the input process given the collected data.
- *Step 3 (parameter estimation)*: to estimate parameters for the distribution.
- *Step 4 (distribution fitness test)*: to evaluate the selected distribution and parameters via goodness-of-fit tests.

Details of the above steps will be discussed in the following subsections.

9.3.1 Data collection

Data collection is crucial in the development of an input model. The data is collected to solve a variety of problems in a real system. The data collection process is based on the principle of garbage-in–garbage-out, which is used to describe the nonsense data output, i.e., "garbage," caused by faulty data collection.

The data collection can be enhanced and facilitated by employing the following:

- *Plan ahead*: to predict circumstances via either practice or pre-observing session, especially keeping track of unusual ones.

- *Analyze data promptly*: to evaluate the data as long as it is observed to check for its appropriateness and accuracy.
- *Combine data*: to combine data sets of the same types over successive time periods.
- *Examine data*: to perform data censoring over the entire collected data set.
- *Check variables*: to validate the relationship between variables in addition to autocorrelation checking.

9.3.2 Probability distribution identification

Given the collected data, this step identifies the probability distribution of the data set by developing a histogram of the data. The histogram represents the frequency distribution of the data, through which the shape of the distribution can be determined. In order to have an appropriate shape, the data distribution depends on the sample size, i.e., the number of observations, and also the dispersion of the data.

Depending on data types, the histogram can be verified through the theoretical distribution as follows:

- *Continuous data*: The shape of histogram should match the probability density function.
- *Discrete data*: The shape of histogram should match the probability mass function.

Based on the shape of the histogram and the context of the input variables, the distribution for random variables can be determined. Some well-known family of distributions will be discussed in detail in Chapters 12 and 13 for discrete and continuous random variables, respectively. Specifically, the physical basis of the distribution can be exploited as a guide to select the appropriate distribution or an approximated distribution. For instance, the following distributions can be identified through their properties and functionalities:

- *Binomial distribution*: to represent the number of successes in a specified number of trials.
- *Poisson distribution*: to represent the number of events that occur in a specified amount of time or space.
- *Normal distribution*: to represent the sum of a number of processes of various components.
- *Exponential distribution*: to represent the time between independent/memoryless events.
- *Weibull distribution*: to represent the time to failure of components.
- *Discrete or continuous uniform distribution*: to represent equally likely outcomes within a specified range.
- *Triangular distribution*: to represent a random process knowing only its minimum, most likely and maximum values.
- *Empirical distribution*: to sample the observed data.

Note that the goal of the distribution identification step to find a good approximation of the distribution function given the physical characteristics of the random process. In order to evaluate the fitness of a distribution, a quantile–quantile (Q–Q) plot can be exploited. The Q–Q plot is a graphical method which allows us to compare two probability distributions through the shape of their quantiles. Here, the quantiles are defined as the value sets of a variable which divide its histogram into contiguous intervals with equal probabilities.

The quantile of a random variable can be defined by Definition 9.5.

Definition 9.5. *If X is a random variable having cdf $F_X(x)$, then the q-quantile, $q \in [0, 1]$, of X is the α such that*

$$F_X(\alpha) \triangleq \Pr[X \le \alpha] = q. \tag{9.19}$$

If $F_X(x)$ is invertible, then α can be determined as

$$\alpha = F_X^{-1}(q). \tag{9.20}$$

Let x_i, $i = 1, 2, \dots, n$, denote a sample of data from the random variable X and y_j, and $j = 1, 2, \dots, n$, denote the observations of the real data in ascending order. It can be approximated that

$$y_j \approx F_X^{-1}\left(\frac{j - 0.5}{n}\right) \tag{9.21}$$

where j is called the ranking or order number. This accordingly means that if X is an appropriate random variable to represent the observed data, then the plot of $\{y_j\}$ versus $\left\{F_X^{-1}\left((j - 0.5)/n\right)\right\}$ is approximately a straight line. A further observation is that if the line has a slope of one then the parameter values of the distribution are appropriate.

Besides the check for appropriateness of a distribution, the Q–Q plot can be employed to check the homogeneity between two sample sets by comparing their plots of the order values. If they are homogeneous, then a distribution can represent both of them.

9.3.3 Parameter estimation

The next step after selecting an appropriate family of distribution is to estimate parameters of the distributions. Specifically, mean and variance are two parameters that are crucial in any distribution. Given a number of observed data, a.k.a. samples, their sample mean and variance are defined by Definition 9.7.

Definition 9.6. *Given a sampled data set of size N, i.e., $\{X_1, X_2, \dots, X_N\}$, its sample mean and variance are determined by*

$$\bar{X} \triangleq \frac{\sum_{i=1}^{N} X_i}{N}, \tag{9.22}$$

$$Var[X] \triangleq \frac{\sum_{i=1}^{N} X_i^2 - N\bar{X}^2}{N - 1}. \tag{9.23}$$

Definition 9.7. *If the sampled data set are grouped with a frequency distribution f_i for the observed value X_i, the sample mean and variance can be given by*

$$\bar{X} = \frac{\sum_{i=1}^{G} f_i X_i}{\sum_{i=1}^{G} f_i}, \tag{9.24}$$

$$Var[X] = \frac{\sum_{i=1}^{G} f_i X_i^2 - \sum_{i=1}^{G} f_i \bar{X}^2}{\sum_{i=1}^{G} f_i - 1}, \tag{9.25}$$

where G is the number of groups in the data set.

Through statistical analysis over a number of observed values, the sample mean and variance are regarded as two statistics of the distribution.

9.3.4 Distribution fitness test

The last step in the development of an input model is to evaluate the selected distribution and the estimated parameters. The appropriateness of the input data distribution can be examined by conducting hypothesis testing, a.k.a. goodness-of-fit test, which is defined as a statistical model used to determine how a distribution fits a sampled data set. Some goodness-of-fit tests can be employed, such as Kolmogorov–Smirnov (K–S) test, chi-squared test for large sample size, Fisher test for small sample size, etc. Note that there is no single correct distribution in a real application, especially when there are numerous data samples available.

Intuitively, a hypothesis test can be carried out by comparing the histogram of the data samples with the shape of the probability mass function for discrete data or the probability density function for continuous data. Specifically, the K–S test is based on the comparison of the distance between the empirical distribution function of the sample and the cdf of the reference distribution to evaluate the fitness of the candidate distribution.

9.4 Related works

For more information, the readers are suggested to refer to the following selected references:

Statistical Models Davison in 2003 [210], Cox in 2006 [211], Rissanen in 2007 [212], Wasserman in 2010 [213], and Kroese and Chan in 2013 [214].

Probability Distribution Papoulis in 2002 [215], Forbes *et al.* in 2010 [216], Montgomery and Runger in 2013 [217], and Ross in 2014 [218].

Poisson Point Process Ross in 1995 [219], Nelson in 1995 [220], Kulkarni in 1995 [202], Papoulis in 2002 [215], Daley and Vere-Jones in 2003 and 2008 [221,222], Illian *et al.* in 2008 [223], Bass in 2011 [224], and Snyder and Miller in 2012 [225].

Input modeling Jain in 1991 [114], Banks *et al.* in 2010 [185], Choi and Kang in 2013 [205], and Obaidat in 2015 [226].

Parameter Estimation Bos in 2007 [227], Levy in 2008 [228], and Aster *et al.* in 2012 [229].

Distribution Fitness Test Massey in 1951 [230], D'Agostino in 1986 [231], Greenwood and Nikulin in 1996 [232], Fan in 1997 [233], and Justel *et al.* in 1997 [234].

9.5 Review questions

[Q9.1] Provide some statistical models which are commonly used in queuing systems.

[Q9.2] How to model packet interarrival time, service time, and failure time in a queuing model?

[Q9.3] What is a Poisson point process (PPP)?

[Q9.4] What is the probability of observing a number of events x within a specified interval in a PPP having an event rate of λ?

[Q9.5] Differentiate between stationary PPP and nonstationary PPP.

[Q9.6] What is the probability function of stationary PPP and nonstationary PPP within a specific interval t?

[Q9.7] Explain how to generate different PPPs with superposition/pooling and splitting/thinning operations.

[Q9.8] Describe the steps for developing an input model in network simulation. Illustrate them with a flowchart.

[Q9.9] How to verify the appropriateness of the collected data?

[Q9.10] Explain how to identify a probability distribution of the collected data.

[Q9.11] Provide some typical distributions that can be easily identified through their properties.

[Q9.12] What are quantiles of a random variable?

[Q9.13] How to determine the q-quantile of a random variable X?

[Q9.14] Explain how to employ Q–Q plot to evaluate the fitness of a distribution of a random variable.

[Q9.15] What are sample mean and sample variance of a data set and how to determine them?

[Q9.16] What is distribution fitness test and its principle in examining a data distribution?

9.6 Problems

[P9.1] Show that the interarrival time in both stationary and nonstationary PPP follows an exponential distribution.

[P9.2] Does the probability function increase or decrease if the rate of the PPP increases? Prove this.

[P9.3] Consider packet arrivals at a server following a PPP with a rate of 25 packets per minute.

- Derive the probability function of the PPP.
- Find the probability that there are 30 packet arrivals in a minute.
- Assume the rate of the PPP is doubled, i.e., 50 packets per minute, find the probability that there are 30 packet arrivals in a minute.

[P9.4] E-mails arrive at a server following a PPP with a rate of 1,000 e-mails per day during the weekday, i.e., from Monday to Friday, and 200 e-mails per day during the weekend. Assume a week starts on Monday, i.e., $t = 0$, and ends on Sunday, i.e., $t = 6$.

- Derive the probability function of the number of e-mails over time $[0, t]$ where $0 \leq t \leq 6$.
- Derive the probability function of the number of e-mails received from Wednesday to Saturday.

[P9.5] Provide two examples to illustrate the pooling and thinning of PPP.

[P9.6] Provide an example to show that Q–Q plot can be used to check the homogeneity between two sample sets.

[P9.7] Do the research on goodness-of-fit tests. Discuss the principle and usage of Kolmogorov–Smirnov test and Fisher test for various sample sets.

Chapter 10
Probabilities in performance analysis

When talking about the randomness of an event in statistical models, it always goes together with the probability concept. This chapter is devoted to introduce the probabilities in the context of network performance analysis. For some people who may have learned and are familiar with the general concept of the probability, this chapter will help them review these and present the terminology and concepts in the communication networks.

In this chapter, I will first introduce the basic concepts and terminology of the probability in Section 10.1. Section 10.2 will present axioms and properties of the probability. Considering the relationship between events, Sections 10.3 and 10.4 will discuss the conditional probability and independence of the events along with their applications in network-performance evaluation.

10.1 Basic concepts and terminology of probability

In statistical theory, some well-known concepts and terminology of the probability can be defined as follows:

Definition 10.1. *An experiment is an examination or a trial run conducted through a procedure of finding, an observation of fact(s) and a model of data.*

Example 10.1. *An experiment could consist of the following:*

- *A procedure of finding a server for data storage among a number of available servers.*
- *An observation of which servers are appropriate or inappropriate.*
- *A model of selecting an appropriate server after a specified number of trials given various constraints on the criteria of server selection.*

Definition 10.2. *An outcome is any possible observation of an experiment.*

Definition 10.3. *A sample space is a set of all possible outcomes of an experiment satisfying that the set is finest grain, mutually exclusive, and collectively exhaustive.*

In order to understand clearly Definition 10.3, in-depth knowledge of set theory is required. The concept "finest grain," "mutually exclusive," and "collectively exhaustive" can be simply understood as follows:

- *Finest grain set*: is a set consisting of all distinguishable outcomes.
- *Mutually exclusive set*: A set $\mathfrak{S}$ of N elements x_i, $i = 1, 2, \ldots, N$, is mutually exclusive when

  $$x_i \neq x_j, \qquad i \neq j. \tag{10.1}$$

  If the set $\mathfrak{S}$ has K subsets $\mathfrak{S}_1, \mathfrak{S}_2, \ldots, \mathfrak{S}_K \subset \mathfrak{S}$, then $\mathfrak{S}$ is a mutually exclusive set when

  $$\mathfrak{S}_i \cap \mathfrak{S}_j = \varnothing, \qquad i \neq j, \tag{10.2}$$

  where $\varnothing$ denotes a null set which has no elements and $\cap$ is an intersection operation of sets corresponding to logical AND operation (denoted by $\wedge$). By definition,

  $$x \in \mathfrak{A} \cap \mathfrak{B} \Leftrightarrow x \in \mathfrak{A} \wedge x \in \mathfrak{B}. \tag{10.3}$$

- *Collectively exhaustive set*: is a set consisting of every outcomes, i.e.,

  $$\mathfrak{S} = \mathfrak{S}_1 \cup \mathfrak{S}_2 \cup \cdots \cup \mathfrak{S}_K \triangleq \bigcup_{i=1}^{K} \mathfrak{S}_i, \tag{10.4}$$

  where $\cup$ is a union operation of sets corresponding to logical OR operation (denoted by $\vee$). By definition,

  $$x \in \mathfrak{A} \cup \mathfrak{B} \Leftrightarrow x \in \mathfrak{A} \vee x \in \mathfrak{B}. \tag{10.5}$$

Remark 10.1. *A set is mutually exclusive and collectively exhaustive when it satisfies both* (10.2) *and* (10.4).

Depending on data types, a sample space can be either a *discrete* or *continuous* set.

Example 10.2. *The sample space when decoding three received data packets is*

$$\mathfrak{S} = \{s_1s_2s_3, s_1s_2f_3, s_1f_2s_3, f_1s_2s_3, s_1f_2f_3, f_1s_2f_3, f_1f_2s_3, f_1f_2f_3\},$$

where s_i and f_i, $i = 1, 2, 3$, are the successful and failed outcomes, respectively, of the ith packet.

Definition 10.4. *An event is a specified set/subset of outcomes in an experiment. The corresponding event space is a mutually exclusive and collectively exhaustive set of all the outcomes of the event.*

An illustration of event in an experiment is shown in Figure 10.1 where the sample space of the experiment has eight outcomes, and a specified event consisting of two outcomes is selected.

Example 10.3. *Suppose we examine three decoded data packets. Let s_i and f_i, $i = 1, 2, 3$, denote the successful packet and failed packet, respectively, of the ith*

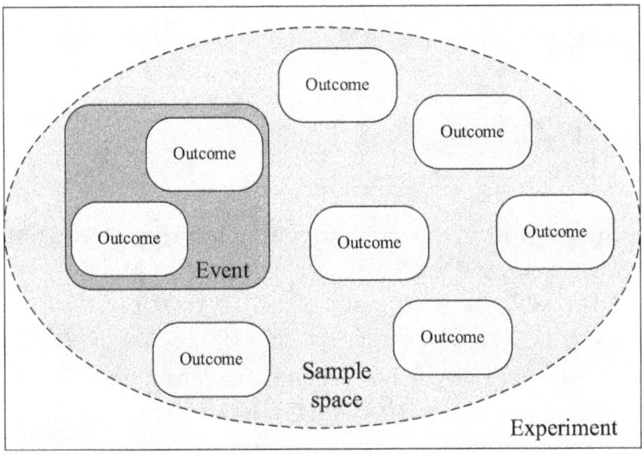

Figure 10.1 Event in an experiment

packet. An event E consists of all possible outcomes which have at least two successful packets. The corresponding event space $\mathfrak{E}$ of the event E is

$$\mathfrak{E} = \{s_1s_2s_3, s_1s_2f_3, s_1f_2s_3, f_1s_2s_3\},$$

Definition 10.5. *An experiment is random if its outcome cannot be predicted with certainty or determined with a probability.*

Most experiments in practice are random. For instance, the communication channels in wireless networks suffer from random fading and noise due to the randomness of the communication media.

10.2 Axioms and properties of probability

In a sample space, each event is mapped to a real number called a probability measure. The probability of an event E, denoted by $\Pr[E]$, also means the likelihood of outcomes in that event.

Remark 10.2. *Two well-known axioms of the probability are*

$$0 \le \Pr[E] \le 1 \tag{10.6}$$

$$\Pr[\mathfrak{S}] = \sum_{i=1}^{K} \Pr[\mathfrak{S}_i] = 1, \tag{10.7}$$

where $\mathfrak{S}$ is a sample space having $\mathfrak{S} = \bigcup_{i=1}^{K} \mathfrak{S}_i$ and $\mathfrak{S}_i \cap \mathfrak{S}_j = \varnothing \forall i \neq j$.

Equation (10.7) can also be applied for any sets having a number of subsets as in Theorem 10.1.

Theorem 10.1. *If a set $\mathfrak{A}$ has M collectively exhaustive and mutually exclusive subsets $\{\mathfrak{A}_1, \mathfrak{A}_2, \ldots, \mathfrak{A}_M\}$, then*

$$\Pr[\mathfrak{A}] = \Pr\left[\bigcup_{i=1}^{M} \mathfrak{A}_i\right] = \sum_{i=1}^{M} \Pr[\mathfrak{A}_i]. \tag{10.8}$$

Proof. The proof follows from the definition of union and intersection operations of collectively exhaustive set and mutually exclusive set, respectively, with $\mathfrak{A} = \mathfrak{A}_1 \cup \mathfrak{A}_2 \cup \cdots \cup \mathfrak{A}_M$ and $\mathfrak{A}_i \cap \mathfrak{A}_j = \varnothing, \forall i \neq j$. □

A special case of Theorem 10.1 is when considering an event with a specified number of outcomes as in Lemma 10.1 and Corollary 10.1.

Lemma 10.1. *If an event E has M outcomes, i.e., $E = \{e_1, e_2, \ldots, e_M\}$, then*

$$\Pr[E] = \sum_{i=1}^{M} \Pr[e_i]. \tag{10.9}$$

Proof. The proof follows from Theorem 10.1 where each outcome e_i, $i = 1, 2, \ldots, M$, is regarded as a subset of a set E. □

Corollary 10.1. *If each outcome e_i, $i = 1, 2, \ldots, M$, of a sample space $\mathfrak{S}$ is equally likely, i.e., $\Pr[e_1] = \Pr[e_2] = \cdots = \Pr[e_M]$, then*

$$\Pr[e_i] = \frac{1}{M}. \tag{10.10}$$

Proof. The proof follows from the probability axiom $\Pr[\mathfrak{S}] = 1$ and from Lemma 10.1 with $\Pr[\mathfrak{S}] = \sum_{i=1}^{M} \Pr[e_i] = M \Pr[e_i]$. □

Example 10.4. *A data symbol consists of 2 b. There are four different data symbols which can be generated from these 2 b. Assume that they are equally likely. Applying Corollary 10.1, the probability of each outcome is* $1/4$.

Some other properties of probability can be obtained from the set theory concepts as in Theorems 10.2 and 10.3.

Theorem 10.2. *A null set $\varnothing$ has*

$$\Pr[\varnothing] = 0. \tag{10.11}$$

Proof. The proof follows from the fact that there is nothing in the null set. □

Theorem 10.3. *A complement of set $\mathfrak{A}$, denoted by $\overline{\mathfrak{A}}$, has*

$$\Pr[\overline{\mathfrak{A}}] = 1 - \Pr[\mathfrak{A}]. \tag{10.12}$$

Proof. The proof follows from the definition of the complement set, i.e.,

$$x \in \overline{\mathfrak{A}} \Leftrightarrow x \notin \mathfrak{A}. \tag{10.13}$$

Equivalently, we have

$$\Pr[\mathfrak{A}] + \Pr[\overline{\mathfrak{A}}] = 1. \tag{10.14}$$

□

Example 10.5. *A data packet is received correctly with a probability of* 0.2. *Then, the probability of the packet to be corrupted or lost is* $1 - 0.2 = 0.8$.

Theorem 10.4. *A subset* $\mathfrak{A}_i$ *of a set* $\mathfrak{A}$ *has*

$$\Pr[\mathfrak{A}_i] \leq \Pr[\mathfrak{A}]. \tag{10.15}$$

Proof. The proof follows from the definition of a subset, i.e., if $x \in \mathfrak{A}_i$, then $x \in \mathfrak{A}$. □

Example 10.6. *A data symbol consists of 2 b* 0 *and* 1. *Four different data symbols generated from these 2 b are assumed to be equally likely. Let A denote the event that the data symbol consists of at least 1 b;* 1 *and B denote the event that the data symbol consists of exact 2 b* 1. *Events A and B can have the following possible outcomes:*

$$A = \{01, 10, 11\}$$

$$B = \{11\}$$

It can be seen that $B \subset A$, $\Pr[B] = 1/4$, *and* $\Pr[A] = 3/4$. *This verifies that* $\Pr[B] \leq \Pr[A]$ *as shown in Theorem 10.4.*

Theorem 10.5. *Given any two sets* $\mathfrak{A}$ *and* $\mathfrak{B}$, *we have*

$$\Pr[\mathfrak{A} \cup \mathfrak{B}] = \Pr[\mathfrak{A}] + \Pr[\mathfrak{B}] - \Pr[\mathfrak{A} \cap \mathfrak{B}] \tag{10.16}$$

Proof. The proof follows from the definition of union set. The union set of $\mathfrak{A}$ and $\mathfrak{B}$ is equal to set $\mathfrak{A}$ plus set $\mathfrak{B}$ minus the intersection set between $\mathfrak{A}$ and $\mathfrak{B}$, i.e., $\mathfrak{A} \cap \mathfrak{B}$. □

Note that the probability of intersection set $\mathfrak{A} \cap \mathfrak{B}$ can be simply written as $\Pr[\mathfrak{A}\mathfrak{B}]$.

Example 10.7. *Suppose that a source* $\mathcal{S}$ *wants to send data packets to both destinations* $\mathcal{D}_1$ *and* $\mathcal{D}_2$. *The probability that the packet is received correctly: (i) at* $\mathcal{D}_1$ *is* 0.3, *(ii) at* $\mathcal{D}_2$ *is* 0.4, *and (iii) at both* $\mathcal{D}_1$ *and* $\mathcal{D}_2$ *is* 0.1. *Then, by employing* (10.16) *in Theorem 10.5, the probability that the packet is received correctly at either* $\mathcal{D}_1$, $\mathcal{D}_2$, *or both can be found as* $0.3 + 0.4 - 0.1 = 0.6$.

Theorem 10.6. *Given probabilities of mutually exclusive intersection sets* $\mathfrak{A} \cap \mathfrak{B}_1, \mathfrak{A} \cap \mathfrak{B}_2, \ldots, \mathfrak{A} \cap \mathfrak{B}_M$, *the probability of set* $\mathfrak{A}$ *can be determined by*

$$\Pr[\mathfrak{A}] = \sum_{i=1}^{M} \Pr[\mathfrak{A}\mathfrak{B}_i]. \tag{10.17}$$

Proof. The proof follows from the fact of union and intersection sets that

$$\mathfrak{A} = (\mathfrak{A} \cap \mathfrak{B}_1) \cup (\mathfrak{A} \cap \mathfrak{B}_2) \cup \cdots \cup (\mathfrak{A} \cap \mathfrak{B}_M) = \bigcup_{i=1}^{M} (\mathfrak{A} \cap \mathfrak{B}_i), \tag{10.18}$$

and from the computation of the probability of the union set as in Theorem 10.1. □

Example 10.8. *A model of mobile phone usage consists of voice (V) and data (D). Each voice or data call can be classified as either long (L) or short (S). The probability of a long voice call, i.e.,* $\Pr[LV]$, *is* 0.2, *of a short voice call, i.e.,* $\Pr[SV]$, *is* 0.3, *of a long data call, i.e.,* $\Pr[LD]$, *is* 0.4, *and of a short data call, i.e.,* $\Pr[SD]$, *is* 0.1. *Applying Theorem 10.6, we can obtain the probability of a long call, i.e.,* $\Pr[L]$, *and of a voice call, i.e.,* $\Pr[V]$, *as follows:*

$$\begin{aligned} \Pr[L] &= \Pr[LV] + \Pr[LD] = 0.2 + 0.4 = 0.6 \\ \Pr[V] &= \Pr[LV] + \Pr[SV] = 0.2 + 0.3 = 0.5. \end{aligned}$$

10.3 Conditional probability

In a practical system, events generally occur in sequence. The probability that an event occurs given that another event has occurred is called a conditional probability. This section will introduce the concept of conditional probability.

Definition 10.6. *The conditional probability of an event A given an event B is denoted by* $\Pr[A|B]$ *which can be determined by*

$$\Pr[A|B] \triangleq \frac{\Pr[AB]}{\Pr[B]}. \tag{10.19}$$

Remark 10.3. *Definition 10.6 also means that in order for both events A and B to occur, event B can occur first and then event A occurs given that B has already occurred. Equation* (10.19) *can be equivalently written as*

$$\Pr[AB] = \Pr[A|B]\Pr[B]. \tag{10.20}$$

Remark 10.4. *A notice in* (10.19) *is that there is an implicit condition of* $\Pr[B] > 0$. *Indeed, if* $\Pr[B] = 0$, *i.e., event B does not exist, then* $\Pr[A|B]$ *has no meaning.*

Example 10.9. *The validation of a communication channel between two devices is performed with two error checks. 20% of the data packets have errors that can be* *detected in both checks, while 40% of the data packets can only be detected for errors in the first check.*

Let A and B denote the events that the packets can be detected for errors in the first check and in the second check, respectively. We have $\Pr[AB] = 0.2$ *and* $\Pr[A] = 0.4$.

The probability of the packets that can be detected for errors in the second check if they can also be detected in the first check is

$$\Pr[B|A] = \frac{\Pr[AB]}{\Pr[A]} = \frac{0.2}{0.4} = 0.5.$$

Remark 10.5. *Similar to the axioms of the probability, we have some well-known axioms of the conditional probability as follows:*

$$0 \leq \Pr[A|B] \leq 1, \tag{10.21}$$

$$\Pr[A|A] = 1. \tag{10.22}$$

Some other properties of the conditional probability are shown in the following theorems:

Theorem 10.7. *If a set* $\mathfrak{A}$ *has M collectively exhaustive and mutually exclusive subsets* $\{\mathfrak{A}_1, \mathfrak{A}_2, \ldots, \mathfrak{A}_M\}$*, then*

$$\Pr[\mathfrak{A}|\mathfrak{B}] = \Pr\left[\bigcup_{i=1}^{M} \mathfrak{A}_i|\mathfrak{B}\right] = \sum_{i=1}^{M} \Pr[\mathfrak{A}_i|\mathfrak{B}]. \tag{10.23}$$

Proof. The proof follows from Definition 10.6 and Theorem 10.1, i.e.,

$$\begin{aligned} \Pr[\mathfrak{A}|\mathfrak{B}] &= \frac{\Pr[\mathfrak{A}\mathfrak{B}]}{\Pr[\mathfrak{B}]} \\ &= \frac{\Pr\left[\bigcup_{i=1}^{M} \mathfrak{A}_i\mathfrak{B}\right]}{\Pr[\mathfrak{B}]} \\ &= \Pr\left[\bigcup_{i=1}^{M} \mathfrak{A}_i|\mathfrak{B}\right] \\ &= \sum_{i=1}^{M} \Pr[\mathfrak{A}_i|\mathfrak{B}]. \end{aligned}$$

□

Theorem 10.8 (Bayes' theorem). *The relationship between conditional probabilities of two events A and B is*

$$\Pr[A|B] = \frac{\Pr[B|A]\Pr[A]}{\Pr[B]}. \tag{10.24}$$

Proof. From Definition 10.6 of the conditional probability, $\Pr[B|A]$ is given by

$$\Pr[B|A] = \frac{\Pr[AB]}{\Pr[A]}$$

or

$$\Pr[AB] = \Pr[B|A]\Pr[A]. \tag{10.25}$$

Substituting (10.25) into (10.19) completes the proof. □

Example 10.10. *A server carries out two tests on every received data packets. 25% of the packets pass both tests, while 40% of the packets pass the first test, and 30% of the packets pass the second test.*

Let A and B denote the event that the packets pass the first test and the second test, respectively. We have $\Pr[AB] = 0.25$, $\Pr[A] = 0.4$, *and* $\Pr[B] = 0.3$

The probability of the packets that pass the second test if they also pass the first test is

$$\Pr[B|A] = \frac{\Pr[AB]}{\Pr[A]} = \frac{0.25}{0.4} = 0.625.$$

The probability of the packets that pass the first test if they also pass the second test is

$$\Pr[A|B] = \frac{\Pr[AB]}{\Pr[B]} = \frac{0.25}{0.3} = 0.833.$$

$\Pr[A|B]$ *can also be interpreted as the probability that the accepted data packets in the second test can pass the first test.*

Remark 10.6. *It can be noted in example that* $\Pr[A|B] \neq \Pr[B|A]$. *These two conditional probabilities also have different meanings.*

A special case is when two events A and B are equally likely as in the following lemma.

Lemma 10.2. *If two events A and B are equally likely, then*

$$\Pr[A|B] = \Pr[B|A]. \tag{10.26}$$

Proof. The proof follows from Theorem 10.8 when $\Pr[A] = \Pr[B]$. □

Considering the intersection of different events, we have a law of total probability in Theorem 10.9.

Theorem 10.9. *Given probabilities of M events $\{B_1, B_2, \ldots, B_M\}$ and conditional probabilities of event A given B_i, $i = 1, 2, \ldots, M$, the probability of event A can be determined by*

$$\Pr[A] = \sum_{i=1}^{M} \Pr[A|B_i] \Pr[B_i]. \tag{10.27}$$

Proof. The proof follows from Theorem 10.6 of the intersection sets and Definition 10.6 of conditional probability

$$\begin{aligned}\Pr[A] &= \sum_{i=1}^{M} \Pr[AB_i] \\ &= \sum_{i=1}^{M} \Pr[A|B_i] \Pr[B_i].\end{aligned}$$

□

Example 10.11. *A cloud computing data center has three servers, denoted by S_1, S_2, and S_3, to process the data. 20% of the data is processed at the first server, 30% of data at the second server, and 50% of the data at the third server.*

Let A denote the event that the data is successfully processed at the data center.

Observing the successful data process from the three servers, we have the following conditional probabilities: $\Pr[A|S_1] = 0.8$, $\Pr[A|S_2] = 0.7$, *and* $\Pr[A|S_3] = 0.9$.

The probability that the data is successfully processed can be determined by the law of total probability in Theorem 10.9 as

$$\begin{aligned}\Pr[A] &= \Pr[A|S_1] \Pr[S_1] + \Pr[A|S_2] \Pr[S_2] + \Pr[A|S_3] \Pr[S_3] \\ &= 0.8 \times 0.2 + 0.7 \times 0.3 + 0.9 \times 0.5 = 0.82.\end{aligned}$$

By applying the Bayes' theorem (cf. Theorem 10.8), the probabilities that the successfully processed data comes from the first, second, and third server are, respectively, given by

$$\begin{aligned}\Pr[S_1|A] &= \frac{\Pr[A|S_1] \Pr[S_1]}{\Pr[A]} = \frac{0.8 \times 0.2}{0.82} \approx 0.195, \\ \Pr[S_2|A] &= \frac{\Pr[A|S_2] \Pr[S_2]}{\Pr[A]} = \frac{0.7 \times 0.3}{0.82} \approx 0.256, \\ \Pr[S_3|A] &= \frac{\Pr[A|S_3] \Pr[S_3]}{\Pr[A]} = \frac{0.9 \times 0.5}{0.82} \approx 0.549.\end{aligned}$$

It can be noticed that

$$\Pr[S_1|A] + \Pr[S_2|A] + \Pr[S_3|A] = 1.$$

This accordingly means that the successfully processed data has to be from either one of three servers, and this is always true. However,

$$\Pr[A|S_1] + \Pr[A|S_2] + \Pr[A|S_3] \neq 1.$$

From the observation in Example 10.11, we have the following lemma.

Lemma 10.3. *Given M events* $\{B_1, B_2, \ldots, B_M\}$, *the conditional probabilities of event* B_i, $i = 1, 2, \ldots, M$, *given an event A should satisfy*

$$\sum_{i=1}^{M} \Pr[B_i|A] = 1. \tag{10.28}$$

Proof. The proof follows from the law of total probability in Theorem 10.9 and Bayes' theorem, i.e., Theorem 10.8, as

$$\begin{aligned}\sum_{i=1}^{M} \Pr[B_i|A] &= \sum_{i=1}^{M} \frac{\Pr[A|B_i]\Pr[B_i]}{\Pr[A]} \\ &= \frac{1}{\Pr[A]} \sum_{i=1}^{M} \Pr[A|B_i]\Pr[B_i] \\ &= \frac{1}{\Pr[A]} \Pr[A] = 1.\end{aligned}$$

□

10.4 Independence of events

Events in practical systems can be either dependent or independent of each other. The independence of two events can be defined as follows:

Definition 10.7. *Two events A and B are independent if and only if the probability of their intersection set equals to the product of their individual probabilities, i.e.,*

$$\Pr[A \cap B] = \Pr[AB] = \Pr[A]\Pr[B]. \tag{10.29}$$

Example 10.12. *Consider two independent checks of data packets arriving at a server. The acceptance rates in the first check and the second check are* 0.5 *and* 0.6, *respectively.*

Let A and B denote the events that the packets pass the first check and the second check, respectively. We have $\Pr[A] = 0.5$ *and* $\Pr[B] = 0.6$.

Since A and B are independent, the probability that the packets pass both checks is

$$\Pr[AB] = \Pr[A]\Pr[B] = 0.5 \times 0.6 = 0.3.$$

Intuitively, if two events A and B are independent, then the occurrence of them does not affect each other. This can be shown in Theorem 10.10.

Theorem 10.10. *If two events A and B are independent, then*

$$\Pr[A] = \Pr[A|B], \tag{10.30}$$

$$\Pr[B] = \Pr[B|A]. \tag{10.31}$$

Proof. The proof follows from Definition 10.6 of the conditional probability and Definition 10.7 of the independent events, i.e.,

$$\begin{aligned}\Pr[B] &= \frac{\Pr[AB]}{\Pr[A|B]} \\ &= \frac{\Pr[A]\Pr[B]}{\Pr[A|B]}.\end{aligned}$$

We thus deduce (10.30) and similarly obtain (10.31), which completes the proof. □

The above definition of two independent events can be extended to the general case of more than two events as follows:

Definition 10.8. *M events $\{A_1, A_2, \ldots, A_M\}$ are independent if and only if every set of $(M-1)$ sets is independent and*

$$\Pr\left[\prod_{i=1}^{M} A_i\right] = \prod_{i=1}^{M} \Pr[A_i]. \tag{10.32}$$

Considering conditionally independent events, we have the following property:

Theorem 10.11. *If two events A and B are conditionally independent given an event C, then*

$$\Pr[AB|C] = \Pr[A|C]\Pr[B|C] \tag{10.33}$$

and

$$\Pr[A|BC] = \Pr[A|C]. \tag{10.34}$$

Proof. The proof of (10.33) follows from the Bayes' theorem (cf. Theorem 10.8) and Definition 10.6 as

$$\begin{aligned}\Pr[AB|C] &= \frac{\Pr[AC]}{\Pr[C]}\frac{\Pr[BC]}{\Pr[C]} \\ &= \Pr[A|C]\Pr[B|C]\end{aligned}$$

The proof of (10.34) follows from Theorem 10.10, i.e.,

$$\begin{aligned}\Pr[A|BC] &= \Pr[A|B|C] \\ &= \Pr[A|C].\end{aligned}$$

□

Remark 10.7. *The independence concept is also related to the RMA analysis in Chapter 5 (cf. Section 5.9).*

Let S denote the event that the whole system consisting of N independent devices is successful and D_n, $n = 1, 2, \ldots, N$, denote the event that the nth device is successful. If these devices are connected in series, it can be deduced that

$$\Pr[S] = \Pr\left[\prod_{n=1}^{N} D_n\right] = \prod_{n=1}^{N} \Pr[D_n]. \tag{10.35}$$

This accordingly verifies (5.13). Similarly, if N independent devices are connected in parallel, we can arrive at (5.16), i.e.,

$$\Pr[S] = 1 - \prod_{n=1}^{N} (1 - \Pr[D_n]). \tag{10.36}$$

10.5 Related works

For more information, the readers are suggested to refer to the following selected references: Feller in 1968 [235], Nelson in 1995 [220], Papoulis in 2002 [215], Yates and Goodman in [236], Gubner in 2006 [237], Leon-Garcia in 2007 [238], Durrett in 2010 [239], Montgomery and Runger in 2013 [217], Ross in 2014 [218], Gallager in 2014 [240], Steyer and Nagel in 2017 [241], and Mitzenmacher and Upfal in 2017 [242].

10.6 Review questions

[Q10.1] What is an experiment? Give an example of an experiment in wireless communications and explain its possible outcomes.

[Q10.2] How to determine sample space of an experiment?

[Q10.3] Differentiate between mutually exclusive set and collectively exhaustive set.

[Q10.4] What is an event in an experiment? Give an example of an event in the experiment in Question [Q10.1].

[Q10.5] What is the probability of an event having M outcomes each of which occurs with a probability of P_i, $i = 1, 2, \ldots, M$?

[Q10.6] If all M outcomes of a sample space of an experiment are equally likely, what is the probability of each outcome?

[Q10.7] What is a complement set and how to determine its probability?

[Q10.8] How to compute the probability of a union of two sets $\mathfrak{A}$ and $\mathfrak{B}$?

[Q10.9] Determine the probability of a set $\mathfrak{A}$ given the probabilities of mutually exclusive intersection sets $\mathfrak{A} \cap \mathfrak{B}_1, \mathfrak{A} \cap \mathfrak{B}_2, \ldots, \mathfrak{A} \cap \mathfrak{B}_M$.

[Q10.10] What is the conditional probability of an event A given an event B?

[Q10.11] Describe Bayes' theorem and prove it.

[Q10.12] Determine the probability of an event A given the probabilities of M events and the conditional probabilities of A given each of these M events.

[Q10.13] Prove that the summation of all conditional probabilities of event B_i, $i = 1, 2, \ldots, M$, given an event A is always one.

[Q10.14] What is the condition of two independent events?

10.7 Problems

[P10.1] Monitor three consecutive phone calls going through a telephone switching office. Classify each one as a voice call (v) if someone is speaking or a data call (d) if the call is carrying a modem or fax signal. Your observation is a sequence of three letters (each letter is either v or d). For example, two voice calls followed by one data call corresponds to vvd. Write the elements of the following sets:

- $A_1 = \{\text{first call is a voice call}\}$
- $A_2 = \{\text{second call is a voice call}\}$
- $A_3 = \{\text{all calls are the same}\}$
- $A_4 = \{\text{one or more voice calls}\}$
- $B_1 = \{\text{first call is a data call}\}$
- $B_2 = \{\text{second call is a data call}\}$
- $B_3 = \{\text{voice and data alternate}\}$
- $B_4 = \{\text{two or more data calls}\}$

For each pair of events A_1 and B_1, A_2 and B_2, and so on, identify whether the pair of events is either mutually exclusive or collectively exhaustive or both.

[P10.2] Differentiate between mutually exclusive events and independent events.

[P10.3] Derive the probability of the union of two events A and B for the following cases:

- A and B are mutually exclusive.
- A and B are independent.

[P10.4] Monitor a phone call. Classify the call as a voice call (V) if someone is speaking or a data call (D) if the call is carrying a modem or fax signal. Classify the call as long (L) if the call lasts for more than 3 min; otherwise, classify the call as brief (B). Based on data collected by the telephone company, we use the following probability model: $P[V] = 0.7$, $P[L] = 0.6$, $P[VL] = 0.35$. Find the following probabilities:

- $\Pr[DL]$
- $\Pr[D \cup L]$
- $\Pr[VB]$
- $\Pr[V \cup L]$
- $\Pr[V \cup D]$
- $\Pr[LB]$

[P10.5] Monitor three consecutive phone calls going through a telephone switching office. Classify each one as a voice call (v) if someone is speaking or a data call (d) if the call is carrying a modem or fax signal. Your observation is a sequence of three letters (each one is either v or d). For example, three voice calls corresponds to vvv. The outcomes vvv and ddd have probability 0.2, whereas each of the other outcomes vvd, vdv, vdd, dvv, dvd, and ddv has probability 0.1. Count the number of voice calls N_V in the three calls you have observed. Consider the four events $N_V = 0$, $N_V = 1$, $N_V = 2$, $N_V = 3$. Describe in words and calculate the following probabilities:

- $\Pr[N_V = 2]$
- $\Pr[N_V \geq 1]$
- $\Pr[\{vvd\}|N_V = 2]$
- $\Pr[\{ddv|\}|N_V = 2]$
- $\Pr[N_V = 2|N_V \geq 1]$
- $\Pr[N_V \geq 1|N_V = 2]$

Determine whether the pair of events $\{N_V = 2\}$ and $\{N_V \geq 1\}$ is independent.

[P10.6] In a cellular phone system, a mobile phone must be paged to receive a phone call. However, paging attempts do not always succeed because the mobile phone may not receive the paging signal clearly. Consequently, the system will page a phone up to three times before giving up. If a single paging attempt succeeds with probability 0.8, sketch a probability tree for this experiment and find the probability $P[F]$ that the phone is found.

[P10.7] A machine produces routers in pairs. Tests show that the first router is acceptable with probability 70%. If the first router is acceptable, then the second router will be acceptable with probability 90%. Otherwise, if the first router is defective, then the second router will be acceptable with probability 30%.

- Find the probability that exactly one router of a pair is acceptable.
- Find the probability that both routers in a pair are acceptable.
- Suppose the quality of each pair of routers is independent of the quality of every other pair of routers. Find the probability of at least one acceptable router out of two pairs of routers.
- Find the minimum number of pair of routers the machine must produce to reach a probability of 99% that there will be at least one acceptable router.

[P10.8] Consider a binary code with 4 b (0 or 1) in each code word. An example of a code word is 0110.

- Find the number of code words that can be generated.
- Find the number of code words that have exactly two zeroes.
- Find the number of code words that begin with a zero.
- In a constant-ratio binary code, each code word has N bits. In every word, M of the N bits are 1 and the other $N - M$ bits are 0. Find the number of code words in the code when $N = 8$ and $M = 3$.

[P10.9] Consider 64-quadrature amplitude modulation (64-QAM) with equally likely data symbols. Let A and B denote the events that the data symbol consists of at least 4 b 1 and at least 4 b 0, respectively.

- Find the number of bits in each 64-QAM symbol.
- List all possible data symbols in A and B.
- Find $\Pr[A]$ and $\Pr[B]$.
- Find $\Pr[A \cap B]$ and $\Pr[A \cup B]$.
- Verify if A and B are either mutually exclusive, collectively exhaustive, or independent.
- Find $\Pr[A|B]$ and $\Pr[B|A]$.

[P10.10] A server performs two tests on N data packets. Suppose that M packets pass both tests, M_1 packets pass only the first test and M_2 packets pass only the second test. Let A and B denote the events that the packets pass the first test and pass the second test, respectively.

- Find $\Pr[A \cap B]$ and $\Pr[A \cup B]$
- Find $\Pr[A|B]$ and $\Pr[B|A]$.
- Find the conditions of N, M, M_1, and M_2 that A and B are independent.

[P10.11] A data center has K servers to process the data. Assume load balancing where each server processes the same amount of data. Observing the data processing at each server, we notice the kth server, $k = 1, 2, \ldots, K$, has a success rate of p_k.

- Find the probability that the data is successfully processed at the data center.
- Find the probability that the successful processed data comes from the kth server where $k = 1, 2, \ldots, K$.

[P10.12] A memory module consists of M chips. The device is designed with redundancy so that it works even if one of its chips is defective. Each chip contains N transistors and functions properly if all of its transistors work. A transistor works with a probability p independent of any other transistor. Find the probability that a chip works and the probability that a memory module works.

Chapter 11
Random variables in network modeling and simulation

In network modeling and simulation, a real network can be represented via statistical models. Due to the randomness of the phenomenon, possible outcomes of the event can be assigned by a real number of random quantity which is known as a random variable. The random variable is specified by a probability distribution. Depending on the data type of the random quantity, we have two kinds of random variables including discrete and continuous random variables.

In this chapter, I will first introduce the basic concepts of random variables in Section 11.1. Discrete and continuous random variables will be sequentially presented in Sections 11.2 and 11.3. Specifically, distribution functions, mean, variance, and moments will be discussed for each random variable type.

11.1 Introduction of random variables

In an experiment, a probability measure is employed for every random event over a sample space $\mathfrak{S}$. A random variable (cf. Figure 11.1) can be defined as follows:

Definition 11.1. *A random variable X is a mapping function that assigns a real number to each outcome of the experiment, i.e.,*

$$X : \mathfrak{S} \to \mathbb{R}. \tag{11.1}$$

where $\mathfrak{S}$ is the sample space consisting of all possible outcomes in the experiment. Let ω denote a sample point in $\mathfrak{S}$, then $X(\omega) \in \mathbb{R}$ is the measured value of ω.

Example 11.1. *Consider a device-to-device communication channel between two mobile devices A and B. The device A wants to share two files with B. Due to the long distance between them, this sharing can be either successful (s) or failed (f). The set of all possible outcomes is $\mathfrak{S} = \{ss, sf, fs, ff\}$.*

Let a random variable X represent the number of files that are successfully received and ω denote a sample point in $\mathfrak{S}$. We have

$$X(\omega) = \begin{cases} 0 & \text{if } \omega = ff, \\ 1 & \text{if } \omega = sf, fs, \\ 2 & \text{if } \omega = ss. \end{cases}$$

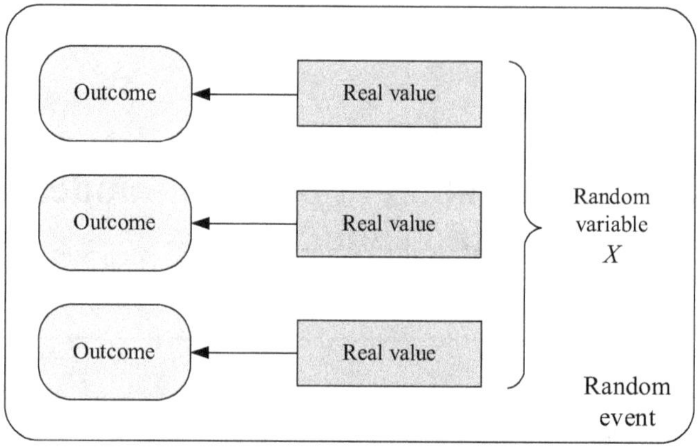

Figure 11.1 Random variable for representing a random event

A random variable is represented by a probability value corresponding to the probability of an event. For instance, the probability that X has a value of x is[1] written as $\Pr[X(\omega) = x]$ or simply $\Pr[X = x]$. Aggregating all probability values of a random variable constructs a probability distribution of that random variable.

Basically, there are two types of random variables with respect to its data type which could be either discrete or continuous random variables. These random variables have different distribution functions. In the following sections, I will discuss in detail these two types of random variables along with their distribution functions.

11.2 Discrete random variables

Definition 11.2. *A random variable $X: \mathfrak{S} \rightarrow \mathbb{R}$ is a discrete random variable if the range of X is a countably infinite set, i.e.,*

$$\mathfrak{S}_X = \{x_1, x_2, \ldots\} \tag{11.2}$$

or a finite set of N elements, i.e.,

$$\mathfrak{S}_X = \{x_1, x_2, \ldots, x_N\}. \tag{11.3}$$

[1]Note that the upper case letter is used to name the random variable, while the lower case letter is for its value.

11.2.1 Probability mass function

A discrete random variable is described by a probability mass function (pmf) which can be defined as follows:

Definition 11.3. *The pmf $f_X(x)$ of a discrete random variable X assigns a probability to each measured value x of the outcome, i.e.,*

$$f_X(x) \triangleq \Pr[X = x]. \tag{11.4}$$

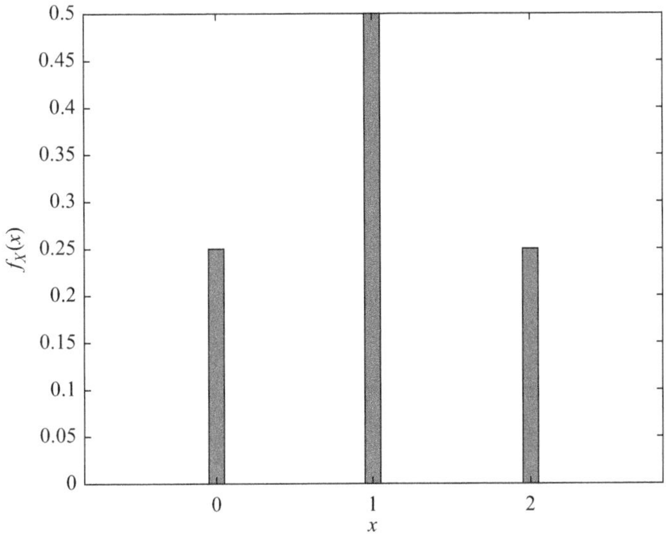

Figure 11.2 pmf of random variable in Example 11.2

Example 11.2. *In Example 11.1, it is assumed that the success and fail of file sharing between two devices are equally likely. The outcomes in the sample space $\mathfrak{S} = \{ss, sf, fs, ff\}$ are thus equally likely, each of which has a probability of 1/4. The pmf of the number of files that are successfully received, i.e., random variable X (cf. Figure 11.2), is given by*

$$f_X(x) = \begin{cases} 1/4 & \text{if } x = 0, 2, \\ 1/2 & \text{if } x = 1, \\ 0 & \text{otherwise.} \end{cases}$$

Remark 11.1. *From Definition 11.3, it can be noticed that the pmf of a discrete random variable X has the following properties:*

$$0 \le f_X(x) \le 1 \quad \forall x \in \mathfrak{S}_X, \tag{11.5}$$

$$\sum_{x \in \mathfrak{S}_X} f_X(x) = 1. \tag{11.6}$$

Similar to conditional probability, we have conditional pmf which is defined as follows:

Definition 11.4. *The conditional pmf of a random variable X given an event B is denoted by* $f_{X|B}(x)$ *which can be determined by*

$$f_{X|B}(x) \triangleq \Pr[X = x|B] \tag{11.7}$$

$$= \frac{\Pr[X = x, B]}{\Pr[B]}. \tag{11.8}$$

We also have the law of total pmf as follows:

Theorem 11.1. *Given probabilities of M events* $\{B_1, B_2, \ldots, B_M\}$ *and conditional pmf of a random variable X given* B_i, $i = 1, 2, \ldots, M$, *the pmf of X can be determined by*

$$f_X(x) = \sum_{i=1}^{M} f_{X|B}(x) \Pr[B_i]. \tag{11.9}$$

Proof. The proof is similar to that of Theorem 10.9. □

Considering a derived random variable Y which is a function of X, i.e., $Y = g(X)$, its pmf can be determined by the following theorem:

Theorem 11.2. *The pmf of a derived discrete random variable* $Y = g(X)$ *is given by*

$$f_Y(y) = \sum_{x:x=g^{-1}(y)} f_X(g^{-1}(y)). \tag{11.10}$$

Proof. The proof follows from Definition 11.3, i.e.,

$$\begin{aligned} f_Y(y) &= \Pr[Y = y] \\ &= \Pr[g(X) = y] \\ &= \sum_{x:x=g^{-1}(y)} \Pr[X = g^{-1}(y)] \\ &= \sum_{x:x=g^{-1}(y)} f_X(g^{-1}(y)) \end{aligned}$$

□

11.2.2 Cumulative distribution function

In addition to pmf, in order to represent a discrete random variable, there is a related function, namely, cumulative distribution function (cdf) which is defined as follows:

Definition 11.5. *The cdf* $F_X(x)$ *of a discrete random variable X is the probability that X will take a value less than or equal to x, i.e.,*

$$F_X(x) \triangleq \Pr[X \le x] \tag{11.11}$$

$$= \sum_{x' \le x \in \mathfrak{S}_X} f_X(x'). \tag{11.12}$$

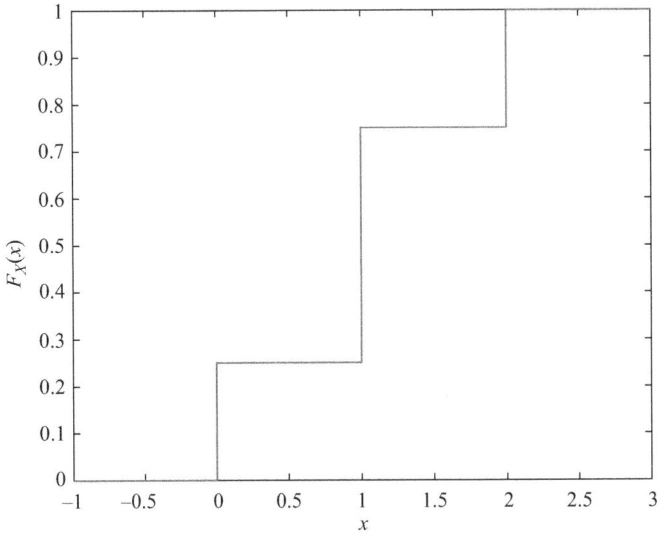

Figure 11.3 cdf of random variable in Example 11.2

Example 11.3. *In Example 11.1, the cdf of the number of files that are successfully received, i.e., random variable X (cf. Figure 11.3), is given by*

$$F_X(x) = \begin{cases} 0 & \text{if } x < 0, \\ 1/4 & \text{if } 0 \le x < 1, \\ 3/4 & \text{if } 1 \le x < 2, \\ 1 & \text{if } x \ge 2. \end{cases}$$

Remark 11.2. *From Definition 11.5, it can be noticed that the cdf of a discrete random variable X has the following properties:*

$$0 \le F_X(x) \le 1 \quad \forall x \in \mathfrak{S}_X, \tag{11.13}$$

$$F_X(-\infty) = 0, \tag{11.14}$$

$$F_X(\infty) = 1. \tag{11.15}$$

Theorem 11.3. *The cdf of a random variable X is an increasing function over the whole range of x in the sample space, i.e.,*

$$F_X(x_1) \le F_X(x_2) \quad \forall x_1 \le x_2 \in \mathfrak{S}_X. \tag{11.16}$$

Proof. The proof follows from Definition 11.5 with $x_1 \le x_2$, i.e.,

$$F_X(x_1) = \Pr[X \le x_1] \le \Pr[X \le x_2] = F_X(x_2)]. \qquad \square$$

Theorem 11.4. *The probability that a discrete random variable X is within a specified interval can be determined via its cdf and pmf as follows:*

$$\Pr[a < X \leq b] = F_X(b) - F_X(a), \tag{11.17}$$

$$\Pr[a \leq X \leq b] = F_X(b) - F_X(a) + f_X(a), \tag{11.18}$$

$$\Pr[a < X < b] = F_X(b) - F_X(a) - f_X(b), \tag{11.19}$$

$$\Pr[a \leq X < b] = F_X(b) - F_X(a) + f_X(a) - f_X(b). \tag{11.20}$$

Proof. The proof follows from Definitions 11.5 and 11.3 as

$$\begin{aligned}\Pr[a < X \leq b] &= \Pr[X \leq b] - \Pr[X \leq a] \\ &= F_X(b) - F_X(a),\end{aligned}$$

$$\begin{aligned}\Pr[a \leq X \leq b] &= \Pr[X \leq b] - \Pr[X \leq a] + \Pr[X = a] \\ &= F_X(b) - F_X(a) + f_X(a),\end{aligned}$$

$$\begin{aligned}\Pr[a < X < b] &= \Pr[X \leq b] - \Pr[X \leq a] - \Pr[X = b] \\ &= F_X(b) - F_X(a) - f_X(b),\end{aligned}$$

$$\begin{aligned}\Pr[a \leq X < b] &= \Pr[X \leq b] - \Pr[X \leq a] + \Pr[X = a] - \Pr[X = b] \\ &= F_X(b) - F_X(a) + f_X(a) - f_X(b).\end{aligned}$$ □

11.2.3 Expected value

The expected value, a.k.a. mean, of a discrete random variable can be simply understood as its average value over a number of tests or trials. Statistically, it is defined as follows:

Definition 11.6. *The expected value of a discrete random variable X is a statistical probability-weighted average value of all possible outcomes, i.e.,*

$$E[X] \triangleq \sum_{x \in \mathfrak{S}_X} x f_X(x), \tag{11.21}$$

where $E[\cdot]$ denotes the expectation operator, and $f_X(x)$ is the pmf of X given by (11.4).

In statistics, the expected value can also be represented by μ_X. For consistency, the expected value is denoted by $E[X]$ throughout.

Example 11.4. *Consider Example 11.1 with the pmf of random variable X given by*

$$f_X(x) = \begin{cases} 1/4 & \text{if } x = 0, 2, \\ 1/2 & \text{if } x = 1, \\ 0 & \text{otherwise.} \end{cases}$$

The expected value of X is calculated by using (11.21) *as*

$$E[X] = \sum_{x\in\{0,1,2\}} xf_X(x) = 0 \times 1/4 + 1 \times 1/2 + 2 \times 1/4 = 1.$$

Theorem 11.5. *The expected value of a derived discrete random variable $Y = g(X)$ is given by*

$$E[Y] = \sum_{x\in\mathfrak{S}_X} g(x)f_X(x), \tag{11.22}$$

where $f_X(x)$ is the pmf of X given by (11.4).

Proof. The proof follows from Definition 11.6 and Theorem 11.2 as

$$\begin{aligned} E[Y] &= \sum_{y\in\mathfrak{S}_Y} yf_Y(y) \\ &= \sum_{y\in\mathfrak{S}_Y} g(x) \sum_{x:x=g^{-1}(y)} f_X(g^{-1}(y)) \\ &= \sum_{y\in\mathfrak{S}_Y} \sum_{x:x=g^{-1}(y)} g(x)f_X(x) \\ &= \sum_{x\in\mathfrak{S}_X} g(x)f_X(x). \end{aligned}$$

□

Lemma 11.1. *The expected value of a random variable $Y = aX + b$ is given by*

$$E[Y] = E[aX + b] = aE[X] + b. \tag{11.23}$$

Proof. From Theorem 11.5, we have

$$\begin{aligned} E[Y] &= \sum_{x\in\mathfrak{S}_X} g(x)f_X(x) \\ &= \sum_{x\in\mathfrak{S}_X} (ax + b)f_X(x) \\ &= a\sum_{x\in\mathfrak{S}_X} xf_X(x) + b\sum_{x\in\mathfrak{S}_X} f_X(x) \\ &= aE[X] + b, \end{aligned}$$

where the last equal sign is due to (11.6), i.e., $\sum_{x\in\mathfrak{S}_X} f_X(x) = 1$, and Definition 11.6. □

Considering multiple random variables, we have the following theorems:

Theorem 11.6. *The expected value of a linear combination of N arbitrary random variables* $\{X_1, X_2, \ldots, X_N\}$ *is given by*

$$E[Y] = E\left[\sum_{i=1}^{N} a_i X_i\right] = \sum_{i=1}^{N} a_i E[X_i]. \tag{11.24}$$

Proof. The proof follows from Definition 11.6 and Theorem 11.5 as

$$\begin{aligned} E[Y] &= \sum_{i=1}^{N} \sum_{x_i \in \mathfrak{S}_{X_i}} a_i f_{X_i}(x_i) \\ &= \sum_{i=1}^{N} a_i E[X_i]. \end{aligned}$$

□

Theorem 11.7. *If two random variables X and Y are independent, then*

$$E[XY] = E[X]E[Y]. \tag{11.25}$$

Proof. From Definition 11.6, we have

$$\begin{aligned} E[XY] &= \sum_{x,y \in \mathfrak{S}_{X,Y}} xy \Pr[X = x, Y = y] \\ &= \sum_{x \in \mathfrak{S}_X} x \Pr[X = x] \sum_{y \in \mathfrak{S}_Y} y \Pr[Y = y] \\ &= E[X]E[Y]. \end{aligned}$$

where the second equal sign is based on Definition 10.7 of two independent events. □

11.2.4 Variance and standard deviation

Besides expected value, it is vital to determine variance of a discrete random variable. The variance of a random variable is used to show how far the outcomes are stretched from the average value. Statistically, it is defined as follows:

Definition 11.7. *The variance of a discrete random variable X is a statistical average value of the squared distance between the outcomes and their expected value, i.e.,*

$$V[X] \triangleq E\left[(X - E[X])^2\right] \tag{11.26}$$

$$= \sum_{x \in \mathfrak{S}_X} (x - E[X])^2 f_X(x), \tag{11.27}$$

where $V[X]$ *denotes the variance of X,* $E[X]$ *is the expected value of X given by* (11.21)*, and* $f_X(x)$ *is the pmf of X given by* (11.4).

In statistics, the expected value can also be represented by σ_X^2. For consistency, the variance value is denoted by $V[X]$ throughout.

Example 11.5. *Consider Example 11.1 with the pmf of random variable X given by*

$$f_X(x) = \begin{cases} 1/4 & \text{if } x = 0, 2, \\ 1/2 & \text{if } x = 1, \\ 0 & \text{otherwise.} \end{cases}$$

and the expected value of X is $E[X] = 1$. The variance of X is calculated by using (11.27) *as*

$$\begin{aligned} V[X] &= \sum_{x\in\{0,1,2\}} (x - E[X])^2 f_X(x) \\ &= (0-1)^2 1/4 + (1-1)^2 1/2 + (2-1)^2 1/4 = 1/2. \end{aligned}$$

Theorem 11.8. *The variance of a random variable X can be determined by*

$$V[X] = E[X^2] - E^2[X], \tag{11.28}$$

where $E[X]$ is the expected value of X given by (11.21) *and $E[X^2]$ is given by*

$$E[X^2] = \sum_{x\in\mathfrak{S}_X} x^2 f_X(x). \tag{11.29}$$

Proof. From (11.26) in Definition 11.7, we have

$$\begin{aligned} V[X] &= E\left[(X - E[X])^2\right] \\ &= E\left[X^2 - 2XE[X] + E^2[X]\right] \\ &\overset{(a)}{=} E[X^2] - 2E[X]E[X] + E^2[X] \\ &= E[X^2] - E^2[X], \end{aligned}$$

where (a) follows from Lemma 11.1. □

Note that $E[X^2]$ is different from $E^2[X]$.

Theorem 11.9. *The variance of a derived discrete random variable $Y = g(X)$ is given by*

$$V[Y] = \sum_{x\in\mathfrak{S}_X} (g(x) - E[g(X)])^2 f_X(x), \tag{11.30}$$

where $f_X(x)$ is the pmf of X given by (11.4).

Proof. The proof is similar to that of Theorem 11.5. □

Lemma 11.2. *The variance of a random variable $Y = aX + b$ is given by*

$$V[Y] = V[aX + b] = a^2 V[X]. \tag{11.31}$$

Proof. From Theorem 11.9, we have

$$\begin{aligned} V[Y] &= \sum_{x \in \mathfrak{S}_X} (g(x) - E[g(X)])^2 f_X(x) \\ &\overset{(a)}{=} \sum_{x \in \mathfrak{S}_X} (ax + b - (aE[X] + b))^2 f_X(x) \\ &= a^2 \sum_{x \in \mathfrak{S}_X} (x - E[X])^2 f_X(x) \\ &\overset{(b)}{=} a^2 V[X], \end{aligned}$$

where (a) follows from Lemma 11.1 and (b) follows from Definition 11.7. □

The variance of a random variable can also be represented via a standard deviation which is simply defined as the square root of its variance.

Definition 11.8. *The standard deviation of a random variable X is the square root of its variance, i.e.,*

$$\sigma_X \triangleq \sqrt{V[X]}, \tag{11.32}$$

where σ_X denotes the standard deviation of X, $V[X]$ is the variance of X given by either (11.26) *or* (11.28).

Example 11.6. *In Example 11.5, the variance of random variable X is $V[X] = 1/2$, so the standard deviation of X is $\sigma_X = 1/\sqrt{2}$.*

Considering multiple random variables, there is a new term called covariance to reflect the relationship between two random variables. The covariance is defined as follows:

Definition 11.9. *The covariance of two random variables X and Y, denoted by $cov[X, Y]$, is defined as*

$$cov[X, Y] \triangleq E[(X - E[X])(Y - E[Y])]. \tag{11.33}$$

Theorem 11.10. *The covariance of two random variables X and Y can be determined by*

$$cov[X, Y] = E[XY] - E[X]E[Y]. \tag{11.34}$$

Proof. The proof follows from Definition 11.9 and Theorem 11.6 as

$$\begin{aligned} cov[X, Y] &= E[(X - E[X])(Y - E[Y])] \\ &= E[XY - XE[Y] - YE[X] + E[X]E[Y]] \\ &= E[XY] - E[X]E[Y]. \end{aligned}$$

□

Lemma 11.3. *If X and Y are independent, then $cov[X, Y] = 0$*

Proof. The proof follows from Theorems 11.7 and 11.10. □

Theorem 11.11. *The variance of a linear combination of N arbitrary random variables $\{X_1, X_2, \ldots, X_N\}$ is given by*

$$V[Y] = V\left[\sum_{i=1}^{N} a_i X_i\right] = \sum_{i=1}^{N} a_i^2 V[X_i] + 2\sum_{i=1}^{N} \sum_{j \neq i \in [1,N]} a_i a_j cov[X_i, X_j]. \quad (11.35)$$

Proof. The proof follows from Definitions 11.7 and 11.9 as

$$\begin{aligned} V[Y] &= V\left[\sum_{i=1}^{N} a_i X_i\right] \\ &= E\left[\left(\sum_{i=1}^{N} a_i X_i - \left(\sum_{i=1}^{N} a_i E[X_i]\right)\right)^2\right] \\ &= E\left[\sum_{i=1}^{N} a_i^2 (X_i - E[X_i])^2\right] \\ &\quad + 2\sum_{i=1}^{N} \sum_{j \neq i \in [1,N]} a_i a_j E[(X_i - E[X_i])(X_j - E[X_j])] \\ &= \sum_{i=1}^{N} a_i^2 V[X_i] + 2\sum_{i=1}^{N} \sum_{j \neq i \in [1,N]} a_i a_j cov[X_i, X_j]. \end{aligned}$$

□

Lemma 11.4. *If random variables $\{X_1, X_2, \ldots, X_N\}$ are independent, then*

$$V[Y] = V\left[\sum_{i=1}^{N} a_i X_i\right] = \sum_{i=1}^{N} a_i^2 V[X_i]. \quad (11.36)$$

Proof. The proof follows from Theorem 11.11, and the fact in Lemma 11.3 that $cov[X_i, X_j] = 0$ if X_i and X_j are independent. □

11.2.5 Moments

In order to model a generalized change of a probability distribution of a random variable about its average value, a new term, namely, central moment, is widely used in statistics. The central moment and its variant forms of a discrete random variable are defined as follows:

Definition 11.10. *The nth central moment of a discrete random variable X is*

$$E\left[(X - E[X])^n\right] = \sum_{x \in S_X} (x - E[X])^n f_X(x), \tag{11.37}$$

where $f_X(x)$ is the pmf of X given by (11.4).

It can be noticed in (11.37) that the second central moment, i.e., $n = 2$, is the variance of X.

Definition 11.11. *The nth moment, a.k.a. raw moment, of a discrete random variable X is*

$$E[X^n] = \sum_{x \in S_X} x^n f_X(x), \tag{11.38}$$

where $f_X(x)$ is the pmf of X given by (11.4).

The raw moment is actually the nth central moment about zero. It can be noticed in (11.38) that the first moment, i.e., $n = 1$, is the expected value of X.

The moment is also represented in a normalized form, namely, normalized moment, which is defined as follows:

Definition 11.12. *The nth normalized moment, a.k.a. standardized moment, of a discrete random variable X is*

$$\frac{E\left[(X - E[X])^n\right]}{\sigma_X^n} = \frac{\sum_{x \in S_X} (x - E[X])^n f_X(x)}{\left(\sqrt{\sum_{x \in S_X} (x - E[X])^2 f_X(x)}\right)^n}, \tag{11.39}$$

where σ_X is the standard deviation of X given by (11.32) *and $f_X(x)$ is the pmf of X given by* (11.4).

Specifically, the normalized moment has been applied in analyzing the shape of a probability distribution. For instance,

- When $n = 3$, the third normalized moment is called skewness which is a measure of the asymmetry of the probability distribution of a random variable about its expected value.
- When $n = 4$, the fourth normalized moment is called kurtosis which is a measure of the heaviness of the tail of the probability distribution of a random variable.

11.3 Continuous random variables

Definition 11.13. *A random variable $X : \mathfrak{S} \to \mathbb{R}$ is a continuous random variable if the range of X is an uncountable set of numerical values in a specified interval either closed, half-bounded, or open, i.e.,*

$$\mathfrak{S}_X = (x_1, x_2) \text{ or } (x_1, x_2] \text{ or } [x_1, x_2) \text{ or } [x_1, x_2]. \tag{11.40}$$

Note that the probability of any individual outcome of a continuous random variable is zero, i.e., $\Pr[X = x] = 0, \forall x \in \mathfrak{S}_X$. This means that there is no pmf for continuous random variables. Instead, a continuous random variable is represented by a probability density function (pdf) and a cdf.

11.3.1 Cumulative distribution function

The cdf of a continuous random variable is defined as follows:

Definition 11.14. *The cdf $F_X(x)$ of a continuous random variable X is*

$$F_X(x) \triangleq \Pr[X \leq x]. \tag{11.41}$$

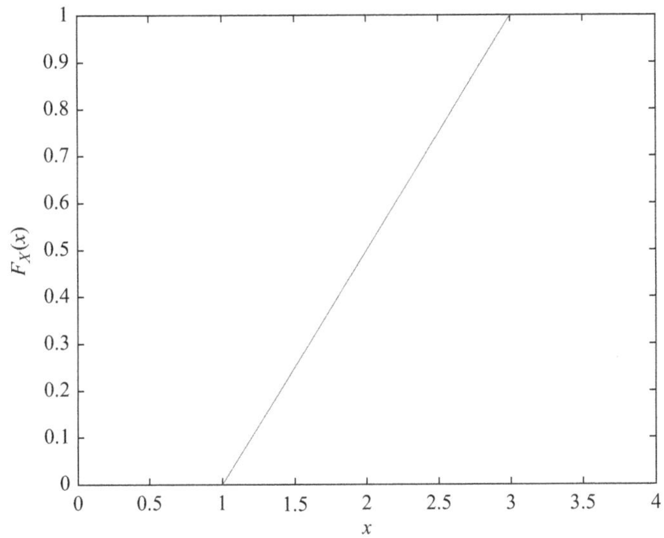

Figure 11.4 cdf of a continuous uniform random variable in $[1, 3]$

Example 11.7. *cdf of a continuous uniform random variable in the range* $[1, 3]$ *as shown in Figure 11.4 is given by*

$$F_X(x) = \begin{cases} 0 & \text{if } x \leq 1, \\ \dfrac{x-1}{2} & \text{if } 1 \leq x \leq 3, \\ 1 & \text{if } x > 3. \end{cases}$$

Remark 11.3. *Note that the cdf given by* (11.41) *is identical to that for discrete random variable as in* (11.11)*; however, it is not computed by* (11.12) *which is only feasible for a finite set. Due to the continuity of the real values in the sample space, the cdf of a continuous random variable is also a continuous function.*

Remark 11.4. *It can be shown that the cdf of a continuous random variable also has the properties of the discrete random variables, i.e.,*

$$0 \leq F_X(x) \leq 1 \quad \forall x \in \mathfrak{S}_X, \tag{11.42}$$

$$F_X(-\infty) = 0, \tag{11.43}$$

$$F_X(\infty) = 1, \tag{11.44}$$

$$F_X(x_1) \leq F_X(x_2) \quad \forall x_1 \leq x_2 \in \mathfrak{S}_X. \tag{11.45}$$

11.3.2 Probability density function

In order to describe the density of a continuous random variable, the pdf is defined as follows:

Definition 11.15. *The pdf $f_X(x)$ of a continuous random variable X is*

$$f_X(x) = \frac{dF_X(x)}{dx}, \tag{11.46}$$

where $F_X(x)$ is the cdf of X determined by (11.41).

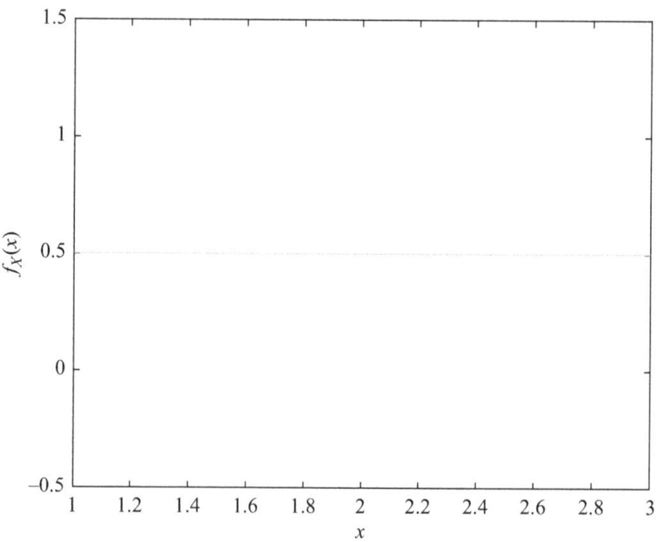

Figure 11.5 pdf of a continuous uniform random variable in $[1, 3]$

Example 11.8. *pdf of a continuous uniform random variable in the range* $[1, 3]$ *as shown in Figure 11.5 is determined by*

$$f_X(x) = \begin{cases} \frac{1}{2} & \text{if } 1 \leq x \leq 3, \\ 0 & \text{otherwise.} \end{cases}$$

Remark 11.5. *From Definition 11.15, it can be noticed that the pdf of a continuous random variable X has the following properties:*

$$f_X(x) \geq 0 \quad \forall x \in \mathfrak{S}_X, \tag{11.47}$$

$$\int_{-\infty}^{\infty} f_X(x)dx = 1, \tag{11.48}$$

Theorem 11.12. *The cdf of a continuous random variable X can be determined via its pdf as*

$$F_X(x) = \int_{-\infty}^{x} f_X(t)dt, \tag{11.49}$$

where $f_X(t)$ is the pdf of X.

Proof. The proof follows from Definitions 11.5 and 11.15, i.e.,

$$F_X(x) = \Pr[X \leq x] = \int_{-\infty}^{x} f_X(t)dt.$$

□

Theorem 11.13. *The probability that a continuous random variable X is within a specified interval can be determined via its cdf and pdf as follows:*

$$\Pr[a < X \leq b] = F_X(b) - F_X(a) \tag{11.50}$$

$$= \int_{a}^{b} f_X(x)dx. \tag{11.51}$$

Proof. The proof of (11.50) is similar to that of Theorem 11.4, and the proof of (11.51) follows from (11.49) in Theorem 11.12, i.e.,

$$\begin{aligned}\Pr[a < X \leq b] &= F_X(b) - F_X(a)\\ &= \int_{-\infty}^{b} f_X(x)dx - \int_{-\infty}^{a} f_X(x)dx\\ &= \int_{a}^{b} f_X(x)dx.\end{aligned}$$

□

Remark 11.6. *Note that the interval between a and b in Theorem 11.13 can be either left-bounded, right-bounded, closed, or open since* $\Pr[X = x] = 0, \forall x \in \mathfrak{S}_X$, *i.e.,*

$$\Pr[a \le X < b] = \Pr[a \le X \le b] = \Pr[a < X < b] = F_X(b) - F_X(a). \quad (11.52)$$

11.3.3 Expected value

The expected value, a.k.a. mean, of a continuous random variable is defined as follows:

Definition 11.16. *The expected value of a continuous random variable X is determined by*

$$E[X] \triangleq \int_{-\infty}^{\infty} x f_X(x) dx, \quad (11.53)$$

where $f_X(x)$ *is the pdf of X given by* (11.46).

Theorem 11.14. *The expected value of a derived continuous random variable* $Y = g(X)$ *is given by*

$$E[Y] \triangleq \int_{-\infty}^{\infty} g(x) f_X(x) dx, \quad (11.54)$$

where $f_X(x)$ *is the pdf of X given by* (11.46).

Proof. The proof is similar to that of Theorem 11.5. □

11.3.4 Variance and standard deviation

The variance and standard deviation of a continuous random variable are defined as follows:

Definition 11.17. *The variance of a continuous random variable X is a statistical average value of the squared distance between the outcomes and their expected value, i.e.,*

$$V[X] \triangleq E\left[(X - E[X])^2\right] \quad (11.55)$$

$$= \int_{-\infty}^{\infty} (x - E[X])^2 f_X(x) dx, \quad (11.56)$$

where $E[X]$ *is the expected value of X given by* (11.53) *and* $f_X(x)$ *is the pdf of X given by* (11.46).

Theorem 11.15. *The variance of a continuous random variable X can be determined by*

$$V[X] = E[X^2] - E^2[X], \quad (11.57)$$

where $E[X]$ is the expected value of X given by (11.53) *and $E[X^2]$ is given by*

$$E[X^2] = \int_{-\infty}^{\infty} x^2 f_X(x)dx. \tag{11.58}$$

Proof. The proof is similar to that of Theorem 11.8. □

Note that $E[X^2]$ is different from $E^2[X]$.

Theorem 11.16. *The variance of a derived continuous random variable $Y = g(X)$ is given by*

$$V[Y] = \int_{-\infty}^{\infty} (g(x) - E[g(X)])^2 f_X(x)dx, \tag{11.59}$$

where $f_X(x)$ is the pdf of X given by (11.46).

Proof. The proof is similar to that of Theorem 11.9. □

Similarly, the standard deviation of a continuous random variable is defined as the square root of its variance.

Definition 11.18. *The standard deviation of a continuous random variable X is the square root of its variance, i.e.,*

$$\sigma_X \triangleq \sqrt{V[X]}, \tag{11.60}$$

where σ_X denotes the standard deviation of X and $V[X]$ is the variance of X given by either (11.56) *or* (11.57).

11.3.5 Moments

The central moment and its variant forms of a continuous random variable are defined as follows:

Definition 11.19. *The nth central moment of a continuous random variable X is*

$$E\left[(X - E[X])^n\right] = \int_{-\infty}^{\infty} (x - E[X])^n f_X(x)dx, \tag{11.61}$$

where $f_X(x)$ is the pdf of X given by (11.46).

Definition 11.20. *The nth moment, a.k.a. raw moment, of a continuous random variable X is*

$$E[X^n] = \int_{-\infty}^{\infty} x^n f_X(x)dx, \tag{11.62}$$

where $f_X(x)$ is the pdf of X given by (11.46).

Definition 11.21. *The nth normalized moment, a.k.a. standardized moment, of a continuous random variable X is*

$$\frac{E\left[(X - E[X])^n\right]}{\sigma_X^n} = \frac{\int_{-\infty}^{\infty} (x - E[X])^n f_X(x)dx}{\left(\sqrt{\int_{-\infty}^{\infty} (x - E[X])^2 f_X(x)dx}\right)^n}, \tag{11.63}$$

where σ_X is the standard deviation of X and $f_X(x)$ is the pdf of X given by (11.46).

Remark 11.7. *Similar to the discrete random variable, the first raw moment, the second central moment, the third normalized moment, and the fourth normalized moment of the continuous random variable stand for its variance, expected value, skewness, and kurtosis, respectively.*

11.4 Related works

For more information, the readers are suggested to refer to the following selected references: Feller in 1968 [235], Prohorov and Rozanov in 1969 [243], Papoulis in 2002 [215], Yates and Goodman in 2004 [236], Johnson *et al.* in 1994 and 2005 [244,245], Gubner in 2006 [237], Leon-Garcia in 2007 [238], Stewart in 2009 [246], [247], Forbes *et al.* in 2010 [216], and Mitzenmacher and Upfal in 2017 [242].

11.5 Review questions

[Q11.1] What is a random variable in an experiment?

[Q11.2] Provide two examples to differentiate between discrete and continuous random variables.

[Q11.3] What are the probability mass function (pmf) and cumulative distribution function (cdf) of a discrete random variable?

[Q11.4] What are the properties of pmf and cdf of a discrete random variable?

[Q11.5] Show that the cdf of a random variable X is an increasing function over the whole range of values in the sample space.

[Q11.6] Derive the formula to compute the probability that a discrete random variable is within a specified interval.

[Q11.7] Derive the pmf of a derived discrete random variable $Y = g(X)$.

[Q11.8] What is the meaning of expected value, variance, and standard deviation of a random variable?

[Q11.9] How to determine the expected value and variance of a discrete random variable given its pmf?

[Q11.10] Derive the expected value of a linear sum of N random variables.

[Q11.11] Derive the expected value of a product of N independent random variables.

[Q11.12] Prove that the variance of a random variable X can be computed through the first and the second moments of X, i.e., $V[X] = E[X^2] - E^2[X]$.

[Q11.13] Derive the expected value and variance of a derived random variable $Y = g(X)$.

[Q11.14] What is covariance of two random variables and its meaning? Draw the condition of two independent random variables.

[Q11.15] Derive the variance of a linear sum of N independent random variables.

[Q11.16] What are the nth central moment, the nth moment, and the nth normalized moment of a discrete random variable?

[Q11.17] What is the probability density function (pdf) of a continuous random variable and its relation with the cdf?

[Q11.18] Differentiate between the pdf and the pmf. Explain why the continuous random variable cannot be represented by the pmf.

[Q11.19] Derive the probability that a continuous random variable is within a specified interval via its pdf and cdf.

[Q11.20] How to determine the expected value and variance of a discrete random variable given its pdf?

[Q11.21] Demonstrate the differences between the continuous and discrete random variables in calculating the expected value, variance, and moments.

11.6 Problems

[P11.1] A random variable Z is given by $Z = \sum_{i=1}^{N} a_i X_i + \sum_{j=1}^{M} b_j Y_j + c$. The expected value and variance of $X_i, i = 1, 2, \ldots, N$, and $Y_j, j = 1, 2, \ldots, M$, are $E[X_i] = \mu_{X_i}$, $V[X_i] = \sigma_{X_i}^2$, $E[Y_j] = \mu_{Y_j}$, and $V[Y_j] = \sigma_{Y_j}^2$.

- Derive the general expressions for the expected value and variance of Z.
- Find $E[Z]$ and $V[Z]$ for the following special cases:
  - $cov[X_i, X_j] = 0$, $\forall i \neq j$, $\{i,j\} \in \{1, 2, \ldots, N\}$.
  - $cov[Y_i, Y_j] = 0$, $\forall i \neq j$, $\{i,j\} \in \{1, 2, \ldots, M\}$.
  - $cov[X_i, Y_j] = 0$, $\forall i \in \{1, 2, \ldots, N\}$ and $j \in \{1, 2, \ldots, M\}$.

[P11.2] Consider the data transmission between two devices A and B. Assume that A transmits three packets, each of which can be either successful (s) or failed (f) to receive at B with the same probability.

- Find the sample space of this experiment.
- Let X denote the number of packets that are successfully received at B. Derive the probability mass function (pmf) and the cumulative distribution function (cdf) of X.
- Find the probability that there is at least one packet successfully received.
- Find the probability that there is either one or two packets successfully received.

- Find the expected value, variance, and standard deviation of X.
- Find the skewness and kurtosis of X.

[P11.3] A discrete random variable X has the following pmf:

$$f_X(x) = \begin{cases} c/x & \text{if } x = 1, 2, 3, \\ 0 & \text{otherwise,} \end{cases}$$

where c is a constant.

- Find the value of constant c.
- Find $\Pr[X \geq 2]$.
- Find the expected value and variance of X.

[P11.4] A discrete random variable X has the following pmf:

$$f_X(x) = \begin{cases} 1/8 & \text{if } x = 0, 2, \\ 3/8 & \text{if } x = 1, 3, \\ 0 & \text{otherwise.} \end{cases}$$

- Derive the cdf of X.
- Find $\Pr[0 \leq X \leq 2]$.
- Find the expected value and variance of X.
- Let $Y = 3X + 1$. Derive the pmf and cdf of Y.
- Find the expected value and variance of Y.
- Find the covariance of X and Y.

[P11.5] The probability that a call is a voice call is $\Pr[V] = 0.7$. The probability of a data call is $\Pr[D] = 0.3$. Voice calls cost 25 pence each and data calls cost 40 pence each. Let C denote the cost (in pence) of one telephone call.

- Derive the pmf of C.
- Find the expected value and variance of C.

[P11.6] Monitor three phone calls and observe whether each one is a voice call or a data call. The number of voice calls N has the following pmf

$$f_N(n) = \begin{cases} 0.1 & \text{if } n = 0, \\ 0.3 & \text{if } n = 1, 2, 3, \\ 0 & \text{otherwise.} \end{cases}$$

Voice calls cost 25 pence each and data calls cost 40 pence each. Let T (pence) denote the cost of the three telephone calls monitored in the experiment.

- Find the expected value and variance of N.
- Express T as a function of N.
- Derive the pmf of T.
- Find the expected value and variance of T based on those of N.

[P11.7] In an experiment to monitor two calls, the pmf of the number of voice calls, i.e., N, is

$$f_N(n) = \begin{cases} 0.1 & \text{if } n = 0, \\ 0.4 & \text{if } n = 1, \\ 0.5 & \text{if } n = 2, \\ 0 & \text{otherwise.} \end{cases}$$

- Find the expected value and variance of N.
- Find the skewness and kurtosis of N.

[P11.8] The number of memory chips M needed in a personal computer depends on how many application programs, A, the owner wants to run simultaneously. The number of chips M and the number of application programs A are described by

$$M = \begin{cases} 4 & \text{if } A = 1, 2, \\ 6 & \text{if } A = 3, \\ 8 & \text{if } A = 4. \end{cases}$$

$$f_A(a) = \begin{cases} 0.1(5 - a) & \text{if } a = 1, 2, 3, 4, \\ 0 & \text{otherwise.} \end{cases}$$

- Find the expected number of programs.
- Express M as a function of A.
- Find the expected number of memory chips.

[P11.9] The cdf of a continuous random variable X is given by

$$F_X(x) = \begin{cases} 0 & \text{if } x < 0, \\ cx & \text{if } 0 \le x \le 4, \\ 1 & \text{if } x > 4. \end{cases}$$

- Find the constant c.
- Find $\Pr[X \le 1]$.
- Find $\Pr[2 < X \le 3]$.
- Find $\Pr[X > 3/2]$.
- Derive the probability density function (pdf) of X.
- Find the expected value, variance, and standard deviation of X.

[P11.10] A continuous random variable X has the following pdf

$$f_X(x) = \begin{cases} cxe^{-x/2} & \text{if } x \ge 0, \\ 0 & \text{otherwise.} \end{cases}$$

- Find the constant c.
- Derive the cdf of X.
- Find $\Pr[0 \le X \le 4]$.
- Find $\Pr[-2 \le X \le 2]$.
- Find the expected value, variance, and standard deviation of X.

[P11.11] The pdf of a random variable X is

$$f_X(x) = \begin{cases} 3x^2/2 & \text{if } -1 \leq x \leq 1, \\ 0 & \text{otherwise.} \end{cases}$$

- Find the expected value, variance, and standard deviation of X.
- Find $\Pr[0 \leq X \leq 1]$ using the pdf.
- Derive the cdf of X. Verify $\Pr[0 \leq X \leq 1]$ with the derived cdf.

[P11.12] A triangular distributed random variable X has the following pdf:

$$f_X(x) = \begin{cases} x - 1 & \text{if } 1 \leq x \leq 2, \\ 3 - x & \text{if } 2 < x \leqslant 3, \\ 0 & \text{otherwise.} \end{cases}$$

- Derive the cdf of X.
- Find the expected value and variance of X.

Chapter 12

Discrete random variables and their applications

Discrete random variables, as introduced in Chapter 11, have been well developed to model a number of practical network phenomena which are vital for network simulation. This chapter will discuss various families of discrete random variables and their applications in modeling and validating random events.

In this chapter, I will first provide a review of formulas related to discrete random variables in Section 12.1. Then, various kinds of discrete random variables, including Bernoulli, binomial, geometric, Pascal, Poisson, and discrete uniform random variables, will be sequentially presented in Sections 12.2–12.7. Specifically, the probability mass function (pmf), expected value, variance, and properties of each of these random variables will be derived and discussed in detail.

12.1 Review of discrete random variables

A discrete random variable $X\colon \mathfrak{S} \to \mathbb{R}$ has a countably infinite sample space

$$\mathfrak{S}_X = \{x_1, x_2, \ldots\}$$

or a finite sample space

$$\mathfrak{S}_X = \{x_1, x_2, \ldots, x_N\}.$$

Its distribution functions, properties, and related formulas are summarized in Table 12.1, where $f_X(x)$, $F_X(x)$, $E[X]$, $V[X]$, and σ_X denote the pmf, cumulative distribution function, expected value, variance, and standard deviation, respectively, of the random variable X.

12.2 Bernoulli random variable

Bernoulli random variable is named for a random variable with only two possible outcomes. Let us consider the following example.

Example 12.1. *Perform a test of a network device. There are only two possible outcomes which could be either a success or a failure. This is also known as boolean or binary outcomes. The test is successful with probability p and failed with probability $1 - p$. Let a random variable X represent the number of successful devices in a test, where X has the value of 1 if the test is successful and 0 if the test is failed.*

Table 12.1 Formulas of discrete random variables

pmf	$f_X(x) \triangleq \Pr[X = x]$ $0 \leq f_X(x) \leq 1$ $\sum_{x \in \mathfrak{S}_X} f_X(x) = 1$ $f_{X\|B}(x) \triangleq \Pr[X = x\|B]$ $f_X(x) = \sum_{i=1}^{M} f_{X\|B}(x) \Pr[B_i]$
cdf	$F_X(x) \triangleq \Pr[X \leq x] = \sum_{x' \leq x \in \mathfrak{S}_X} f_X(x')$ $0 \leq F_X(x) \leq 1, F_X(-\infty) = 0, F_X(\infty) = 1$ $F_X(x_1) \leq F_X(x_2) \quad \forall x_1 \leq x_2$ $\Pr[a < X \leq b] = F_X(b) - F_X(a) \quad \forall a < b$
$E[X]$	$E[X] \triangleq \sum_{x \in \mathfrak{S}_X} x f_X(x)$
$V[X]$	$V[X] \triangleq E\left[(X - E[X])^2\right] = \sum_{x \in \mathfrak{S}_X} (x - E[X])^2 f_X(x)$ $V[X] = E[X^2] - E^2[X]$
σ_X	$\sigma_X \triangleq \sqrt{V[X]}$
$Y = g(X)$	$f_Y(y) = \sum_{x:x=g^{-1}(y)} f_X(g^{-1}(y))$ $E[Y] = \sum_{x \in \mathfrak{S}_X} g(x) f_X(x)$ $E[aX + b] = aE[X] + b$ $V[Y] = \sum_{x \in \mathfrak{S}_X} (g(x) - E[g(X)])^2 f_X(x)$ $V[aX + b] = a^2 V[X]$
Central moment	$E[(X - E[X])^n] = \sum_{x \in S_X} (x - E[X])^n f_X(x)$
Raw moment	$E[X^n] = \sum_{x \in S_X} x^n f_X(x)$
Normalized moment	$(E[(X - E[X])^n])/\sigma_X^n = \left(\sum_{x \in S_X} (x - E[X])^n f_X(x)\right) /$ $\left(\sqrt{\sum_{x \in S_X} (x - E[X])^2 f_X(x)}\right)^n$

In Example 12.1, the random variable X is called Bernoulli random variable, and the experiment is well known as Bernoulli trial. For simplicity, it is written as $X \sim Ber(p)$, and its distribution function is defined as follows:

Definition 12.1. *The pmf of a Bernoulli random variable $X \sim Ber(p)$, $0 < p < 1$, is*

$$f_X(x) = \begin{cases} p & \text{if } x = 1, \\ 1 - p & \text{if } x = 0, \\ 0, & \text{otherwise.} \end{cases} \tag{12.1}$$

An illustration of a Bernoulli random variable $X \sim Ber(0.6)$ is shown in Figure 12.1.

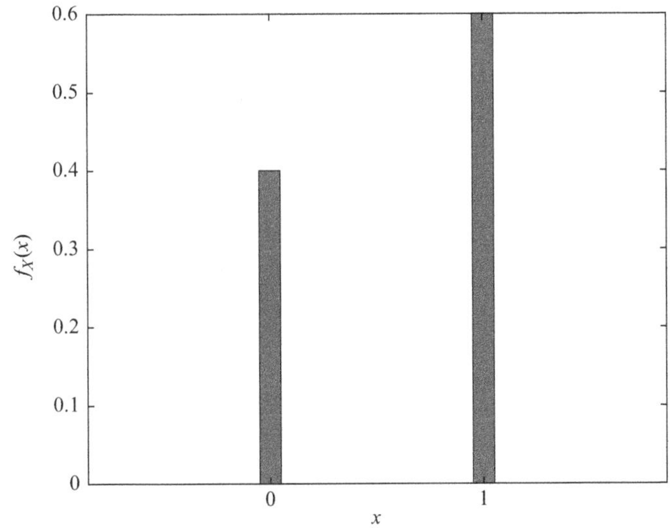

Figure 12.1 pmf of $X \sim Ber(0.6)$

Remark 12.1. *It can be observed in* (12.1) *that* $\sum_{x \in \mathfrak{S}_X} f_X(x) = 1$. *This reflects the fact that the total pmf should be equal to 1 (cf. (11.6)). Note that this fact applies to all discrete random variables including those covered in this chapter.*

Theorem 12.1. *The expected value and variance of a Bernoulli random variable* $X \sim Ber(p)$ *are respectively given by*

$$E[X] = p, \tag{12.2}$$

$$V[X] = p(1 - p). \tag{12.3}$$

Proof. The proof follows from Definitions 11.6 and 11.7 as

$$\begin{aligned} E[X] &= 1 \times f_X(1) + 0 \times f_X(0) \\ &= p, \\ V[X] &= (1 - p)^2 f_X(1) + (0 - p)^2 f_X(0) \\ &= (1 - p)^2 p + p^2(1 - p) \\ &= p(1 - p), \end{aligned}$$

□

12.3 Binomial random variable

As a generalization of Bernoulli random variable, binomial random variable represents either the number of successes or the number of failures in a sequence of independent Bernoulli trials. Let us consider the following example.

Example 12.2. *Perform the test of N network devices, each of which is successful with probability p and failed with probability $(1-p)$. The test result of a device is independent of those of other devices. Let a random variable X represent the number of successful devices in the N tests.*

In Example 12.2, each individual test is a Bernoulli trial as in Example 12.1, and the random variable X is named binomial random variable. For simplicity, it is written as $X \sim Bin(N,p)$.

Definition 12.2. *The pmf of a binomial random variable $X \sim Bin(N,p)$, $0 < p < 1$, $N \in \mathbb{Z}^+$, is*

$$f_X(x) = \begin{cases} \binom{N}{x} p^x (1-p)^{N-x} & \text{if } x = 0, 1, 2, \ldots, N, \\ 0 & \text{otherwise,} \end{cases} \tag{12.4}$$

where $\binom{N}{x}$ denotes the binomial coefficient given by

$$\binom{N}{x} \triangleq \frac{N!}{x!(N-x)!}, \tag{12.5}$$

and $k!$ denotes the factorial of a nonnegative integer k computed by

$$k! \triangleq \prod_{i=1}^{k} i. \tag{12.6}$$

An illustration of a binomial random variable $X \sim Bin(20, 0.6)$ is shown in Figure 12.2.

Remark 12.2. *In (12.4), the binomial coefficient $\binom{N}{x}$ is indeed the number of ways to select x items out of N items, and thus it is read as "N choose x." In this case, this is the number of successful items out of N tests, each of which has a successful probability of p.*

Remark 12.3. *A binomial random variable is a Bernoulli random variable when $N = 1$.*

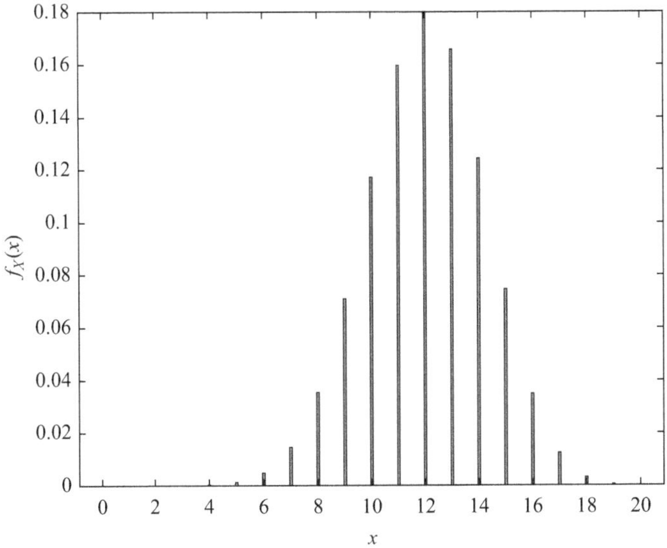

Figure 12.2 pmf of $X \sim Bin(20, 0.6)$

Remark 12.4. *Note that* $0! = 1$, *and thus* $\binom{N}{0} = \binom{N}{N} = 1$, *which means* $f_X(x)$ *still exists when* $x = 0$ *and* $x = N$. *In these two special cases,* $f_X(0) = (1-p)^N$ *and* $f_X(N) = p^N$.

Considering the total pmf, we have the following theorem:

Theorem 12.2. *For* $0 < p < 1$ *and* $N \in \mathbb{Z}^+$,

$$\sum_{x=0}^{N} \binom{N}{x} p^x (1-p)^{N-x} = 1. \tag{12.7}$$

Proof. The proof follows from the well-known binomial formula for the expansion of $(x+y)^n$ based on the Pascal's triangle, i.e.,

$$(x+y)^n = \sum_{k=0}^{n} \binom{n}{k} x^{n-k} y^k.$$

Applying to the pmf given by (12.4), we have

$$\sum_{x=0}^{N} \binom{N}{x} p^x (1-p)^{N-x} = (1-p+p)^N = 1.$$

□

Theorem 12.3. *The expected value and variance of a binomial random variable $X \sim Bin(N,p)$ are respectively given by*

$$E[X] = Np, \tag{12.8}$$

$$V[X] = Np(1-p). \tag{12.9}$$

Proof. The proof follows from the fact that the binomial random variable can be regarded as the summation of N Bernoulli random variables, i.e.,

$$X = \sum_{i=1}^{N} X_i,$$

where X_i, $i = 1, 2, \ldots, N$, is a Bernoulli random variable.

From Theorems 11.11 and 12.1 with $E[X_i] = p$, $i = 1, 2, \ldots, N$, the expected value of X can be obtained by

$$\begin{aligned} E[X] &= \sum_{i=1}^{N} E[X_i] \\ &= Np. \end{aligned}$$

Since $\{X_1, X_2, \ldots, X_N\}$ are independent, from Lemma 11.4 with the variance of each X_i, $i = 1, 2, \ldots, N$, given by $V[X_i] = p(1-p)$ (cf. Theorem 12.1), the variance of X can be determined by

$$\begin{aligned} V[X] &= \sum_{i=1}^{N} V[X_i] \\ &= Np(1-p). \end{aligned}$$

□

12.4 Geometric random variable

A geometric random variable is used to represent the number of tests required to obtain the first success.[1] Let us consider the following example.

Example 12.3. *Perform a test of network devices, each of which is successful with probability p and failed with probability* $(1-p)$*. Let X be the number of tests up to and including the first successful test.*

In Example 12.3, the random variable X is called geometric random variable due to the fact that its function distribution has the form of a geometric progression

[1]The geometric random variable may be defined differently in some materials as the number of tests until having the first success or before the first success. It is also defined as the number of failures prior to having the first success. In either case, the distribution function can be similarly obtained by shifting the random variable.

as shown later in Definition 12.3. For simplicity, the geometric random variable is written as $X \sim Geo(p)$.

Definition 12.3. *The pmf of a geometric random variable* $X \sim Geo(p)$, $0 < p < 1$, *is*

$$f_X(x) = \begin{cases} p(1-p)^{x-1} & \text{if } x = 1, 2, \ldots, \\ 0 & \text{otherwise.} \end{cases} \tag{12.10}$$

An illustration of a geometric random variable $X \sim Geo(0.6)$ is shown in Figure 12.3.

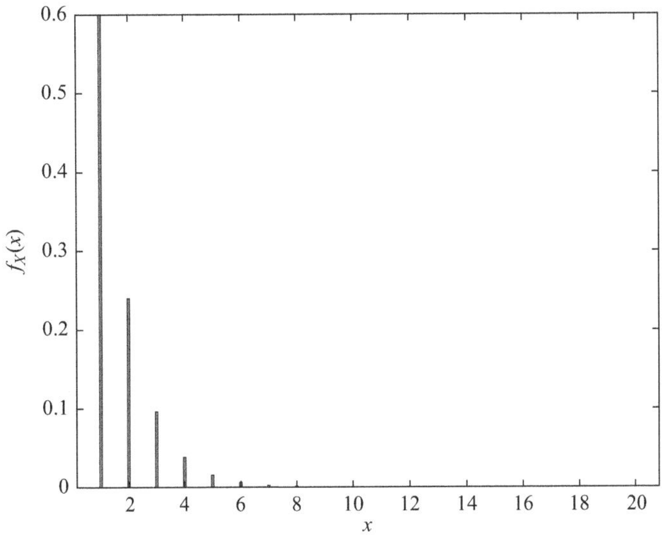

Figure 12.3 pmf of $X \sim Geo(0.6)$

Remark 12.5. *In* (12.10), *it can be easily seen that there is only one success with probability p, while the remaining* $(x-1)$ *tests are failed with probability* $(1-p)$.

Remark 12.6. *If we define a random variable Y as the number of failures until having the first success, i.e.,* $Y = X - 1$, *then the pmf of Y should be*

$$f_Y(y) = \begin{cases} p(1-p)^{y} & \text{if } y = 0, 1, 2, \ldots, \\ 0 & \text{otherwise.} \end{cases} \tag{12.11}$$

In this case, there is y failures with probability $(1-p)$ *and one success with probability p. Notice that the range of y is from* 0 *for the case of no failure, which is different from the range of random variable X in Definition 12.3 starting from* 1.

Considering the total pmf, we have the following theorem

Theorem 12.4. *For* $0 < p < 1$,

$$\sum_{x=1}^{\infty} p(1-p)^{x-1} = 1. \tag{12.12}$$

Proof. We have

$$\begin{aligned}
\sum_{x=1}^{\infty} p(1-p)^{x-1} &= p\sum_{x=1}^{\infty}(1-p)^{x-1} \\
&= p\sum_{x=0}^{\infty}(1-p)^{x} \\
&\overset{(a)}{=} p\frac{1}{1-(1-p)} \\
&= 1,
\end{aligned}$$

where (a) follows from the infinite sum of geometric series, i.e.,

$$\sum_{x=0}^{\infty} q^x = \frac{1}{1-q}, \quad \forall |q| < 1.$$

□

Theorem 12.5. *The expected value and variance of a geometric random variable* $X \sim Geo(p)$ *are respectively given by*

$$E[X] = \frac{1}{p}, \tag{12.13}$$

$$V[X] = \frac{1-p}{p^2}. \tag{12.14}$$

Proof. From Definition 11.6, the expected value of X is given by

$$\begin{aligned}
E[X] &= \sum_{x \in \mathfrak{S}_X} x f_X(x) \\
&\overset{(a)}{=} \sum_{x=1}^{\infty} xp(1-p)^{x-1} \\
&= p\sum_{x=1}^{\infty} x(1-p)^{x-1} \\
&\overset{(b)}{=} p\frac{1}{(1-(1-p))^2} \\
&= \frac{1}{p}.
\end{aligned}$$

where (a) follows from the pmf of X in (12.3) and (b) follows from the derivative of geometric progression, i.e.,

$$\sum_{x=1}^{\infty} nx^{n-1} = \frac{1}{(1-x)^2}, \quad \forall |x| < 1.$$

In order to derive the variance of X, we need to first compute $E[X^2]$ as

$$\begin{aligned} E[X^2] &= \sum_{x \in \mathfrak{S}_X} x^2 f_X(x) \\ &= \sum_{x=1}^{\infty} x^2 p(1-p)^{x-1} \\ &= p\sum_{x=1}^{\infty} x(x+1)(1-p)^{x-1} - p\sum_{x=1}^{\infty} x(1-p)^{x-1} \\ &\overset{(a)}{=} p\frac{2}{(1-(1-p))^3} - \frac{1}{p} \\ &= \frac{2}{p^2} - \frac{1}{p}, \end{aligned}$$

where (a) follows from the derivative of geometric progression, i.e.,

$$\sum_{x=1}^{\infty} n(n+1)x^{n-1} = \frac{2}{(1-x)^3}, \quad \forall |x| < 1.$$

From Theorem 11.8, we can obtain the variance of X as

$$\begin{aligned} V[X] &= E[X^2] - E^2[X] \\ &= \frac{2}{p^2} - \frac{1}{p} - \frac{1}{p^2} \\ &= \frac{1-p}{p^2}. \end{aligned}$$

□

12.5 Pascal random variable

A Pascal random variable is generalized from geometric random variable to represent the number of tests required to obtain a specified number of successes. Let us consider the following example.

Example 12.4. *Perform a test of network devices, each of which is successful with probability p and failed with probability* $(1-p)$*. Let X be the number of tests up to and including the k successful tests.*

In Example 12.4, the random variable X is called Pascal random variable and, for simplicity, it is written as $X \sim Pas(k,p)$.

Definition 12.4. *The pmf of a Pascal random variable $X \sim Pas(k,p)$, $0 < p < 1$, $k \in \mathbb{Z}^+$, is*

$$f_X(x) = \begin{cases} \binom{x-1}{k-1} p^k (1-p)^{x-k} & \text{if } x = k, k+1, \ldots, \\ 0 & \text{otherwise.} \end{cases} \tag{12.15}$$

An illustration of a Pascal random variable $X \sim Pas(4, 0.6)$ is shown in Figure 12.4.

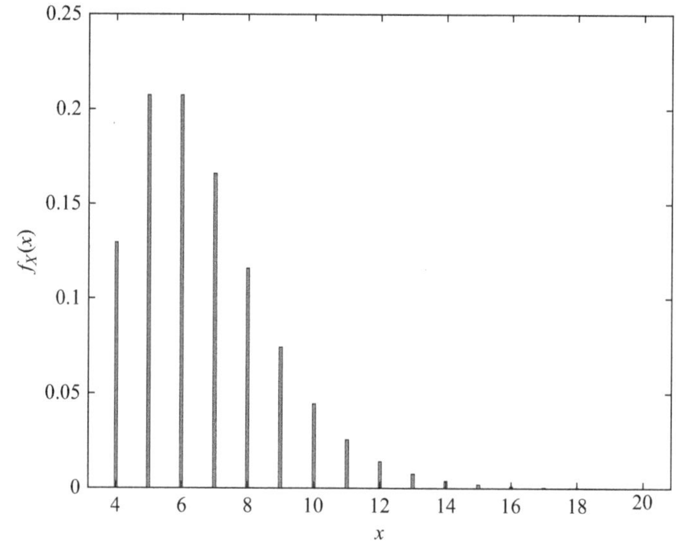

Figure 12.4 pmf of $X \sim Pas(4, 0.6)$

Remark 12.7. *In (12.15), the binomial coefficient $\binom{x-1}{k-1}$ is the number of possibilities to have $(k-1)$ successful tests out of $(x-1)$ tests given the fact that the last test is successful, i.e., the kth successful test.*

Remark 12.8. *It can be seen that the binomial coefficient in the pmf of Pascal random variable has a reversed form of that of the binomial random variable where the number of tests is fixed. Therefore, the Pascal random variable is also called as "negative" binomial random variable in some materials.*

Remark 12.9. *A Pascal random variable is a geometric random variable when $k = 1$.*

Considering the total pmf, we have the following theorem

Theorem 12.6. *For* $0 < p < 1$ *and* $N \in \mathbb{Z}^+$,

$$\sum_{x=k}^{\infty} \binom{x-1}{k-1} p^k (1-p)^{x-k} = 1. \tag{12.16}$$

Proof.

$$\begin{aligned}
\sum_{x=k}^{\infty} \binom{x-1}{k-1} p^k (1-p)^{x-k} &= p^k \sum_{x=k}^{\infty} \binom{x-1}{k-1} (1-p)^{x-k} \\
&= p^k \frac{1}{(1-p)^{k-1}} \sum_{x=k}^{\infty} \binom{x-1}{k-1} (1-p)^{x-1} \\
&\overset{(a)}{=} p^k \frac{1}{(1-p)^{k-1}} \frac{(1-p)^{k-1}}{(1-(1-p))^k} \\
&= 1,
\end{aligned}$$

where (a) follows from the ordinary generating function of binomial coefficients, i.e.,

$$\sum_{n=m}^{\infty} \binom{n}{m} x^n = \frac{x^m}{(1-x)^{m+1}}$$

when $n = x - 1$, $m = k - 1$, and $x = 1 - p$. □

Theorem 12.7. *The expected value and variance of a Pascal random variable* $X \sim Pas(k, p)$ *are respectively given by*

$$E[X] = \frac{k}{p}, \tag{12.17}$$

$$V[X] = \frac{k(1-p)}{p^2}. \tag{12.18}$$

Proof. The proof follows from the fact that the Pascal random variable can be regarded as the summation of k geometric random variables, i.e.,

$$X = \sum_{i=1}^{k} X_i,$$

where X_i, $i = 1, 2, \ldots, k$, is a geometric random variable.

From Theorems 11.11 and 12.5 with $E[X_i] = 1/p$, $i = 1, 2, \dots, N$, the expected value of X can be obtained by

$$E[X] = \sum_{i=1}^{k} E[X_i]$$
$$= \frac{k}{p}.$$

Since $\{X_1, X_2, \dots, X_k\}$ are independent, from Lemma 11.4 with the variance of each X_i, $i = 1, 2, \dots, N$, given by $V[X_i] = (1-p)/p^2$ (cf. Theorem 12.5), the variance of X can be determined by

$$V[X] = \sum_{i=1}^{k} V[X_i]$$
$$= \frac{k(1-p)}{p^2}.$$

□

12.6 Poisson random variable

A Poisson random variable is associated with Poisson point process which was described in Chapter 9 (cf. Section 9.2). In this section, the Poisson random variable is discussed in detail showing its pmf, expected value, and variance.

A Poisson random variable X is used to represent the number of events occurring in a specified time interval. For simplicity, X is written as $X \sim Poi(\lambda)$ where λ is a Poisson parameter. As shown later, λ is also the average number of events, and thus it is also known as event rate.

Definition 12.5. *The pmf of a Poisson random variable $X \sim Poi(\lambda)$, $\lambda > 0$ is*

$$f_X(x) = \begin{cases} \dfrac{\lambda^x e^{-\lambda}}{x!} & \text{if } x = 0, 1, 2, \dots, \\ 0 & \text{otherwise.} \end{cases} \tag{12.19}$$

An illustration of a Poisson random variable $X \sim Poi(5)$ is shown in Figure 12.5. Considering the total pmf, we have the following theorem:

Theorem 12.8. *For $\lambda > 0$,*

$$\sum_{x=0}^{\infty} \frac{\lambda^x e^{-\lambda}}{x!} = 1. \tag{12.20}$$

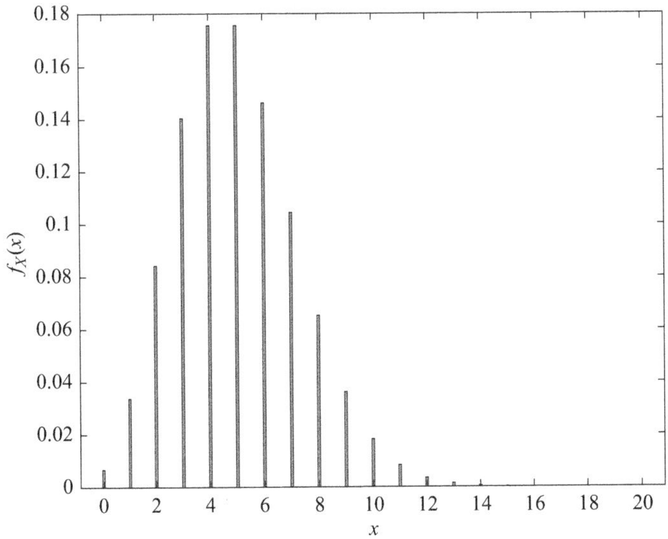

Figure 12.5 pmf of $X \sim Poi(5)$

Proof. We have

$$\begin{aligned}\sum_{x=0}^{\infty} \frac{\lambda^x e^{-\lambda}}{x!} &= e^{-\lambda} \sum_{x=0}^{\infty} \frac{\lambda^x}{x!} \\ &\stackrel{(a)}{=} e^{-\lambda} e^{\lambda} \\ &= 1,\end{aligned}$$

where (a) follows from the Maclaurin series of exponential function, i.e.,

$$e^x = \sum_{n=0}^{\infty} \frac{x^n}{n!}.$$

□

Theorem 12.9. *The expected value and variance of a Poisson random variable $X \sim Poi(\lambda)$ are respectively given by*

$$E[X] = \lambda, \tag{12.21}$$

$$V[X] = \lambda. \tag{12.22}$$

Proof. From Definition 11.6, the expected value of X is given by

$$\begin{aligned}
E[X] &= \sum_{x \in \mathfrak{S}_X} x f_X(x) \\
&\overset{(a)}{=} \sum_{x=0}^{\infty} x \frac{\lambda^x e^{-\lambda}}{x!} \\
&= \sum_{x=1}^{\infty} x \frac{\lambda^x e^{-\lambda}}{x!} \\
&= \lambda e^{-\lambda} \sum_{x=1}^{\infty} \frac{\lambda^{x-1}}{(x-1)!} \\
&= \lambda e^{-\lambda} \sum_{x=0}^{\infty} \frac{\lambda^x}{x!} \\
&\overset{(b)}{=} \lambda e^{-\lambda} e^{\lambda} \\
&= \lambda
\end{aligned}$$

where (a) follows from the pmf of X given by (12.19) and (b) follows from the Maclaurin series of exponential function, i.e.,

$$e^x = \sum_{n=0}^{\infty} \frac{x^n}{n!}.$$

From Theorem 11.8, the variance of X is determined by

$$\begin{aligned}
V[X] &= E[X^2] - E^2[X] \\
&= \sum_{x=0}^{\infty} x^2 \frac{\lambda^x e^{-\lambda}}{x!} - \lambda^2 \\
&= \sum_{x=0}^{\infty} x \frac{\lambda^x e^{-\lambda}}{(x-1)!} - \lambda^2 \\
&= \lambda e^{-\lambda} \sum_{x=0}^{\infty} x \frac{\lambda^{x-1}}{(x-1)!} - \lambda^2 \\
&= \lambda e^{-\lambda} \sum_{x=1}^{\infty} x \frac{\lambda^{x-1}}{(x-1)!} - \lambda^2 \\
&= \lambda e^{-\lambda} \left(\sum_{x=1}^{\infty} (x-1) \frac{\lambda^{x-1}}{(x-1)!} + \sum_{x=1}^{\infty} \frac{\lambda^{x-1}}{(x-1)!} \right) - \lambda^2 \\
&= \lambda e^{-\lambda} \left(\lambda \sum_{x=2}^{\infty} \frac{\lambda^{x-2}}{(x-2)!} + \sum_{x=1}^{\infty} \frac{\lambda^{x-1}}{(x-1)!} \right) - \lambda^2 \\
&\overset{(a)}{=} \lambda e^{-\lambda} (\lambda e^{\lambda} + e^{\lambda}) - \lambda^2 \\
&= \lambda,
\end{aligned}$$

where (a) follows the same Maclaurin series of exponential function as in the proof of the expected value of X. □

12.7 Discrete uniform random variable

A discrete uniform random variable is used to describe an experiment with equiprobable outcomes within a finite range of integer numbers.

Let X denote the discrete uniform random variable with the sample space $\mathfrak{S}_X = \{a, a+1, \ldots, b\}$, where $a, b \in \mathbb{Z}$ and $a < b$. The random variable X can be simply written as $X \sim U(a, b)$.

Definition 12.6. *The pmf of a discrete uniform random variable* $X \sim U(a, b)$*,* $a, b \in \mathbb{Z}$*,* $a < b$*, is*

$$f_X(x) = \begin{cases} \dfrac{1}{b-a+1} & \text{if } x = a, a+1, a+2, \ldots, b, \\ 0 & \text{otherwise.} \end{cases} \tag{12.23}$$

An illustration of a discrete uniform random variable $X \sim U(1, 5)$ is shown in Figure 12.6.

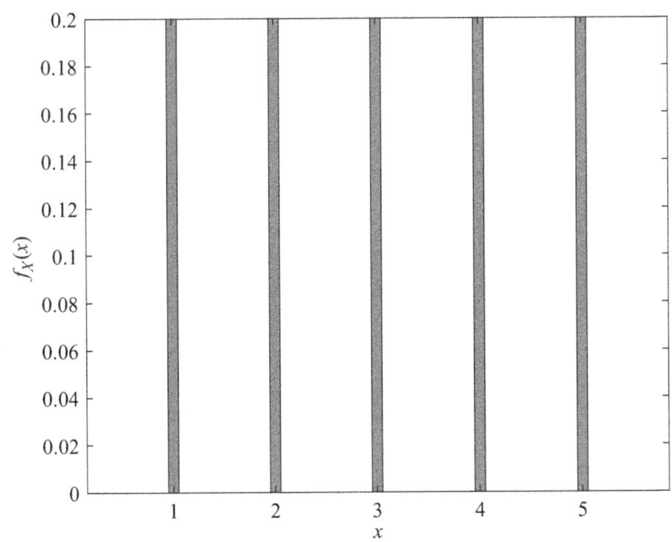

Figure 12.6 pmf of $X \sim U(1, 5)$

Remark 12.10. *It can be observed in* (12.23) *that*

$$\sum_{x=a}^{b} \frac{1}{b-a+1} = 1.$$

This verifies that the total pmf should be equal to 1 (cf. (11.6)).

Theorem 12.10. *The expected value and variance of a discrete uniform random variable* $X \sim U(a,b)$ *is*

$$E[X] = \frac{a+b}{2}, \tag{12.24}$$

$$V[X] = \frac{(b-a+1)^2 - 1}{12}. \tag{12.25}$$

Proof. From Definition 11.6, the expected value of X is given by

$$\begin{aligned} E[X] &= \sum_{x \in \mathfrak{S}_X} x f_X(x) \\ &\overset{(a)}{=} \sum_{x=a}^{b} x \frac{1}{b-a+1} \\ &= \frac{1}{b-a+1} \sum_{x=a}^{b} x \\ &\overset{(b)}{=} \frac{1}{b-a+1} \frac{(b-a+1)(a+b)}{2} \\ &= \frac{a+b}{2}, \end{aligned}$$

where (a) follows from the pmf of X given by (12.23) and (b) follows from the sum of series that

$$\sum_{i=a}^{b} i = \frac{(b-a+1)(a+b)}{2}.$$

In order to derive the variance of X, for simplicity, let us define another random variable $Y = X - a + 1$. Y is also a uniform random variable, but it is defined within a simple range, i.e., $Y \sim U(1,n)$, where $n = b - a + 1$. The pmf of Y is thus given by

$$f_Y(y) = \begin{cases} \dfrac{1}{n} & \text{if } y = 1, 2, \ldots, n, \\ 0 & \text{otherwise.} \end{cases}$$

The expected value of Y can be obtained as

$$
\begin{aligned}
E[Y] &= \sum_{y \in \mathfrak{S}_Y} y f_Y(y) \\
&= \sum_{y=1}^{n} y \frac{1}{n} \\
&= \frac{1}{n} \sum_{y=1}^{n} y \\
&= \frac{n+1}{2}.
\end{aligned}
$$

Since Y is a linear combination of X, we have another approach to derive $E[X]$ using Theorem 11.6 as

$$
\begin{aligned}
E[X] &= E[Y] + a - 1 \\
&= \frac{b - a + 2}{2} + a - 1 \\
&= \frac{a+b}{2}.
\end{aligned}
$$

In order to derive $V[Y]$, we need to first compute $E[Y^2]$ as follows

$$
\begin{aligned}
E[Y^2] &= \sum_{y \in \mathfrak{S}_Y} y^2 f_Y(x) \\
&= \sum_{y=1}^{n} y^2 \frac{1}{n} \\
&= \frac{1}{n} \sum_{y=1}^{n} y^2 \\
&\overset{(a)}{=} \frac{1}{n} \frac{n(n+1)(2n+1)}{6} \\
&= \frac{(n+1)(2n+1)}{6},
\end{aligned}
$$

where (a) follows from the sum of sequence of squares, i.e.,

$$
\sum_{i=1}^{n} i^2 = \frac{n(n+1)(2n+1)}{6}.
$$

The variance of Y can be derived from Theorem 11.8 as

$$\begin{aligned} V[Y] &= E[Y^2] - E^2[Y] \\ &= \frac{(n+1)(2n+1)}{6} - \frac{(n+1)^2}{4} \\ &= \frac{n^2-1}{12} \\ &= \frac{(b-a+1)^2-1}{12}. \end{aligned}$$

According to Lemma 11.2, we have

$$V[X] = V[Y] = \frac{(b-a+1)^2-1}{12}.$$

□

12.8 Related works

For more information, the readers are suggested to refer to the following selected references:

Bernoulli Distribution Johnson *et al.* in 2005 [245] and Forbes *et al.* in 2010 [216].
Binomial Distribution Miller in 1954 [248], Kaas and Buhrman in 1980 [249], and Wang in 1993 [250].
Geometric Distribution Andrews in 1998 [251] and Spiegel in 2003 [252].
Pascal Distribution Wilf in 2006 [253] and Hilbe in 2011 [254].
Poisson Distribution Haight in 1967 [255], Knusel in 1986 [256], Daley and Vere-Jones in 2003 and 2008 [221,222], Illian *et al.* in 2008 [223], Bass in 2011 [224], and Snyder and Miller in 2012 [225].
Discrete Uniform Distribution Papoulis in 2002 [215], Johnson *et al.* in 2005 [245], and Forbes *et al.* in 2010 [216].

12.9 Review questions

[Q12.1] Give an example for the following discrete random variables:
- Bernoulli random variable.
- Binomial random variable.
- Geometric random variable.
- Pascal random variable.
- Poisson random variable.
- Discrete uniform random variable.

[Q12.2] For each of the above random variables, describe its probability mass function (pmf).

[Q12.3] Derive the expected value and variance of the above random variables.

[Q12.4] What is the relationship between binomial random variable and Bernoulli random variable?

[Q12.5] What is the relationship between Pascal random variable and geometric random variable?

[Q12.6] What is the physical meaning of the Poisson parameter in defining the distribution function of a Poisson random variable?

12.10 Problems

[P12.1] A Bernoulli test of network devices is represented by a Bernoulli random variable $X \sim Ber(0.4)$. Find its expected value and variance of X.

[P12.2] Perform a test of five devices, each of which follows a Bernoulli distribution with a success probability of 0.3. Assume the tests are independent of each other.

- Derive the probability mass function (pmf) of the number of successful devices denoted by X.
- Find the expected value and variance of X.

[P12.3] A test of devices has a successful probability of 0.6. Let X denote the number of tests up to and including the last successful attempt.

- How to represent X?
- Derive the probability mass function (pmf) of X.
- Find the expected value and variance of X.

[P12.4] Consider a test of network devices. Let X denote the number of failures until the first success.

- Derive the pmf of X.
- Find the expected value and variance of X.
- What is the relationship between X and the geometric random variable?

[P12.5] Each time a modem transmits 1 b, the receiving modem analyzes the signal that arrives and decides whether the transmitted bit is 0 or 1. It makes an error with probability p, independent of whether any other bit is received correctly.

- If the transmission continues until the receiving modem makes its first error, what is the pmf of X, the number of bits transmitted?
- If $p = 0.1$, what is the probability that $X = 10$? What is the probability that $X \geq 10$?
- If the modem transmits 100 b, what is the pmf of Y, the number of errors?
- If $p = 0.01$ and the modem transmits 100 b, what is the probability of $Y = 2$ errors at the receiver? What is the probability that $Y \leq 2$?
- If the transmission continues until the receiving modem makes three errors, what is the pmf of Z, the number of bits transmitted?
- If $p = 0.25$, what is the probability of $Z = 12$ b transmitted?

[P12.6] A device is rejected with probability of $p = 1/4$. Let X represent the number of tests up to and including the first reject.

- Derive the pmf of X.
- Derive the cumulative distribution function (cdf) of X.
- Use the derived cdf to find the following probabilities:
  - $\Pr[X < 1]$
  - $\Pr[X \leq 1]$
  - $\Pr[X > 2]$
  - $\Pr[X \geq 2]$
  - $\Pr[3 < X \leq 8]$
  - $\Pr[X = 1]$
  - $\Pr[X = 3]$

[P12.7] The number of hits at a website in any time interval is a Poisson random variable. A particular site has on average $\lambda = 2$ hits per second.

- Derive the pmf of the Poisson random variable $X \sim Poi(\lambda)$.
- Find the probability that there are no more than two hits in an interval of 1 s.
- Let Y denote the number of hits in 0.25 s. Derive the pmf of Y and find the probability that there is no hit within that interval.
- Find the probability that there are at least three hits in a 2-s interval.

[P12.8] Consider information flow arriving at a router. Suppose that an average of 50 flows per minute arrives at the router following a Poisson process. Let X denote the number of arrivals during the next 5 s.

- Derive the pmf and cdf of X.
- Find the probability that three or more flows will arrive at the router in the next 5 s.

[P12.9] The amplitude V [V] of a sinusoidal signal follows a discrete uniform distribution within a range of value $\{-3, -2, -1, 0, 1, 2, 3\}$.

- Derive the pmf of V.
- Find the expected value and variance of V.
- Let $W = V^2/2$ [W] denote the average power of the transmitted signal.
  - Derive the pmf of W.
  - Find the expected value and variance of W.

Chapter 13

Continuous random variables and their applications

Together with discrete random variables for modeling discrete random events, continuous random variables have also been well developed to represent network phenomena in which the variables can randomly take infinite and uncountable values. This chapter will discuss various families of continuous random variables and their applications in modeling and validating random events.

In this chapter, I will first provide a review of formulas related to continuous random variables in Section 13.1. Then, various kinds of continuous random variables, including uniform, exponential, Erlang, normal, and lognormal random variables, will be sequentially discussed in Sections 13.2–13.6. For each random variable, its probability density function (pdf), cumulative distribution function (cdf), expected value, and variance will be derived along with discussion of their properties and relevant applications.

13.1 Review of continuous random variables

A continuous random variable $X: \mathfrak{S} \to \mathbb{R}$ has an uncountable sample space consisting of numerical values in a specified interval which can be either closed, half-bounded, or open, i.e.,

$$\mathfrak{S}_X = (x_1, x_2) \text{ or } (x_1, x_2] \text{ or } [x_1, x_2) \text{ or } [x_1, x_2].$$

Its distribution functions, properties, and related formulas are summarized in Table 13.1, where $f_X(x)$, $F_X(x)$, $E[X], V[X]$, and σ_X denote the pdf, cdf, expected value, variance, and standard deviation, respectively, of the random variable X.

13.2 Continuous uniform random variable

Similar to discrete uniform random variable in Chapter 12 (cf. Section 12.7), a continuous uniform random variable is used to represent an event with equiprobable outcomes having values within a specified range.

Table 13.1 Formulas of continuous random variables

pdf	$f_X(x) = ((dF_X(x))/dx)$
	$0 \le f_X(x) \le 1$
	$\int_{-\infty}^{\infty} f_X(x) = 1$
cdf	$F_X(x) \triangleq \Pr[X \le x] = \int_{-\infty}^{x} f_X(t)dt$
	$0 \le F_X(x) \le 1, F_X(-\infty) = 0, F_X(\infty) = 1$
	$F_X(x_1) \le F_X(x_2) \quad \forall x_1 \le x_2$
	$\Pr[a < X \le b] = F_X(b) - F_X(a) = \int_a^b f_X(x)dx \quad \forall a < b$
$E[X]$	$E[X] \triangleq \int_{-\infty}^{\infty} x f_X(x)dx$
$V[X]$	$V[X] \triangleq E\left[(X - E[X])^2\right] = \int_{-\infty}^{\infty} (x - E[X])^2 f_X(x)dx$
	$V[X] = E[X^2] - E^2[X]$
σ_X	$\sigma_X \triangleq \sqrt{V[X]}$
$Y = g(X)$	$f_Y(y) = \sum_{x:x=g^{-1}(y)} f_X(g^{-1}(y))$
	$E[Y] = \int_{-\infty}^{\infty} g(x) f_X(x)dx$
	$E[aX + b] = aE[X] + b$
	$V[Y] = \int_{-\infty}^{\infty} (g(x) - E[g(X)])^2 f_X(x)dx$
	$V[aX + b] = a^2 V[X]$
Central moment	$E[(X - E[X])^n] = \int_{-\infty}^{\infty} (x - E[X])^n f_X(x)dx$
Raw moment	$E[X^n] = \int_{-\infty}^{\infty} x^n f_X(x)dx$
Normalized moment	$((E[(X - E[X])^n])/\sigma_X^n)$
	$= \left((\int_{-\infty}^{\infty} (x - E[X])^n f_X(x)dx) / \left(\sqrt{\int_{-\infty}^{\infty} (x - E[X])^2 f_X(x)dx} \right)^n \right)$

Let X denote the continuous uniform random variable with the sample space $\mathfrak{S}_X = [a, b]$, where $a < b$. The random variable X can be simply written as $X \sim U(a, b)$ with a notice that the range of values is continuous.

Definition 13.1. *The pdf of a continuous uniform random variable $X \sim U(a, b)$, $a < b$, is*

$$f_X(x) = \begin{cases} \dfrac{1}{b - a} & \text{if } a \le x \le b, \\ 0 & \text{otherwise} \end{cases} \tag{13.1}$$

Remark 13.1. *It can be shown from* (13.1) *that*

$$\begin{aligned} \int_{-\infty}^{\infty} f_X(x)dx &= \int_a^b \frac{1}{b - a} dx \\ &= \frac{1}{b - a} x|_a^b \\ &= 1. \end{aligned}$$

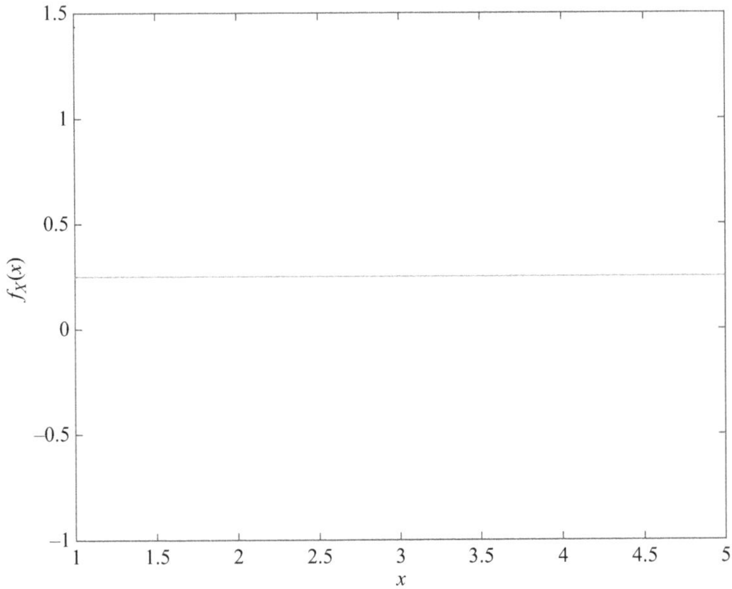

Figure 13.1 pdf of X ~ U(1,5)

This verifies the fact that the total pdf should be equal to 1 (cf. (11.48)). Note that this fact applies to all continuous random variables including those covered in this chapter.

An illustration of a continuous uniform random variable is shown in Figures 13.1 and 13.2 where the pdf and cdf of $X \sim U(1,5)$ are, respectively, plotted.

Theorem 13.1. *The cdf of a continuous uniform random variable $X \sim U(a,b)$, $a < b$, is given by*

$$F_X(x) = \begin{cases} 0 & \text{if } x < a \\ \dfrac{x-a}{b-a} & \text{if } a \leq x \leq b \\ 1 & \text{if } x > b \end{cases} \tag{13.2}$$

Proof. The proof follows from Definition 11.14 and Theorem 11.12 as follows:

- If $x \leq a$, then

$$F_X(x) = \Pr[X \leq x] = 0.$$

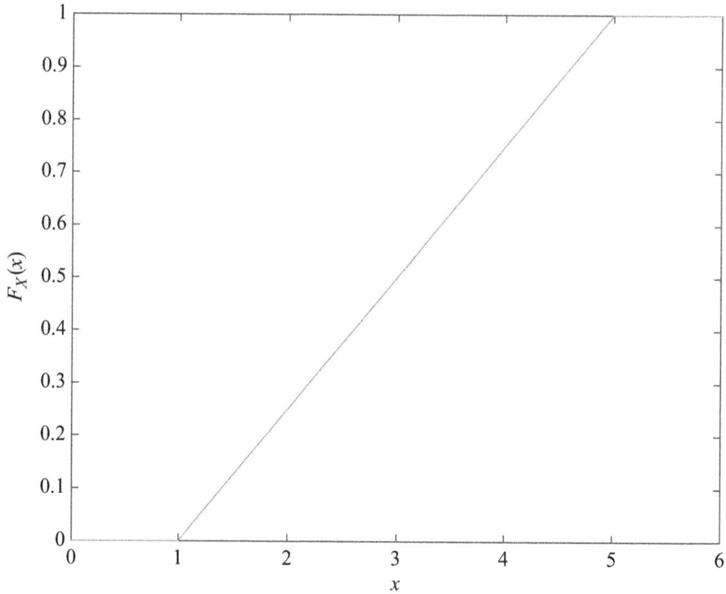

Figure 13.2 cdf of X ~ U(1,5)

- If $a \leq x \leq b$, then

$$\begin{aligned} F_X(x) &= \int_{-\infty}^{x} f_X(t)dt \\ &= \int_{a}^{x} \frac{1}{b-a} f_X(t)dt \\ &= \left(\frac{t}{b-a} \right) \Big|_a^x \\ &= \frac{x-a}{b-a}. \end{aligned}$$

- If $x > b$, then

$$F_X(x) = \Pr[X \leq x] = 1. \qquad \square$$

Theorem 13.2. *The expected value and variance of a continuous uniform random variable* $X \sim U(a, b)$ *are respectively given by*

$$E[X] = \frac{a+b}{2}, \tag{13.3}$$

$$V[X] = \frac{(b-a)^2}{12}. \tag{13.4}$$

Proof. From Definition 11.16, the expected value of X is given by

$$\begin{aligned} E[X] &= \int_{-\infty}^{\infty} x f_X(x)dx \\ &\overset{(a)}{=} \int_a^b \frac{x}{b-a}dx \\ &= \frac{1}{b-a}\left(\frac{x^2}{2}\right)\Bigg|_a^b \\ &= \frac{a+b}{2}, \end{aligned}$$

where (a) follows from (13.1).

From Theorem 11.15, in order to derive the variance of X, we need to calculate $E[X^2]$ which is given by

$$\begin{aligned} E[X^2] &= \int_{-\infty}^{\infty} x^2 f_X(x)dx \\ &= \int_a^b \frac{x^2}{b-a}dx \\ &= \frac{1}{b-a}\left(\frac{x^3}{3}\right)\Bigg|_a^b \\ &= \frac{a^2+ab+b^2}{3}. \end{aligned}$$

The variance is thus obtained by

$$\begin{aligned} V[X] &= E[X^2] - E^2[X] \\ &= \frac{a^2+ab+b^2}{3} - \left(\frac{a+b}{2}\right)^2 \\ &= \frac{(b-a)^2}{12}. \end{aligned}$$

□

13.3 Exponential random variable

An exponential random variable is related to Poisson point process (PPP) which was briefly discussed in Chapter 9 (cf. Section 9.2). As stated in Theorem 9.1, the time difference between events, a.k.a. interarrival time, is exponentially distributed.

The exponential random variable X is represented by a parameter $\lambda > 0$, namely, rate as in PPP. For simplicity, it is generally written as $X \sim Exp(\lambda)$.

Definition 13.2. *The pdf of an exponential random variable $X \sim Exp(\lambda)$, $\lambda > 0$, is*

$$f_X(x) = \begin{cases} \lambda e^{-\lambda x} & \text{if } x \geq 0 \\ 0 & \text{otherwise.} \end{cases} \tag{13.5}$$

Considering the total pdf, we have the following theorem

Theorem 13.3. *For $\lambda > 0$,*

$$\int_0^\infty \lambda e^{-\lambda x} = 1. \tag{13.6}$$

Proof. We have

$$\begin{aligned} \int_0^\infty \lambda e^{-\lambda x} &= \lambda \left(-\frac{1}{\lambda} e^{-\lambda x} \right) \Bigg|_0^\infty \\ &\overset{(a)}{=} \lambda \frac{1}{\lambda} \\ &= 1, \end{aligned}$$

where (a) follows from the fact that

$$\lim_{x \to \infty} e^{-\lambda x} = 0.$$

□

An illustration of an exponential random variable is shown in Figures 13.3 and 13.4 where the pdf and cdf of $X \sim Exp(5)$ are, respectively, plotted.

Theorem 13.4. *The cdf of an exponential random variable $X \sim Exp(\lambda)$, $\lambda > 0$, is given by*

$$F_X(x) = \begin{cases} 1 - e^{-\lambda x} & \text{if } x \geq 0 \\ 0 & \text{otherwise.} \end{cases} \tag{13.7}$$

Proof. The proof follows from Definition 11.14 and Theorem 11.12 as follows:

- If $x < 0$, then

$$F_X(x) = \Pr[X \leq x] = 0.$$

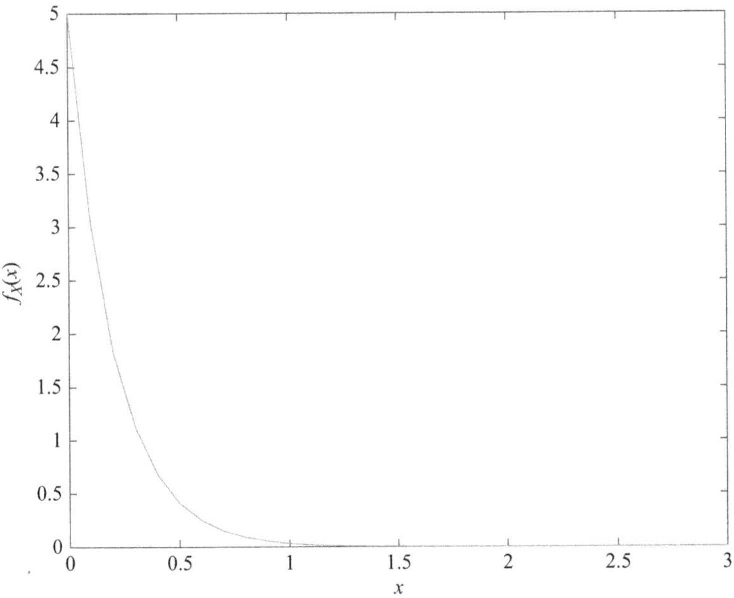

Figure 13.3 pdf of $X \sim Exp(5)$

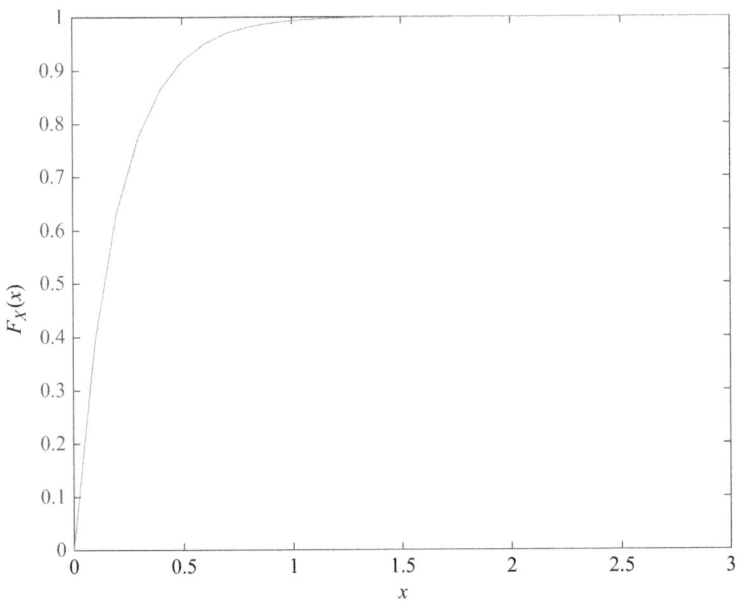

Figure 13.4 cdf of $X \sim Exp(5)$

- If $x \geq 0$, then

$$
\begin{aligned}
F_X(x) &= \int_{-\infty}^{x} f_X(t)dt \\
&= \int_{0}^{x} \lambda e^{-\lambda t} dt \\
&= \left(-e^{-\lambda t}\right)\Big|_0^x \\
&= 1 - e^{-\lambda x}.
\end{aligned}
$$

□

Theorem 13.5. *The expected value and variance of an exponential random variable $X \sim Exp(\lambda)$ are respectively given by*

$$E[X] = \frac{1}{\lambda}, \tag{13.8}$$

$$V[X] = \frac{1}{\lambda^2}. \tag{13.9}$$

Proof. From Definition 11.16, the expected value of X is given by

$$
\begin{aligned}
E[X] &= \int_{-\infty}^{\infty} x f_X(x)dx \\
&\overset{(a)}{=} \int_{0}^{\infty} x\lambda e^{-\lambda x} dx \\
&\overset{(b)}{=} = \left(-xe^{-\lambda x}\right)\Big|_0^\infty + \int_{0}^{\infty} e^{-\lambda x} dx \\
&\overset{(c)}{=} \left(-\frac{1}{\lambda}e^{-\lambda x}\right)\Big|_0^\infty \\
&= \frac{1}{\lambda},
\end{aligned}
$$

where (a) follows from (13.5), (b) follows by performing integration by parts, and (c) follows from the fact that

$$\lim_{x \to \infty} xe^{-\lambda x} = 0.$$

From Theorem 11.15, in order to derive the variance of X, we need to calculate $E[X^2]$ which is given by

$$\begin{aligned}
E[X^2] &= \int_{-\infty}^{\infty} x^2 f_X(x)dx \\
&= \int_{0}^{\infty} x^2 \lambda e^{-\lambda x} dx \\
&\overset{(a)}{=} \left(-x^2 e^{-\lambda x}\right)\Big|_0^{\infty} + \int_0^{\infty} 2xe^{-\lambda x}dx \\
&\overset{(b)}{=} \frac{2}{\lambda}E[X] \\
&\overset{(c)}{=} \frac{2}{\lambda^2},
\end{aligned}$$

where (a) follows by performing integration by parts, and (b) follows from the proof of $E[X]$ with a fact that

$$\lim_{x\to\infty} x^2 e^{-\lambda x} = 0$$

and (c) follows from (13.8).

The variance is thus obtained by

$$\begin{aligned}
V[X] &= E[X^2] - E^2[X] \\
&= \frac{2}{\lambda^2} - \left(\frac{1}{\lambda}\right)^2 \\
&= \frac{1}{\lambda^2}.
\end{aligned}$$

□

Theorem 13.6. *An exponential random variable has a memoryless property, i.e.,*

$$\Pr[X \le a + b | X > a] = \Pr[X \le b] \tag{13.10}$$

Proof. From Definition 10.6 of conditional probability, we have

$$\begin{aligned}
\Pr[X \le a+b|X>a] &= \frac{\Pr[X \le a+b, X>a]}{\Pr[X>a]} \\
&= \frac{\Pr[a<X \le a+b]}{\Pr[X>a]} \\
&\overset{(a)}{=} \frac{F_X(a+b)-F_X(a)}{f-F_X(a)} \\
&\overset{(b)}{=} \frac{1-e^{-\lambda(a+b)}-(1-e^{-\lambda a})}{1-(1-e^{-\lambda a})} \\
&= 1-e^{-\lambda b} \\
&\overset{(c)}{=} F_X(b) \\
&\overset{(d)}{=} \Pr[X \le b],
\end{aligned}$$

where (a) follows from Theorems 11.13 and 10.3, (b) and (c) follow from Theorem 13.4, and (d) follows from Definition 11.14. □

13.4 Erlang random variable

An Erlang random variable has the distribution of a sum of independent exponential variables having the same expected values. It is used to represent the total interarrival time in various random processes such as in queuing systems with PPP (cf. Section 9.2 in Chapter 9).

The Erlang random variable X is represented by two parameters including rate, denoted by $\lambda > 0$, as for exponential random variable and shape, denoted by N, which corresponds the number of exponential random variables. For simplicity, it is generally written as $X \sim Erl(N, \lambda)$.

Definition 13.3. *The pdf of an Erlang random variable $X \sim Erl(N, \lambda)$, $N \in \mathbb{Z}^+$, $\lambda > 0$, is*

$$f_X(x) = \begin{cases} \dfrac{\lambda^N x^{N-1} e^{-\lambda x}}{(N-1)!} & \text{if } x \ge 0, \\ 0 & \text{otherwise.} \end{cases} \tag{13.11}$$

Considering the total pdf, we have the following theorem

Theorem 13.7. *For $N \in \mathbb{Z}^+$ and $\lambda > 0$,*

$$\int_0^\infty \frac{\lambda^N x^{N-1} e^{-\lambda x}}{(N-1)!} dx = 1 \tag{13.12}$$

Proof. We have

$$\begin{aligned}\int_0^\infty \frac{\lambda^N x^{N-1} e^{-\lambda x}}{(N-1)!} dx &= \frac{\lambda^N}{(N-1)!} \int_0^\infty x^{N-1} e^{-\lambda x} dx \\ &= \frac{\lambda^N}{(N-1)!} \frac{(N-1)!}{\lambda^N} \\ &= 1,\end{aligned}$$

where the second equality follows from

$$\int_0^\infty x^m e^{-ax} dx = \frac{m!}{a^{m+1}}$$

when $m = N - 1$ and $a = \lambda$. □

An illustration of an Erlang random variable is shown in Figures 13.5 and 13.6 where the pdf and cdf of $X \sim Erl(3, 5)$ are, respectively, plotted.

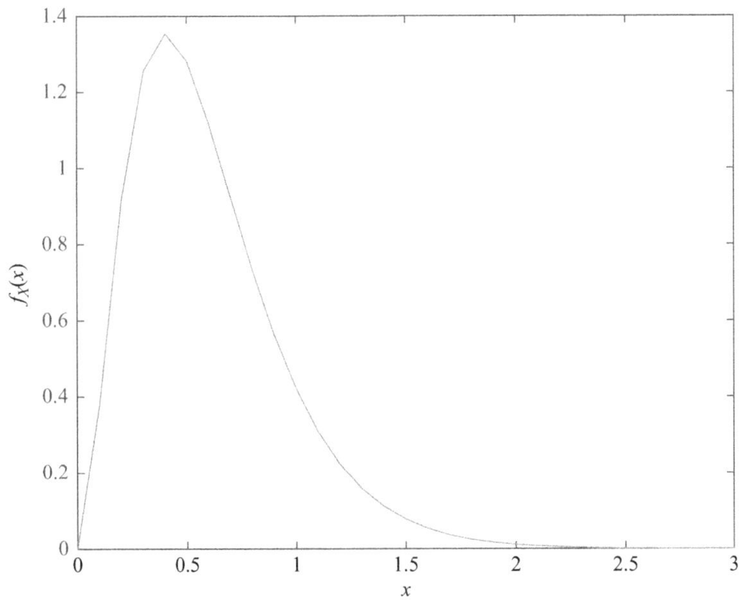

Figure 13.5 pdf of $X \sim Erl(3,5)$

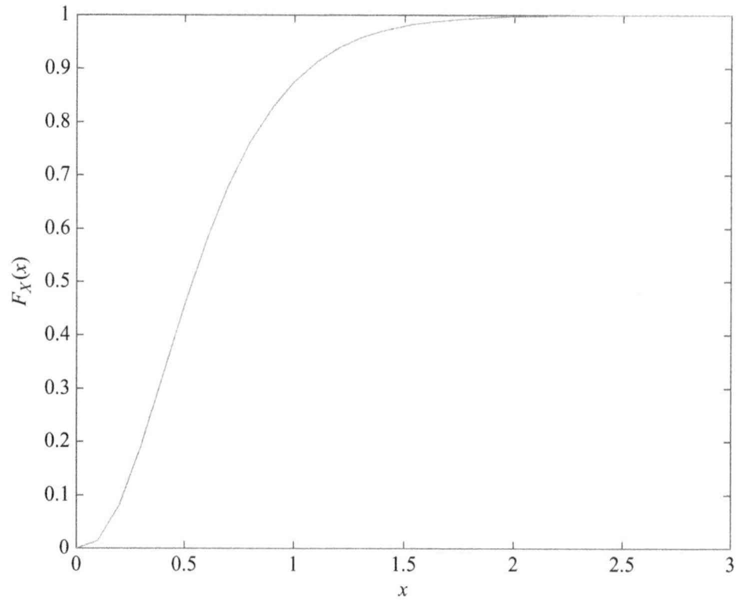

Figure 13.6 cdf of X ∼ Erl(3,5)

Theorem 13.8. *The cdf of an Erlang random variable* $X \sim Erl(N, \lambda)$, $N \in \mathbb{Z}^+$, $\lambda > 0$, *is given by*

$$F_X(x) = 1 - e^{-\lambda x} \sum_{k=0}^{N-1} \frac{(\lambda x)^k}{k!} \tag{13.13}$$

Proof. Let $X_1, X_2, \ldots, X_N$, denote N exponential random variables, each having rate $\lambda > 0$. The Erlang random variable X is thus given by $X = \sum_{i=1}^{N} X_i$.

Related to PPP, the Erlang distribution also means the waiting time, i.e., λx, until the Nth Poisson event occurs. Let Y denote the Poisson random variable, i.e., $Y \sim Poi(\lambda x)$. Therefore, in order to have a waiting time longer than x, all $N - 1$ Poisson random events should occur, i.e.,

$$\begin{aligned}
\Pr[X > x] &= \sum_{k=0}^{N-1} \Pr[Y = k] \\
&\overset{(a)}{=} \sum_{k=0}^{N-1} f_Y(k) \\
&\overset{(b)}{=} \sum_{k=0}^{N-1} \frac{(\lambda x)^k e^{-\lambda x}}{k!} \\
&= e^{-\lambda x} \sum_{k=0}^{N-1} \frac{(\lambda x)^k}{k!},
\end{aligned}$$

where (a) follows from Definition 11.3 with $f_Y(k)$ denoting the probability mass function (pmf) of Y and (b) follows from Definition 12.5 of the pmf of Poisson random variable.

From Definition 11.14, the cdf of X is given by

$$\begin{aligned} F_X(x) &= \Pr[X \le x] \\ &= 1 - \Pr[X > x] \\ &= 1 - e^{-\lambda x} \sum_{k=0}^{N-1} \frac{(\lambda x)^k}{k!}. \end{aligned}$$

□

Theorem 13.9. *The expected value and variance of an Erlang random variable $X \sim Erl(N, \lambda)$ are respectively given by*

$$E[X] = \frac{N}{\lambda}, \tag{13.14}$$

$$V[X] = \frac{N}{\lambda^2}. \tag{13.15}$$

Proof. The proof follows from the fact that the Erlang random variable can be regarded as the summation of N exponential random variables, i.e.,

$$X = \sum_{i=1}^{N} X_i,$$

where X_i, $i = 1, 2, \ldots, N$, is an exponential random variable.

From Theorems 11.11 and 13.5 with $E[X_i] = 1/\lambda$, $i = 1, 2, \ldots, N$, the expected value of X can be obtained by

$$\begin{aligned} E[X] &= \sum_{i=1}^{N} E[X_i] \\ &= \frac{N}{\lambda}. \end{aligned}$$

Since $\{X_1, X_2, \ldots, X_N\}$ are independent, from Lemma 11.4 with the variance of each X_i, $i = 1, 2, \ldots, N$, given by $V[X_i] = 1/\lambda^2$ (cf. Theorem 13.5), the variance of X can be determined by

$$\begin{aligned} V[X] &= \sum_{i=1}^{N} V[X_i] \\ &= \frac{N}{\lambda^2}. \end{aligned}$$

□

13.5 Normal random variable

A normal random variable, a.k.a. Gaussian random variable, is a well-known family of continuous random variables to represent the unknown distributions in general. An example of Gaussian random variable can be found in modeling the random environmental noises in wireless communications, and thus they are also called Gaussian noises. The development of the normal random variable is due to the central limit theorem which states that the summation of various independent distributions converges to the normal distribution having a bell-shaped pdf defined as in Definition 13.4.

The normal random variable X is represented by two parameters including μ and σ^2. As shown later in Theorem 13.12, μ and σ^2 are the expected value and variance, respectively, of X. For simplicity, the normal random variable X is generally written as $X \sim N(\mu, \sigma^2)$.

Definition 13.4. *The pdf of a normal random variable $X \sim N(\mu, \sigma^2)$ is*

$$f_X(x) = \frac{1}{\sqrt{2\pi}\sigma} \exp\left[-\frac{(x-\mu)^2}{2\sigma^2}\right]. \tag{13.16}$$

Considering the total pdf, we have the following theorem

Theorem 13.10. *For all μ and σ,*

$$\int_{-\infty}^{\infty} \frac{1}{\sqrt{2\pi}\sigma} \exp\left[-\frac{(x-\mu)^2}{2\sigma^2}\right] dx = 1. \tag{13.17}$$

Proof. We have

$$\begin{aligned}
\int_{-\infty}^{\infty} \frac{1}{\sqrt{2\pi}\sigma} \exp\left[-\frac{(x-\mu)^2}{2\sigma^2}\right] dx &\overset{(a)}{=} \frac{1}{\sqrt{2\pi}} \int_{-\infty}^{\infty} \exp\left[-\frac{t^2}{2}\right] dt \\
&\overset{(b)}{=} \frac{2}{\sqrt{2\pi}} \int_{0}^{\infty} \exp\left[-\frac{t^2}{2}\right] dt \\
&\overset{(c)}{=} \sqrt{\frac{2}{\pi}} \left(\int_{0}^{\infty} \exp\left[-\frac{u^2}{2}\right] du \int_{0}^{\infty} \exp\left[-\frac{v^2}{2}\right] dv \right)^{1/2} \\
&= \sqrt{\frac{2}{\pi}} \left(\int_{0}^{\infty} \int_{0}^{\infty} \exp\left[-\frac{u^2+v^2}{2}\right] du dv \right)^{1/2}
\end{aligned}$$

$$\stackrel{(d)}{=} \sqrt{\frac{2}{\pi}} \left(\int_0^\infty \int_0^\infty \exp\left[-\frac{u^2 + (wu)^2}{2} \right] u du dw \right)^{1/2}$$

$$= \sqrt{\frac{2}{\pi}} \left(\int_0^\infty \left[-\frac{1}{1+w^2} \exp\left[-\frac{u^2(1+w^2)}{2} \right] \right] \Bigg|_{u=0}^{u\to\infty} dw \right)^{1/2}$$

$$= \sqrt{\frac{2}{\pi}} \left(\int_0^\infty \frac{1}{1+w^2} dw \right)^{1/2}$$

$$= \sqrt{\frac{2}{\pi}} \left(\arctan(w)|_0^\infty \right)^{1/2}$$

$$= \sqrt{\frac{2}{\pi}} \sqrt{\frac{\pi}{2}}$$

$$= 1,$$

where (a) follows by letting $t = (x - \mu)/\sigma$, (b) follows from the fact that $e^{-t^2/2}$ is an even function, (c) follows by duplicating the same integration along with changing variable $t = u$ and $t = v$, and (d) follows by changing variable $v = wu$. □

Remark 13.2. *Denoting Y as the normal random variable with $\mu = 0$ and $\sigma = 1$, we have the simplest form of the pdf given by*

$$f_Y(y) = \frac{1}{\sqrt{2\pi}} \exp\left[-\frac{y^2}{2} \right]. \tag{13.18}$$

The random variable $Y \sim N(0, 1)$ is known as standard normal random variable, and the general normal random variable $X \sim N(\mu, \sigma^2)$ can be written by

$$X = \mu + \sigma Y. \tag{13.19}$$

An illustration of a Gaussian random variable is shown in Figures 13.7 and 13.8 where the pdf and cdf of $X \sim N(1, 9)$ are, respectively, plotted.

Theorem 13.11. *The cdf of a normal random variable $X \sim N(\mu, \sigma^2)$ is given by*

$$F_X(x) = \Phi\left(\frac{x - \mu}{\sigma} \right) \tag{13.20}$$

$$= \frac{1}{2}\left[1 + erf\left(\frac{x - \mu}{\sigma\sqrt{2}} \right) \right], \tag{13.21}$$

where $\Phi(x)$ is the cdf of the standard normal distribution, i.e.,

$$\Phi(x) \triangleq \frac{1}{\sqrt{2\pi}} \int_{-\infty}^{x} \exp\left[-\frac{t^2}{2} \right] dt = \frac{1}{2}\left[1 + erf\left(\frac{x}{\sqrt{2}} \right) \right] \tag{13.22}$$

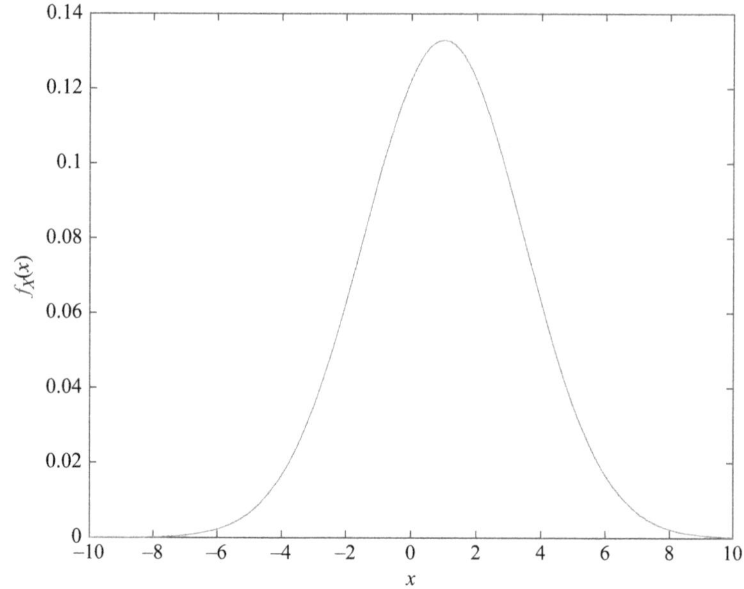

Figure 13.7 pdf of $X \sim N(1,9)$

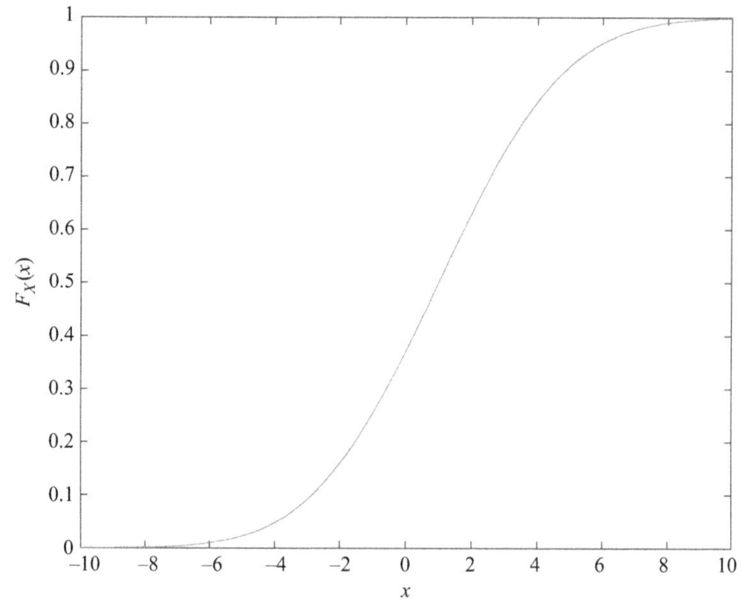

Figure 13.8 cdf of $X \sim N(1,9)$

and erf(x) is the error function defined by

$$erf(x) \triangleq \frac{2}{\sqrt{\pi}} \int_0^x e^{-t^2} dt. \tag{13.23}$$

Proof. The proof follows from Theorem 11.12 as follows:

$$\begin{aligned}
F_X(x) &= \int_{-\infty}^{x} f_X(t)dt \\
&\overset{(a)}{=} \int_{-\infty}^{x} \frac{1}{\sqrt{2\pi}\sigma} \exp\left[-\frac{(t-\mu)^2}{2\sigma^2}\right] dt = \Phi\left(\frac{x-\mu}{\sigma}\right) \\
&\overset{(b)}{=} \frac{1}{2} + \frac{1}{\sqrt{\pi}} \int_0^{((x-\mu)/(\sigma\sqrt{2}))} e^{-u} du \\
&\overset{(c)}{=} \frac{1}{2}\left[1 + erf\left(\frac{x-\mu}{\sigma\sqrt{2}}\right)\right],
\end{aligned}$$

where (a) follows from the pdf of X given by (13.16) and the cdf $\Phi(x)$ defined by (13.22), (b) follows by changing variable $u = (t-\mu)/(\sigma\sqrt{2})$ and the fact that $\int_{-\infty}^{0} f_X(t)dt = 1/2$, and (c) follows from the definition of error function in (13.23). □

Theorem 13.12. *The expected value and variance of a normal random variable $X \sim N(\mu, \sigma^2)$ are respectively given by*

$$E[X] = \mu, \tag{13.24}$$

$$V[X] = \sigma^2. \tag{13.25}$$

Proof. Let us first find the expected value and variance of a standard normal random variable $Y \sim N(0, 1)$ having the pdf given by (13.18).

From Definition 11.16, the expected value of Y is given by

$$\begin{aligned}
E[Y] &= \int_{-\infty}^{\infty} y f_Y(y)dy \\
&= \int_{-\infty}^{\infty} \frac{y}{\sqrt{2\pi}} \exp\left[-\frac{y^2}{2}\right] dy \\
&= 0,
\end{aligned}$$

where the second equality is due to the fact that the integrand is an odd function.

From Theorem 11.15, in order to derive the variance of Y, we need to calculate $E[Y^2]$ which is given by

$$
\begin{aligned}
E[Y^2] &= \int_{-\infty}^{\infty} y^2 f_Y(y)dy \\
&= \int_{-\infty}^{\infty} \frac{y^2}{\sqrt{2\pi}} \exp\left[-\frac{y^2}{2}\right] dy \\
&\overset{(a)}{=} \frac{2}{\sqrt{2\pi}} \int_{0}^{\infty} y^2 \exp\left[-\frac{y^2}{2}\right] dy \\
&\overset{(b)}{=} \sqrt{\frac{2}{\pi}} \frac{\sqrt{\pi}}{(4/(2\sqrt{2}))} \\
&= 1,
\end{aligned}
$$

where (a) follows from the fact that the integrand is an even function and (b) follows from

$$
\int_{0}^{\infty} x^m e^{-ax^2} dx = \frac{\sqrt{\pi}}{4a^{(m+1)/2}}
$$

when $m = 2$ and $a = 1/2$.

The variance of Y is thus obtained by

$$
\begin{aligned}
V[Y] &= E[Y^2] - E^2[Y] \\
&= 1.
\end{aligned}
$$

As stated in Remark 13.2, the random variable X can be represented by $X = \mu + \sigma Y$. From Theorem 11.6 and Lemma 11.4, we have

$$
\begin{aligned}
E[X] &= \mu + \sigma E[Y] \\
&= \mu,
\end{aligned}
$$

$$
\begin{aligned}
V[X] &= \sigma^2 V[Y] \\
&= \sigma^2.
\end{aligned}
$$

□

13.6 Lognormal random variable

A lognormal random variable is a continuous random variable having natural logarithm of its distribution function is a normal distribution. An example of lognormal random variable can be found in modeling the change of received power caused by shadowing in wireless communications.

The lognormal random variable X is also represented by two parameters including μ and σ^2. For simplicity, the lognormal random variable X is written as $X \sim LN(\mu, \sigma^2)$.

Definition 13.5. *The pdf of a lognormal random variable $X \sim LN(\mu, \sigma^2)$ is*

$$f_X(x) = \begin{cases} \dfrac{1}{x\sqrt{2\pi}\sigma} \exp\left[-\dfrac{(\ln x - \mu)^2}{2\sigma^2}\right] & \text{if } x > 0, \\ 0 & \text{otherwise.} \end{cases} \tag{13.26}$$

Considering the total pdf, we have the following theorem

Theorem 13.13. *For all μ and σ,*

$$\int_0^\infty \frac{1}{\sqrt{2\pi}\sigma x} \exp\left[-\frac{(\ln x - \mu)^2}{2\sigma^2}\right] dx = 1. \tag{13.27}$$

Proof. We have

$$\begin{aligned} \int_0^\infty \frac{1}{\sqrt{2\pi}\sigma x} \exp\left[-\frac{(\ln x - \mu)^2}{2\sigma^2}\right] dx &\overset{(a)}{=} \frac{1}{\sqrt{2\pi}} \int_{-\infty}^\infty \exp\left[-\frac{t^2}{2}\right] dt \\ &\overset{(b)}{=} \sqrt{\frac{2}{\pi}}\sqrt{\frac{\pi}{2}} \\ &= 1, \end{aligned}$$

where (a) follows by letting $t = (\ln x - \mu)/\sigma$ and (b) follows the same proof of Theorem 13.10. □

Remark 13.3. *Let Y denote a normal random variable having mean μ and variance σ^2, i.e., $Y \sim N(\mu, \sigma^2)$. By definition, X can be written as*

$$X = e^Y. \tag{13.28}$$

An illustration of a lognormal random variable is shown in Figures 13.9 and 13.10 where the pdf and cdf of $X \sim LN(1, 0.25)$ are, respectively, plotted.

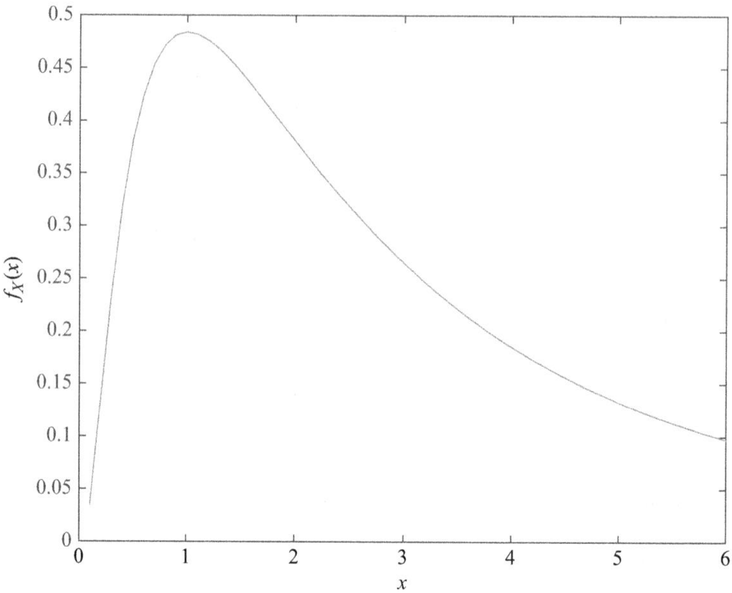

Figure 13.9 pdf of X ~ LN(1,0.25)

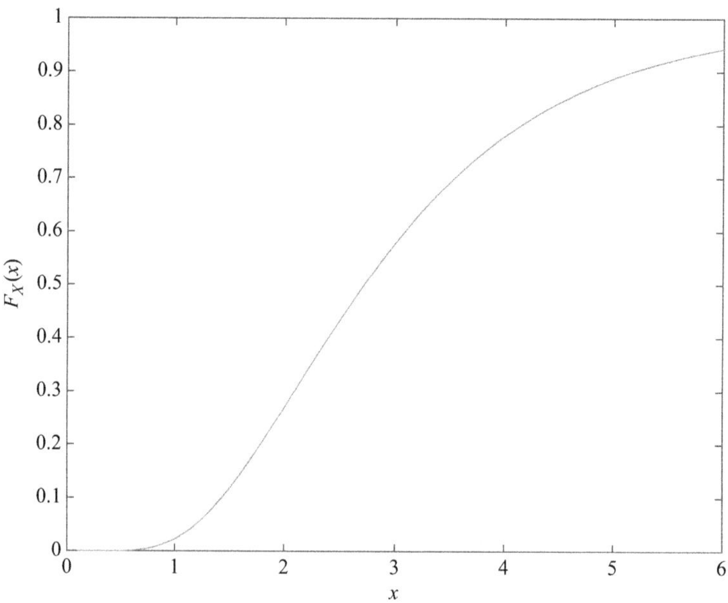

Figure 13.10 cdf of X ~ LN(1,0.25)

Theorem 13.14. *The cdf of a lognormal random variable $X \sim LN(\mu, \sigma^2)$ is given by*

$$F_X(x) = \Phi\left(\frac{\ln x - \mu}{\sigma}\right) \tag{13.29}$$

$$= \frac{1}{2}\left[1 + erf\left(\frac{\ln x - \mu}{\sigma\sqrt{2}}\right)\right], \tag{13.30}$$

where $\Phi(x)$ and $erf(x)$ are given by (13.22) *and* (13.23), *respectively.*

Proof. The proof is similar to that of Theorem 13.11. □

Theorem 13.15. *The expected value and variance of a lognormal random variable $X \sim LN(\mu, \sigma^2)$ are respectively given by*

$$E[X] = \exp\left(\mu + \frac{\sigma^2}{2}\right), \tag{13.31}$$

$$V[X] = e^{2\mu+\sigma^2}(e^{\sigma^2} - 1). \tag{13.32}$$

Proof. As noted in Remark 13.3, X can be represented by $X = e^Y$ where $Y \sim N(\mu, \sigma^2)$. The expected value of X can be derived from the pdf of Y using Theorem 11.14 as

$$\begin{aligned}
E[X] &= \int_{-\infty}^{\infty} e^y f_y(y) dy \\
&\overset{(a)}{=} \int_{-\infty}^{\infty} e^y \frac{1}{\sqrt{2\pi}\sigma} \exp\left[-\frac{(y-\mu)^2}{2\sigma^2}\right] dy \\
&= \exp\left(\mu + \frac{\sigma^2}{2}\right) \int_{-\infty}^{\infty} \frac{1}{\sqrt{2\pi}\sigma} \exp\left[y - \frac{(y-\mu)^2}{2\sigma^2} - \mu - \frac{\sigma^2}{2}\right] dy \\
&= \exp\left(\mu + \frac{\sigma^2}{2}\right) \int_{-\infty}^{\infty} \frac{1}{\sqrt{2\pi}\sigma} \exp\left[-\frac{(y-\mu-\sigma^2)^2}{2\sigma^2}\right] \\
&\overset{(b)}{=} \exp\left(\mu + \frac{\sigma^2}{2}\right).
\end{aligned}$$

where (a) follows from Definition 13.4 and (b) follows from the fact that the integrand is the pdf of a normal random variable $N(\mu + \sigma^2, \sigma^2)$.

From Theorem 11.15, we have

$$\begin{aligned} V[X] &= E[X^2] - E^2[X] \\ &\overset{(a)}{=} E[e^{2Y}] - e^{2\mu+\sigma^2} \\ &\overset{(b)}{=} e^{2\mu+2\sigma^2} - e^{2\mu+\sigma^2} \\ &= e^{2\mu+\sigma^2}(e^{\sigma^2} - 1), \end{aligned}$$

where (a) follows from the proof of the expected value and (b) follows from the fact that $2Y \sim N(2\mu, 4\sigma^2)$. □

13.7 Related works

For more information, the readers are suggested to refer to the following selected references:

Continuous Uniform Distribution Johnson *et al.* in 1994 [244], Papoulis in 2002 [215], Yates and Goodman in 2004 [236], and Forbes *et al.* in 2010 [216].

Exponential Distribution Feller in 1968 [235], Ross in 1995 [219], and Balakrishnan and Basu in 1996 [257].

Erlang Distribution Jambunathan in 1954 [258], Papoulis in 2002 [215], and Forbes *et al.* in 2010 [216].

Normal Distribution Patel and Read in 1996 [259], Papoulis in 2002 [215], Spiegel in 2003 [252], and Forbes *et al.* in 2010 [216].

Lognormal Distribution Aitchison *et al.* in 1957 [260], Crow and Shimizu in 1987 [261], and Forbes *et al.* in 2010 [216].

13.8 Review questions

[Q13.1] Provide an example for the following continuous random variables:
- Continuous uniform random variable.
- Exponential random variable.
- Erlang random variable.
- Normal random variable.
- Lognormal random variable.

[Q13.2] Describe the probability density function, usage, and application of the above random variables.

[Q13.3] Derive the cumulative distribution function, the expected value, and variance for each of the above random variables.

[Q13.4] Show that the exponential random variable has memoryless property.

[Q13.5] What is the relationship between Erlang random variable and exponential random variable?

[Q13.6] What is the relationship between lognormal random variable and normal random variable?

13.9 Problems

[P13.1] A continuous random variable X has $E[X] = 3$ and $V[X] = 9$. Find the probability density function (pdf) and cumulative distribution function (cdf) of X when

- X is a continuous uniform random variable.
- X is an exponential random variable.

[P13.2] The probability that a telephone call lasts no more than t minutes is modeled by the following cdf:

$$F_T(t) = \begin{cases} 1 - e^{-t/3} & \text{if } t \geq 0, \\ 0 & \text{otherwise.} \end{cases}$$

- Derive the pdf of the duration in minutes of a telephone conversation.
- Find the probability that a conversation will last between 2 and 4 min.
- Find the expected duration of a telephone call.
- Find the variance and standard deviation of the call duration.
- Find the probability that a call duration is within ± 1 standard deviation of the expected call duration.

[P13.3] An exponentially distributed random variable X has the following relationship:

$$P[X \leq 1] = 0.9P[X \leq 2].$$

- Find the value of λ.
- Derive the pdf and cdf of X.

[P13.4] The lifetime X, in years, of a networking device is given by the following pdf:

$$f_X(x) = \begin{cases} 0.4e^{-0.4x} & \text{if } x \geq 0, \\ 0 & \text{otherwise.} \end{cases}$$

- Derive the cdf of X.
- Find the expected value and variance of X.
- Find the probability that this device is still alive after 5 years.
- Find the probability that the device dies between 3 and 6 years from the time it is placed in the networking system.

[P13.5] A component has an exponential time-to-failure with a mean of 10,000 h.

- Derive the pdf and cdf of the time-to-failure of the component.
- The component has already been in operation for its mean life. Find the probability that it will fail by 15,000 h.
- At 15,000 h, the component is still in operation. Find the probability that it will operate for another 5,000 h.

[P13.6] Phone company A charges 0.15 per minute for telephone calls. For any fraction of a minute at the end of a call, they charge for a full minute. Phone Company B also charges 0.15 per minute. However, Phone Company B calculates its charge based on the exact duration of a call. Find the expected revenues per call for companies A and B when the duration of a call in minutes is exponentially distributed with a rate $\lambda = 1/3$. (Note that, if X is an exponential (λ) random variable, then $K = \lceil X \rceil$ is a geometric (p) random variable with $p = 1 - e^{-\lambda}$ and $E[K] = 1/p$.)

[P13.7] The time intervals between hits on a web page from remote computers are exponentially distributed, with a mean of 15 s.

- Derive the pdf and cdf of the duration of K hits.
- Find the probability that the third hit connection occurs after 30 s have elapsed.

[P13.8] Consider the validation of a network component. There are four steps to validate the quality of a component. Each step requires an average of 10 min. The validation time for each step is exponentially distributed, and the steps are independent of each other.

- Let X denote the total validation time of all four steps. Derive the pdf and cdf of X.
- Find the expected value and variance of the total validation time.
- Find the probability that the whole validation finishes in 30 min or less.
- Find the probability that the third step of the validation process occurs after 15 min have elapsed.

[P13.9] A standard normal random variable X has expected value $E[X] = 0$ and variance $V[X] = 1$. Let $Y = 2 + 4X$ and $Z = e^Y$.

- Derive the pdf and cdf of Y.
- Derive the pdf and cdf of Z.
- Find the expected value and variance of Z.

Chapter 14
Random variable generation in network simulation

In network modeling and simulation, as introduced in Chapter 11, the randomness of phenomena is represented by various random variables which are specified by probability distribution functions. Specifically, various families of discrete and continuous random variables were discussed in Chapters 12 and 13, respectively. The critical issue in network modeling for performance evaluation is to build a random variable generator to simulate a random event given its probability distribution function. This chapter is devoted to illustrating some widely used techniques for generating random variables.

In this chapter, I will first introduce inverse transform technique in Section 14.1. Specifically, examples of applying the inverse transform technique to generate both discrete and continuous random variables will be provided. Based on the developed random generators, Section 14.2 will present transformation technique to generate related random variables in the same family.

14.1 Inverse transform technique

Given distribution of random variables, inverse transform technique can be used to generate random samples of various discrete and continuous random variables, such as Bernoulli, geometric, discrete uniform, continuous uniform, and exponential random variables.

In order to apply this technique, for discrete random variable, it is required to know the probability mass function (pmf) and/or cumulative distribution function (cdf), while the generation of continuous random variables is developed based on its probability density function (pdf) and/or cdf. Also, as shown later in Theorem 14.1, this technique can only be applied to generate the random variables having invertible cdf.

As a starting point for generating random variables of different families, it is assumed that there exists a basic generator of a uniform random variable R in the range [0, 1], i.e., $R \sim U(0, 1)$.

The inverse transform technique is motivated by noticing that the cdf $F_X(x)$ of a random variable X is monotonically increasing function (cf. Theorem 11.3 in Chapter 11) and $0 \leq F_X(x) \leq 1 \; \forall x \in \mathfrak{S}_X$.

We have Theorem 14.1

Theorem 14.1. *Given a uniform random variable $R \sim U(0,1)$ and a generalized inverse function $F_X^{-1}(r)$, $0 < r < 1$, of $F_X(x)$ defined as*

$$F_X^{-1}(r) = \inf\{x : F_X(x) \geq r\}, \tag{14.1}$$

if a random variable X is determined by

$$X = F_X^{-1}(R), \tag{14.2}$$

then X has the cdf $F_X(x)$.

Proof. We have

$$\begin{aligned} \Pr[X \leq x] &= \Pr[F_X^{-1}(R) \leq x] \\ &= \Pr[R \leq F_X(x)] \\ &= F_X(x), \end{aligned}$$

where the last equality is due to the fact that $\Pr[R \leq r] = r$ if $R \sim U(0,1)$. This according means that $F_X(x)$ is the cdf of X. □

From Theorem 14.1, the inverse transform technique for generating samples of a random variable X given its cdf $F_X(x)$ can be realized via the following steps:

- *Step 1*: Generate samples of a uniform random variable $R \sim U(0,1)$.
- *Step 2*: Determine the generalized inverse function of cdf, i.e., $F_X^{-1}(r)$.
- *Step 3*: For every r, find x such that $x = F_X^{-1}(r)$

Remark 14.1. *Note that the generalized inverse function in* (14.1) *is used when the cdf $F_X(x)$ is not continuous, and thus it can be employed to generate discrete random variables via the following steps:*

- *Step 1: Generate samples of a uniform random variable $R \sim U(0,1)$.*
- *Step 2: For every r, increase x_i from $i = 1$ until $F_X(x_i) > r$, then return $x = x_i$.*

Therefore, all discrete distributions can be generated via inverse-transform technique as long as its pmf is specified.

In the following subsections, some examples of the inverse transform technique will be provided for both discrete and continuous random variables that were studied in Chapters 12 and 13.

14.1.1 Bernoulli random variable generation

Theorem 14.2. *A Bernoulli random variable $X \sim Ber(p)$, $0 < p < 1$, can be generated by*

$$X = \begin{cases} 0 & \text{if } U \leq p, \\ 1 & \text{otherwise.} \end{cases} \tag{14.3}$$

Proof. The proof follows from the pmf of X given by Definition 12.1, i.e.,

$$f_X(x) = \begin{cases} p & \text{if } x = 1, \\ 1 - p & \text{if } x = 0, \\ 0, & \text{otherwise.} \end{cases} \tag{14.4}$$

□

An illustration of generating Bernoulli samples $X \sim Ber(0.6)$ is shown in Figure 14.1.

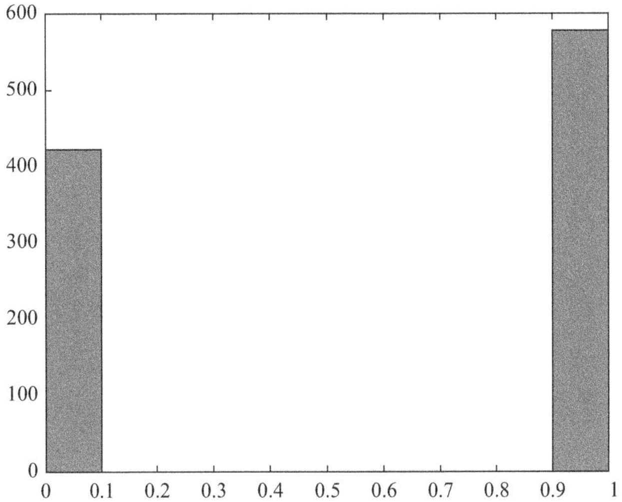

Figure 14.1 Samples of X ~ Ber(0.6)

14.1.2 Geometric random variable generation

Theorem 14.3. *A geometric random variable $X \sim Geo(p)$, $0 < p < 1$, can be generated by*

$$X = \left\lceil \frac{\ln(R)}{\ln(1-p)} \right\rceil, \tag{14.5}$$

where $R \sim U(0, 1)$ and $\lceil \cdot \rceil$ is the ceiling function defined by

$$\lceil x \rceil = \min\{t \in \mathbb{Z} | t \geq x\}. \tag{14.6}$$

Proof. We first derive the cdf of X from its pmf, i.e., $f_X(x)$, defined in Definition 12.3 as

$$f_X(x) = \begin{cases} p(1-p)^{x-1} & \text{if } x = 1, 2, \ldots, \\ 0 & \text{otherwise.} \end{cases} \tag{14.7}$$

From Definition 11.5, the cdf of X is given by

$$\begin{aligned}
F_X(x) &= \Pr[X \leq x] \\
&= 1 - \Pr[X > x] \\
&= 1 - \Pr[X \geq x+1] \\
&= 1 - \sum_{k=x+1}^{\infty} p(1-p)^{k-1} \\
&= 1 - p(1-p)^x \sum_{k=0}^{\infty} (1-p)^k \\
&\overset{(a)}{=} 1 - p(1-p)^x \frac{1}{1-(1-p)} \\
&= 1 - (1-p)^x,
\end{aligned}$$

where (a) follows from the infinite sum of geometric series, i.e.,

$$\sum_{x=0}^{\infty} q^x = \frac{1}{1-q}, \quad \forall |q| < 1.$$

Solving $F_X(X) > 1 - R$ with a notice that X is discrete, the theorem is proved. □

An illustration of generating geometric samples $X \sim Geo(0.6)$ is shown in Figure 14.2.

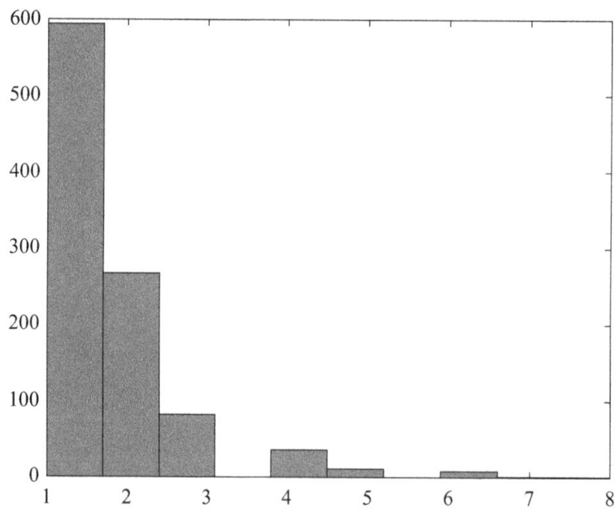

Figure 14.2 Samples of $X \sim Geo(0.6)$

14.1.3 Discrete uniform random variable generation

Theorem 14.4. *A discrete uniform random variable $X \sim U(a,b)$, $a, b \in \mathbb{Z}$, $a < b$, can be generated by*

$$X = a + \lfloor (b-a+1)R \rfloor, \tag{14.8}$$

where $R \sim U(0,1)$ and $\lfloor \cdot \rfloor$ is the floor function defined by

$$\lfloor x \rfloor = \max\{t \in \mathbb{Z} | t \leq x\}. \tag{14.9}$$

Proof. Note that the pdf of X is given in Definition 12.6 as

$$f_X(x) = \begin{cases} \dfrac{1}{b-a+1} & \text{if } x = a, a+1, a+2, \ldots, b, \\ 0 & \text{otherwise.} \end{cases} \tag{14.10}$$

From Definition 11.5, the cdf of X can be derived as

$$F_X(x) = \begin{cases} 1 & \text{if } x > b, \\ \dfrac{x-a+1}{b-a+1} & \text{if } x = a, a+1, a+2, \ldots, b, \\ 0 & \text{if } x < a. \end{cases} \tag{14.11}$$

Solving $F_X(X) > R$ with a notice that X is discrete, the theorem is proved. □

An illustration of generating discrete uniform samples $X \sim U(1,5)$ and $X \in \mathbb{Z}$ is shown in Figure 14.3.

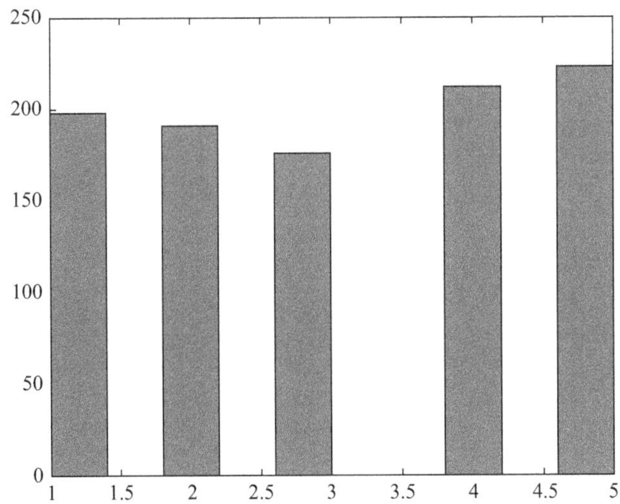

Figure 14.3 Samples of $X \sim U(1,5)$, $X \in \mathbb{Z}$

14.1.4 Continuous uniform random variable generation

Theorem 14.5. *A continuous uniform random variable* $X \sim U(a, b)$, $a, b \in \mathbb{R}$, $a < b$, *can be generated by*

$$X = a + (b - a)R, \tag{14.12}$$

where $R \sim U(0, 1)$.

Proof. The proof follows from Theorem 14.1 where the cdf of X was derived in Theorem 13.1 as

$$F_X(x) = \begin{cases} 0 & \text{if } x \leq a \\ \dfrac{x-a}{b-a} & \text{if } a \leq x \leq b \\ 1 & \text{if } x > b \end{cases} \tag{14.13}$$

Solving $F_X(X) = R$ with $0 < R < 1$ to find X in terms of R, we obtain

$$X = a + (b - a)R.$$ □

An illustration of generating continuous uniform samples $X \sim U(1, 5)$ is shown in Figure 14.4.

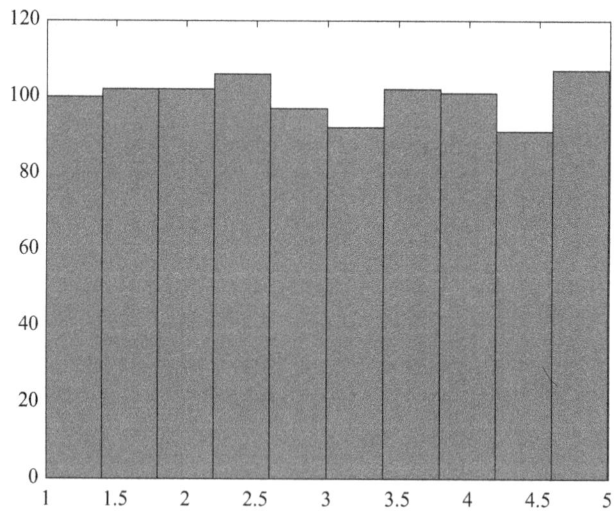

Figure 14.4 Samples of X ~ U(1,5)

14.1.5 Exponential random variable generation

Theorem 14.6. *An exponential random variable $X \sim Exp(\lambda)$, $\lambda > 0$, can be generated by*

$$X = -\frac{1}{\lambda} \ln (1 - R), \tag{14.14}$$

where $R \sim U(0, 1)$.

Proof. The proof follows from Theorem 14.1 where the cdf of X was derived in Theorem 13.4 as

$$F_X(x) = \begin{cases} 1 - e^{-\lambda x} & \text{if } x \geq 0 \\ 0 & \text{otherwise.} \end{cases} \tag{14.15}$$

Solving $F_X(X) = R$ with $0 < R < 1$ to find X in terms of R, we obtain

$$X = -\frac{1}{\lambda} \ln (1 - R).$$

□

An illustration of generating exponential samples $X \sim Exp(5)$ is shown in Figure 14.5.

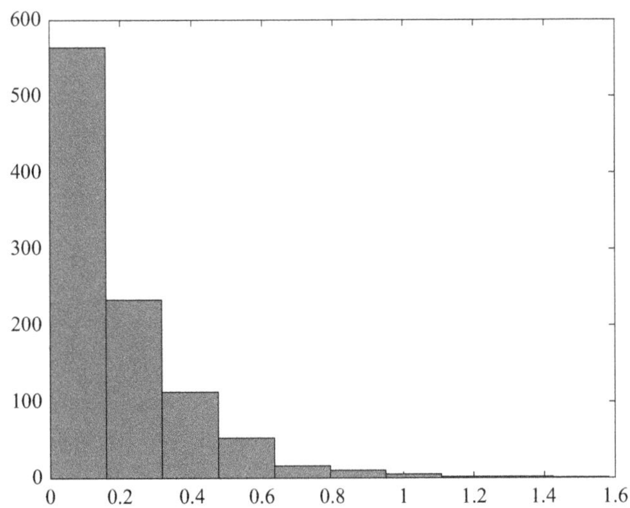

Figure 14.5 Samples of $X \sim Exp(5)$

14.2 Transformation techniques

For some random variables, they can be simply generated based on their features and properties instead of performing the inversion of the cdf. In this section, various transformation techniques will be discussed through the generation of binomial, Pascal, Erlang, and lognormal random variables.

14.2.1 Binomial random variable generation

From the definition of binomial random variable in Section 12.3, it is a generalized form of Bernoulli random variable for a fixed number of tests, i.e., N. The approach to generate the binomial random variable $X \sim Bin(N,p)$, $0 < p < 1, N \in \mathbb{Z}^+$ is known as composition technique which consists of the following steps:

- *Step 1*: Generate N independent uniform random variables $R_1 \sim U(0,1)$, $R_2 \sim U(0,1) \sim U(0,1), \ldots, R_N \sim U(0,1)$.
- *Step 2*: Find X which is the number of random variables R_i satisfying $R_i < p$.

An illustration of generating binomial samples $X \sim Bin(20, 0.6)$ is shown in Figure 14.6.

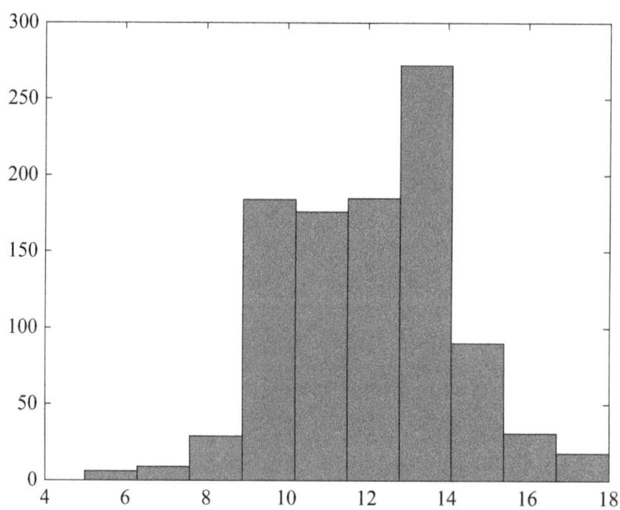

Figure 14.6 Samples of X ~ Bin(20,0.6)

14.2.2 Pascal random variable generation

In Section 12.5, a Pascal random variable $X \sim Pas(k,p)$, $0 < p < 1$, $k \in \mathbb{Z}^+$, is extended from geometric random variable. Specifically, it is the summation of k geometric random variables, i.e.,

$$X = \sum_{i=1}^{k} X_i,$$

where $X_i \sim Geo(p)$, $i = 1, 2, \ldots, k$.

The approach to generate the Pascal random variable $X \sim Pas(k,p)$, $0 < p < 1$, $k \in \mathbb{Z}^+$ is known as convolution method, which can be carried out as follows:

- *Step 1*: Generate k independent geometric random variables $X_1 \sim Geo(p)$, $X_2 \sim Geo(p)$, $\ldots$, $X_k \sim Geo(p)$ (cf. Section 14.1.2).
- *Step 2*: Determine X as

$$X = \sum_{i=1}^{k} X_i.$$

An illustration of generating Pascal samples $X \sim Pas(4, 0.6)$ is shown in Figure 14.7.

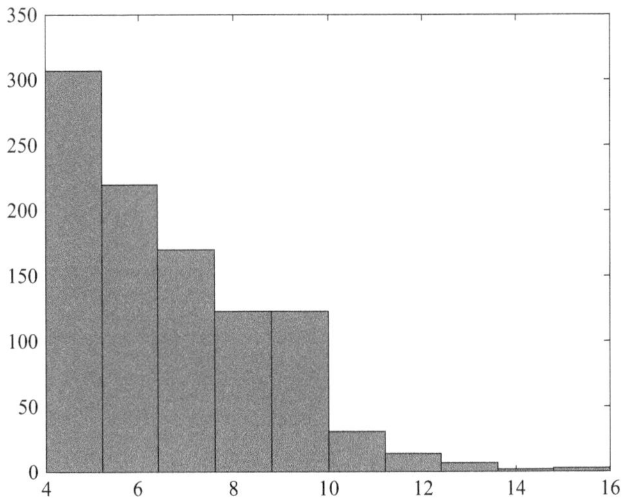

Figure 14.7 Samples of $X \sim Pas(4,0.6)$

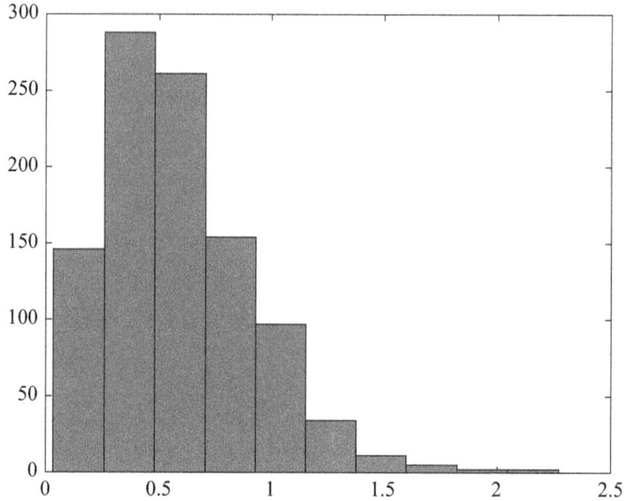

Figure 14.8 Samples of $X \sim Erl(3,5)$

14.2.3 Erlang random variable generation

In Section 13.4, an Erlang random variable $X \sim Erl(N, \lambda)$, $N \in \mathbb{Z}^+$, $\lambda > 0$, is extended from exponential random variable. Specifically, it is the summation of N exponential random variables, i.e.,

$$X = \sum_{i=1}^{N} X_i,$$

where $X_i \sim Exp(\lambda)$, $i = 1, 2, \ldots, k$.

Employing the same approach to generate Pascal random variables, the Erlang random variable $X \sim Erl(N, \lambda)$ can be carried out with convolution technique as follows:

- *Step 1*: Generate N independent exponential random variables $X_1 \sim Exp(\lambda), X_2 \sim Exp(\lambda), \ldots, X_N \sim Exp(\lambda)$ (cf. Section 14.1.5).
- *Step 2*: Determine X as $X = \sum_{i=1}^{N} X_i$.

An illustration of generating Erlang samples $X \sim Erl(3, 5)$ is shown in Figure 14.8.

14.2.4 Lognormal random variable generation

As noted in Section 13.6, a lognormal random variable $X \sim LN(\mu, \sigma^2)$ can be represented as

$$X = e^Y, \tag{14.16}$$

where $Y \sim N(\mu, \sigma^2)$, i.e., a normal random variable with mean μ and variance σ^2.

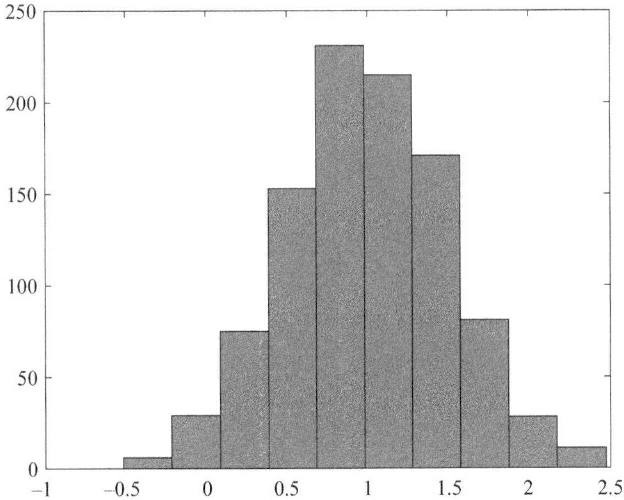

Figure 14.9 Samples of Y ~ N(1,0.25)

Another notice in Section 13.5 is that $Y \sim N(\mu, \sigma^2)$ can be written by

$$Y = \mu + \sigma Z, \tag{14.17}$$

where $Z \sim N(0, 1)$ is a standard normal random variable.

The generation of the lognormal random variable $X \sim LN(\mu, \sigma^2)$ can be realized by employing direct transformation technique as follows:

- *Step 1*: Generate a standard normal random variable $Z \sim N(0, 1)$.
- *Step 2*: Determine Y as $Y = \mu + \sigma Z$.
- *Step 3*: Determine X as $X = e^Y$.

An illustration of generating normal and lognormal samples $Y \sim N(1, 0.25)$ and $X \sim LN(1, 0.25)$ is shown in Figures 14.9 and 14.10, respectively.

14.3 Related works

For more information, the readers are suggested to refer to the following selected references: Box and Muller in 1958 [262], Devroye in 1981 and 1986 [263,264], Barabesi and Pratelli in 2015 [265], Robert and Casella in 2005 [266], Rubinstein and Kroese in 2007 [267], and Gentle in 2009 [247].

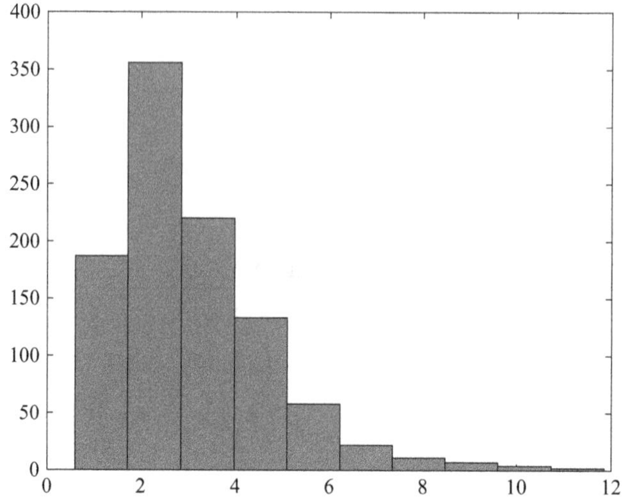

Figure 14.10 Samples of $Y \sim LN(1,0.25)$

14.4 Review questions

[Q14.1] Describe the basic concept of inverse transform technique to generate a random variable for network simulation.

[Q14.2] A random variable X is represented by a cumulative distribution function (cdf) $F_X(x)$. Explain steps to generate X for the following cases:
- X is a continuous random variable.
- X is a discrete random variable.

[Q14.3] Provide steps with proof to generate the following discrete random variables:
- Discrete uniform random variable.
- Bernoulli random variable.
- Geometric random variable.

[Q14.4] Provide steps with proof to generate the following continuous random variables:
- Continuous uniform random variable.
- Exponential random variable.

[Q14.5] Provide some examples of employing transformation technique to generate the following random variables:
- Binomial random variable.
- Pascal random variable.
- Erlang random variable.
- Lognormal random variable.

14.5 Problems

[P14.1] Develop a generator for the following random variables:

- $X \sim Ber(0.5)$.
- $X \sim Geo(0.5)$.
- $X \sim U(2,6), X \in \mathbb{Z}$ (discrete).
- $X \sim U(2,6)$ (continuous).
- $X \sim Exp(3)$.
- $X \sim Bin(10,0.5)$.
- $X \sim Erl(4,3)$.

[P14.2] Develop a random-variate generator for a random variable X having the following cdf:

$$F(x) = \begin{cases} e^{2x}/2 & \text{if } x \le 0, \\ 1 - e^{-2x} & \text{if } x > 0. \end{cases}$$

[P14.3] Develop a generation scheme for the triangular distribution with the following pdf:

$$f(x) = \begin{cases} \dfrac{x-2}{2} & \text{if } 2 \le x \le 3, \\ \dfrac{2 - x/3}{2} & \text{if } 3 < x \le 6, \\ 0 & \text{otherwise.} \end{cases}$$

[P14.4] Develop a generator for the variable having the following cdf:

$$F(x) = \begin{cases} 0 & \text{if } x \le -3, \\ \dfrac{1}{2} + \dfrac{x}{6} & \text{if } -3 < x \le 0, \\ \dfrac{1}{2} + \dfrac{x^2}{32} & \text{if } 0 < x \le 4, \\ 1 & \text{if } x > 4. \end{cases}$$

[P14.5] Develop a generator for a random variable X having the following cdf:

$$F(x) = \begin{cases} 0 & \text{if } x < 0, \\ x^4/16 & \text{if } 0 \le x \le 2, \\ 1 & \text{if } x > 2. \end{cases}$$

[P14.6] Develop a generator for a random variable X having the following cdf:

$$F(x) = \begin{cases} 0 & \text{if } x < 0, \\ x/3 & \text{if } 0 \le x \le 2, \\ 2/3 + (x-2)/24 & \text{if } 2 < x \le 10, \\ 1 & \text{if } x > 10. \end{cases}$$

[P14.7] Develop a generator for a Weibull random variable $X \sim Wei(\lambda, k)$ with the following cdf:

$$F(x) = \begin{cases} 1 - e^{-(x/\lambda)^k} & \text{if } x \geq 0, \\ 0 & \text{if } x < 0, \end{cases}$$

where $\lambda > 0$ and $k > 0$ denote the scale parameter and shape parameter, respectively.

Chapter 15
Queuing theory for network modeling and performance evaluation

As outlined in Chapter 9, statistical models are crucial for network simulation and performance evaluation when most parameters and events in practical networks are randomly distributed and variant over time following different distributions. A typical application of the statistical models can be found in queuing systems to model the waiting experiences. This statistical modeling of the waiting is well known as queuing theory. This chapter is devoted to study the queuing theory in the context of computer and communication networks.

In this chapter, I will first introduce the basic concept of queuing theory in Section 15.1. The application of queuing theory in the form of queuing systems for computer and communication networks will be presented in Section 15.2. As a background of the queuing system, the Poisson point process will be then reviewed in Section 15.3. Various performance measures and their relationship with Little's law will be discussed in Section 15.4. This chapter will be closed by introducing Kendall's notation for specifying queuing systems in Section 15.5.

15.1 An introduction to queuing theory

Queuing theory has been adopted to study the waiting lines in different areas, such as telecommunications, computing, industrial engineering, data management, etc. The queuing theory is shown to be eminently helpful in modeling and validating the performance of a number of facilities and components within computer, communication, manufacturing, and transport systems.

Applying queuing theory to computer and communication systems, a variety of computer processors, disks, terminals, communication links, concentrators, and memories can be modeled as queuing systems. The queuing theory is regarded as a mathematical tool for describing the behavior of queuing or waiting lines in a system. Presented in mathematical terms, the queuing theory allows us to validate various performance measures of interest, such as waiting times, waiting lengths, response times, busy periods, idle periods, blocking probability, etc.

The queuing theory is aimed to model and analyze a system verifying its ability to provide services as demanded, for instance to minimize the response time and delay with respect to various levels of demand. In order to achieve this, the behavior, operation, and properties of all resources as well as relevant parameters within the system need to be understood and considered when developing any queuing models. Indeed, the queuing model is generally subject to constraints of available resources along with contention between them.

As modeling of a waiting line, a queuing model simply represents a service facility to be able to accommodate requests that are waiting to be served. Depending on the number of servers, the queuing model can be categorized into two basic types including the following:

- *Single-server queue*: In a single-server queue, packets are sequentially served one at a time. If there are more than one packet arriving, then there is only one packet to be served, while the remaining packets queue in a buffer. After the service of a packet is complete, it leaves the system and thus the number of packets in the system is reduced by one. The next packet in the queue will be served and the number of packets in the queue is also reduced by one.
- *Multi-server queue*: In a multi-server queue, packets are served by multiple servers in parallel. If the number of packet arrivals is less than the number of servers, then some servers will be idle. Otherwise, the number of packets being served is the number of servers, while the remaining packets will queue in a buffer.

The above two queuing models will be discussed in detail in the subsequent Chapters 16 and 17.

In a practical communication network consisting of multiple nodes, these nodes are generally interconnected via various communication links, and thus they experience multiple queues of different types, namely, a network of queues. In this case, the network of queues can be regarded as an aggregation of various simple queuing models.

15.2 Queuing systems in computer and communication networks

In computer and communication networks, the contention of resources is a critical issue due to their inability to support immediately all requests for different kinds of services. This accordingly results in the development of queues, which cause substantial delays in obtaining the required services. Such delays, in some networks such as telephone or mobile phone networks, may cause rejection or blocking of service requests. Therefore, in queuing systems, there are three major constraints, including reliability, capacity, and delay (cf. Chapter 5 for a brief introduction of these three performance measures).

Basically, a queuing system consists of the following components (cf. Figure 15.1):

- *Source*: is the input of the system which can be either an infinite or a finite source. Although a finite source with a finite queue is more appropriate in practice, the

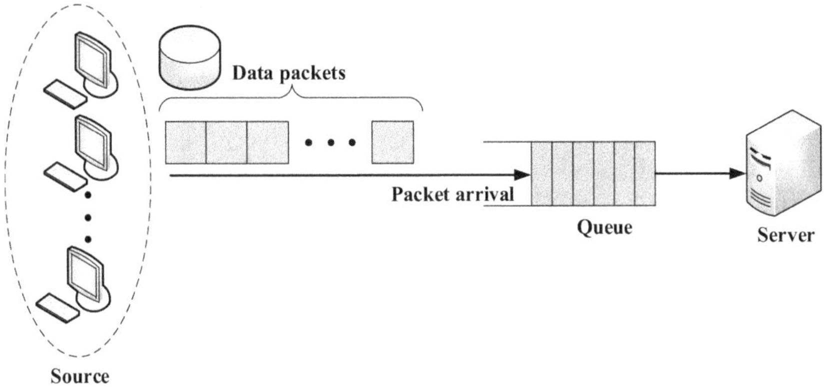

Figure 15.1 Components in a queuing system

source is generally assumed to be very large that can be approximated as an infinite source for simplicity in the analysis.

- *Packet arrivals*: The packets are assumed to arrive at the queuing system randomly over time and may occur according to any arbitrary pattern. Poisson arrivals are shown to be completely random in time, and thus Poisson process is applicable for modeling queuing systems in many applications. The average interarrival time of packets is the reciprocal of the arrival rate.
- *Service mechanism*: depends on the number of servers as categorized above for two basic queues including single-server and multi-server queues. The service is represented by a service rate, which is the reciprocal of the average service time.
- *Queue capacity*: is related to the buffer size which allows the packets to queue in. We therefore have the following:
  - *Zero capacity*: the packets drop when system is busy, i.e., no buffer, and thus it is also known as a loss system.
  - *Infinite capacity*: all packets are allowed to enter the queue with unlimited buffer size.
  - *Positive capacity*: only a limited number of packets are allowed to enter the queue if there is still room.
- *Queue discipline*: defines the rule that the server selects the next packet from the queue, for instance first-in first-out (FIFO) or first-come first-served (FCFS), last-in first-out (LIFO), first-in random-out (FIRO), etc.

15.3 Poisson point process for queuing systems

In this section, Poisson point process (PPP) will be reviewed with a brief summary of related formula that will be used in this chapter and the following Chapters 16 and 17. Details of the Poisson random variables and Poisson point process can be referred to Section 9.2 in Chapter 9 and Section 12.6 in Chapter 12.

The PPP in queuing systems is defined in one dimension to represent a counting process. The PPP is related to Markov process as a special case in which the future probabilities are determined by the current state of the process, but not on how that state was reached. The PPP is also known as a birth–death process where the packet arrival at the queue is "birth" and the service at the server is "death." As shown in Section 9.2, the PPP has the same memoryless property as defined in the Markov process.

Let X denote a Poisson random variable, i.e., $X \sim Poi(\lambda)$, to represent the number of events occurring in a specified time interval with a Poisson parameter $\lambda > 0$. The probability mass function (pmf) of a Poisson random variable X is given by Definition 12.5, i.e.,

$$f_X(x) = \begin{cases} \dfrac{\lambda^x e^{-\lambda}}{x!} & \text{if } x = 0, 1, 2, \ldots, \\ 0 & \text{otherwise.} \end{cases} \tag{15.1}$$

It has been shown in Theorem 12.9 that the expected value and variance of X are

$$E[X] = V[X] = \lambda. \tag{15.2}$$

Considering PPP, let $N(t)$ denote a counting function that represents the number of packet arrivals in $[0, t]$. The packets are assumed to arrive with a rate λ, which, in the scope of this book, is assumed to be a constant, and thus the PPP hereafter is regarded as homogeneous or stationary PPP unless stated otherwise. As defined in Definition 9.3, the packet arrivals in the PPP occur one at a time, $N(t)$ has stationary and independent increments, and $N(t) \sim Poi(\lambda t)$ with the following pmf

$$\Pr[N(t) = n] = \frac{e^{-\lambda t}(\lambda t)^n}{n!}, n \in \mathbb{N} \tag{15.3}$$

Similar to Poisson random variable, $N(t)$ has equal expected value and variance which are given by

$$E[N(t)] = V[N(t)] = \lambda t. \tag{15.4}$$

Due to the stationary increment, the number of packet arrivals in the interval $(t_1, t_2]$, $t_1 \leq t_2$, denoted by $N(t_1, t_2]$, also follows a Poisson distribution with the following pmf:

$$\Pr[N(t_1, t_2] = n] = \frac{e^{-\lambda(t_2 - t_1)}(\lambda(t_2 - t_1))^n}{n!}, \quad n \in \mathbb{N}. \tag{15.5}$$

The average number of events in $(t_1, t_2]$ is thus given by

$$E[N(t_1, t_2]] = \lambda(t_2 - t_1). \tag{15.6}$$

15.4 Performance measures in queuing systems

A queuing system is characterized by various performance measures which are related to probabilistic characteristics of flows and services in the system. This section will briefly introduce popular performance measures in queuing systems, including inter-arrival, service time, waiting time, response time, the number of packets in the system, queue length, traffic intensity, and server utilization. In addition, Little's law will be presented to describe the relationship between these measures. Note that the notation defined in this section will be used hereafter throughout the remaining chapters.

15.4.1 Interarrival and service time in Poisson point process

Definition 15.1. *Interarrival and service time are defined as follows:*

- *Interarrival time is the time difference between packet arrivals.*
- *Service time is the time difference between service completions.*

Let T_I denote the interarrival time between packets in PPP. As shown in Theorem 9.1, T_I is exponentially distributed with an arrival rate of λ, i.e., $T_I \sim Exp(\lambda)$ (cf. Section 13.3 in Chapter 13 for details of exponential random variable). The probability density function (pdf), i.e., $f_{T_I}(t)$, and cumulative distribution function (cdf), i.e., $F_{T_I}(t)$, of T_I are given by Definition 13.2 and Theorem 13.4, respectively, i.e.,

$$f_{T_I}(t) = \lambda e^{-\lambda t}, \tag{15.7}$$

$$F_{T_I}(t) = 1 - e^{-\lambda t}. \tag{15.8}$$

The expected value and variance of T_I are determined in Theorem 13.5 as

$$E[T_I] = \frac{1}{\lambda}, \tag{15.9}$$

$$V[T_I] = \frac{1}{\lambda^2}. \tag{15.10}$$

Similarly, when the service time, denoted by T_S, is exponentially distributed with a service rate μ, i.e., $T_S \sim Exp(\mu)$, the pdf, cdf, expected value, and variance of T_S are, respectively, given by

$$f_{T_S}(t) = \lambda e^{-\mu t}, \tag{15.11}$$

$$F_{T_S}(t) = 1 - e^{-\mu t}, \tag{15.12}$$

$$E[T_S] = \frac{1}{\mu}, \tag{15.13}$$

$$V[T_S] = \frac{1}{\mu^2}. \tag{15.14}$$

15.4.2 Waiting and response time

Definition 15.2. *The waiting time and response time are defined as follows:*

- *Waiting time is the duration the packets stay in the queue.*
- *Response time is the total time the packets spend in the system.*

It can be noticed that the response time involves the waiting time in the queue and the service time at the server.

Let T_W and T_R denote the waiting time and response time, respectively. We then have the following relationship:

$$T_R = T_W + T_S, \tag{15.15}$$

where T_S is the service time defined in Definition 15.1.

15.4.3 The number of packets in the queue and in the system

Definition 15.3. *The number of packets in the queue and in the system is defined as follows:*

- *The number of packets in the queue, a.k.a. queue length, is the number of packets waiting in the queue to be served.*
- *The number of packets in the system is the total number of packets staying in the system including those being served.*

Note that the number of packets being served is the number of servers in the system.

Let n_s, $N_Q(t)$, and $N(t)$ denote the number of servers, the number of packets in the queue, and the number of packets in the system, respectively. We then have

$$N_Q(t) = [N(t) - n_s]^+, \tag{15.16}$$

where $[a]^+ \triangleq \max\{a, 0\}$.

15.4.4 Traffic intensity and server utilization

Traffic intensity and server utilization are defined as follows:

Definition 15.4. *Traffic intensity, denoted by ρ, is defined as a measure of the ability of a server to serve requests, which is determined by*

$$\rho \triangleq \frac{E[T_S]}{E[T_I]}. \tag{15.17}$$

Remark 15.1. *When both T_S and T_I are exponential distributed with $E[T_S] = 1/\mu$ and $E[T_I] = 1/\lambda$, we have*

$$\rho = \frac{\lambda}{\mu}. \tag{15.18}$$

Definition 15.5. *Server utilization, denoted by U_S, is defined as a percentage of time that the server is busy serving requests, i.e.,*

$$U_S = \frac{E[T_B]}{E[T_B] + E[T_0]}, \tag{15.19}$$

where T_B and T_0 are the busy period and idle period, respectively, of the server.

Remark 15.2. *From* (15.19), *the average busy period of the server can be determined by*

$$E[T_B] = \frac{U_S}{1 - U_S} E[T_0]. \tag{15.20}$$

Remark 15.3. *It can be observed that the utilization of server is indeed the probability that the server is busy, and thus it can be obtained by*

$$U_S = 1 - P_0, \tag{15.21}$$

where P_0 *is the steady-state probability that the server is idle. Substituting* (15.21) *into* (15.19), P_0 *can be derived as*

$$P_0 = \frac{E[T_0]}{E[T_B] + E[T_0]}. \tag{15.22}$$

Definition 15.6. *The steady-state distribution of the birth–death process is represented by the following relationship of the steady-state probabilities:*

$$P_n = \frac{\prod_{k=0}^{n-1} \lambda_k}{\prod_{k=1}^{n} \mu_k} P_0, \tag{15.23}$$

where P_n *is the steady-state probability that there are n packets in the system,* λ_k *is the arrival rate of kth packet, and* μ_k *is the service rate of the kth packet.*

Theorem 15.1. *The steady-state probability that the system is idle can be given by*

$$P_0 = \left(1 + \sum_{n=1}^{\infty} \frac{\prod_{k=0}^{n-1} \lambda_k}{\prod_{k=1}^{n} \mu_k}\right)^{-1}. \tag{15.24}$$

Proof. The proof follows from the fact that

$$\begin{aligned} &\sum_{n=0}^{\infty} P_n = 1 \\ \overset{(a)}{\Leftrightarrow}\ & P_0 + \sum_{n=1}^{\infty} \frac{\prod_{k=0}^{n-1} \lambda_k}{\prod_{k=1}^{n} \mu_k} P_0 = 1 \\ \Leftrightarrow\ & P_0 = \left(1 + \sum_{n=1}^{\infty} \frac{\prod_{k=0}^{n-1} \lambda_k}{\prod_{k=1}^{n} \mu_k}\right)^{-1}, \end{aligned}$$

where (a) follows from (15.23). □

15.4.5 Little's law for queuing systems

Figure 15.2 illustrates a simple queuing model with fundamental and explicit performance measures. Their relationship is shown in the following Little's law.

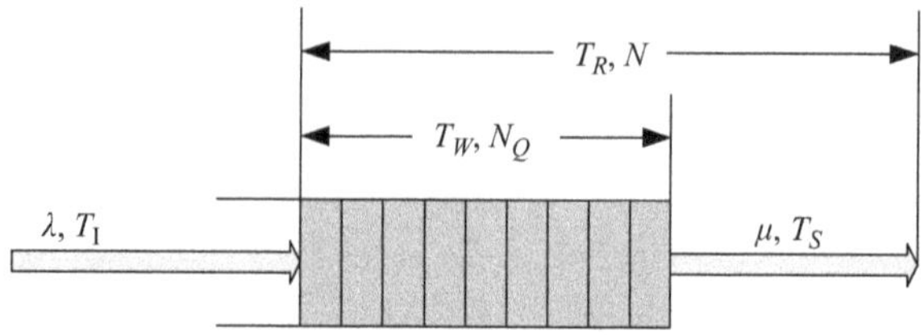

Figure 15.2 Performance measures in a queuing system

Definition 15.7. *Little's law is used to show the relationship between performance measures in queuing systems. We have the following relations:*

$$E[N(t)] = \lambda E[T_R], \tag{15.25}$$

$$E[N_Q(t)] = \lambda E[T_W], \tag{15.26}$$

$$E[T_R] = E[T_W] + E[T_S]. \tag{15.27}$$

Theorem 15.2. *The traffic intensity represents the difference between the average number of packets in the queue and those in the system, i.e.,*

$$\rho = E[N(t)] - E[N_Q(t)]. \tag{15.28}$$

Proof. Substituting (15.27) into (15.25) with (15.26), we obtain

$$\begin{aligned} E[N(t)] &= E[N_Q(t)] + \lambda E[T_S] \\ &\overset{(a)}{=} E[N_Q(t)] + \frac{\lambda}{\mu} \\ &\overset{(b)}{=} E[N_Q(t)] + \rho, \end{aligned}$$

where (a) follows from (15.13), i.e., $E[T_S] = 1/\mu$, and (b) follows from Definition 15.4 of traffic intensity, i.e., $\rho = \lambda/\mu$. This completes the proof. □

15.5 Kendall's notation for queuing systems

Kendall's notation is used to describe queuing systems via six parameters represented as follows:

$$A/B/n_s/n_k/n_p/S,$$

where

- A denotes the distribution of packet interarrivals.
- B denotes the distribution of service time at the server. For communication networks, a server is simply a communication link. In this case, the link transmission time can be regarded as the service time.

- n_s is the number of servers in the system. For communication networks, n_s means the number of communication links between nodes.
- n_k is the system capacity which is the maximum number of packets in the system including all packets in the queue and those being served.
- n_p is the population size, which is defined as the number of sources of arriving packets.
- S denotes the queuing discipline which specifies the order or priorities of packet arrivals at the queue and the packets going out for service. Some commonly used disciplines are FIFO or FCFS, LIFO, FIRO, etc.

In particular, for the first two letters A and B in the Kendall's notation, they are commonly described by one of the following symbols:

- M: means exponential distribution as in Markovian process.
- G: means general distribution which can be defined by either pdf, cdf, or arbitrary distribution.
- D: means deterministic or constant in which all packet interarrivals and service time are invariant.

In the Kendall's notation, if the system capacity is infinity, i.e., $n_k \to \infty$, the population size is infinity, i.e., $n_p \to \infty$, and the service discipline is FIFO, i.e., $S =$ FIFO, then they can be omitted for simplicity.

Example 15.1. *We have some queuing models as follows:*

- $M/M/1$ *queue describes a single-server queue with Poisson arrivals, exponential service time distribution, infinite system capacity, infinity population, and FIFO discipline.*
- $M/G/c/c$ *queue describes a c-server queue with Poisson arrivals, generally distributed service time, finite system capacity with a maximum of c packets in the system, infinite population, and FIFO discipline.*

15.6 Related works

For more information, the readers are suggested to refer to the following selected references:

Queuing Theory Takacs in 1962 [268], Newell in 1971 [269], Nelson in 1995 [220], and Harchol-Balter in 2013 [178].

Queuing Systems Kleinrock in 1975 and 1976 [270,271], Giambene in 2005 [272], and Alfa in 2010 [273].

PPP for Queuing Systems Wolff in 1982 [274], Van Doorn and Regterschot in 1988 [275], and Nelson in 1995 [220].

Queuing Performance Measures Little in 1961 [276], Stidham in 1974 [277], Nelson in 1995 [220], Stewart in 2009 [246], and Harchol-Balter in 2013 [178].

Kendall's Notation Kendall in 1954 [278], Lee in 1966 [279], and Kleinrock in 1975 [270].

15.7 Review questions

[Q15.1] Provide some applications of queuing theory in telecommunications.
[Q15.2] What are the components of a queuing system in computer and communication networks?
[Q15.3] Describe queue capacity and discipline in a queuing system.
[Q15.4] What is birth–death process?
[Q15.5] How to represent a counting process with a Poisson point process (PPP)?
[Q15.6] Describe the probability mass function (pmf) of a Poisson random variable X with arrival rate λ.
[Q15.7] Derive the pmf of a PPP $N(t)$ over time.
[Q15.8] What are interarrival and service time in a queuing system?
[Q15.9] Derive probability density function (pdf) and cumulative distribution function (cdf) of the interarrival and service time in PPP.
[Q15.10] What are waiting time and response time?
[Q15.11] What are traffic intensity and server utilization?
[Q15.12] What is steady-state distribution in a birth–death process?
[Q15.13] Derive the steady-state probability that the system is idle.
[Q15.14] What is Little's law for queuing systems?
[Q15.15] Provide an illustration to show the relationship between performance measures in a simple single-server queuing system.
[Q15.16] Describe Kendall's notation and explain all parameters which are used to represent a queuing system.

15.8 Problems

[P15.1] Show the relationship between the number of packets in the queue and those in the system.
[P15.2] Derive the average busy and idle periods of the server as a function of the server utilization.
[P15.3] Derive the traffic intensity using Little's law.
[P15.4] Interpret the following Kendall's notation for various queuing systems:
- $M/M/1/K$
- $M/M/c/c$
- $M/M/\infty$
- $M/M/c/K/L$
- $M/G/1/K/L$
- $D/D/c/K/K$

[P15.5] A PPP $N(t)$ in a queuing system has an arrival rate $\lambda = 4$ packets/second.
- Derive the pmf and cdf of $N(t)$.
- Find the expected value of $N(t)$ during an interval of a minute.
- Find the probability that there are 100 packets arrived in a minute.

[P15.6] A single-server queuing system has both interarrival time and service time following exponential distribution with an average arrival rate $\lambda = 10$ packets/second and a service rate $\mu = 12$ packets/second, respectively.

- Give the Kendall's notation describing this system.
- Derive the pmf and cdf of both the interarrival time and service time.
- Find the traffic intensity of the system.

Chapter 16

Single-server queues—network behaviors and analysis

Queuing models can be generally classified based on the number of servers in the system. This chapter is devoted to present single-server queues in which packets are sequentially processed by only one server. If there are more than one packet arriving, then they have to wait until the service at the server is complete. Depending on the system capacity and population size of the sources, various models have been developed for single-server queues.

In this chapter, I will first introduce $M/M/1$ queue with infinite system capacity and infinite population in Section 16.1. Variant models of $M/M/1$ queue with finite system capacity and finite population will be sequentially discussed in Sections 16.2 and 16.3. Specifically, network behaviors and their relevant performance measures will be provided in detail for every queuing model.

16.1 *M*/*M*/1 Queue

In $M/M/1$ queue, there is only a server. The system capacity and the population, which by default are not given, mean infinite (cf. Figure 16.1). Both interarrival time, i.e., T_I, and service time, i.e., T_S, are exponentially distributed with $E[T_I] = 1/\lambda$ and $E[T_S] = 1/\mu$.

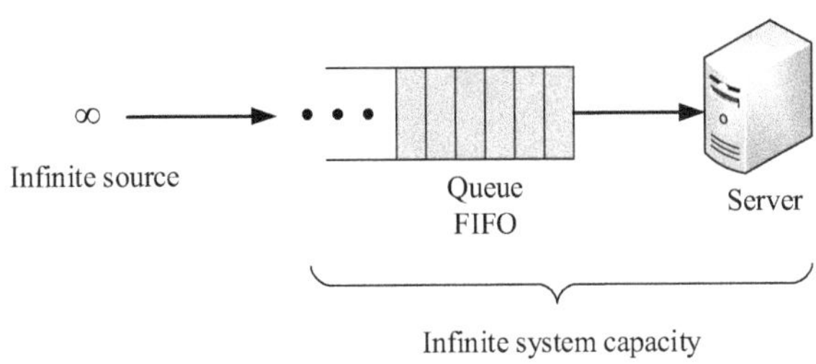

Figure 16.1 $M/M/1$ queue

Remark 16.1. *In single-server queues, in order to prevent from the overload of the queue, it is required that the arrival rate should be less than the service rate, i.e., $\lambda < \mu$. In other terms, the traffic intensity ρ should be*

$$\rho = \frac{\lambda}{\mu} < 1. \tag{16.1}$$

Theorem 16.1. *The steady-state probability that there are n packets, $n = 0, 1, 2, \ldots,$ in the $M/M/1$ queue is given by*

$$P_n = (1-\rho)\rho^n. \tag{16.2}$$

Proof. From Theorem 15.1, we have

$$\begin{aligned} P_0 &= \left(1 + \sum_{n=1}^{\infty} \frac{\prod_{k=0}^{n-1} \lambda_k}{\prod_{k=1}^{n} \mu_k}\right)^{-1} \\ &\overset{(a)}{=} \left(1 + \sum_{n=1}^{\infty} \left(\frac{\lambda}{\mu}\right)^n\right)^{-1} \\ &= \left(\sum_{n=0}^{\infty} \rho^n\right)^{-1} \\ &\overset{(b)}{=} 1-\rho \end{aligned}$$

where (a) follows from the assumption that $\lambda_k = \lambda$ and $\mu_k = \mu \ \forall k$ and (b) follows from the infinite sum of geometric series, i.e.,

$$\sum_{x=0}^{\infty} q^x = \frac{1}{1-q}, \quad \forall |q| < 1.$$

Substituting P_0 into (15.23), we obtain

$$\begin{aligned} P_n &= P_0 \prod_{k=0}^{n-1} \frac{\lambda}{\mu} \\ &= (1-\rho)\rho^n. \end{aligned}$$

□

Theorem 16.2. *The average number of packets in the system and in the queue of $M/M/1$ system is, respectively, given by*

$$E[N] = \frac{\rho}{1-\rho}, \tag{16.3}$$

$$E[N_Q] = \frac{\rho^2}{1-\rho}. \tag{16.4}$$

Proof. We have

$$
\begin{aligned}
E[N] &= \sum_{n=0}^{\infty} nP_n \\
&\overset{(a)}{=} \sum_{n=0}^{\infty} n(1-\rho)\rho^n \\
&= (1-\rho)\rho \sum_{n=1}^{\infty} n\rho^{n-1} \\
&\overset{(b)}{=} (1-\rho)\rho \frac{1}{(1-\rho)^2} \\
&= \frac{\rho}{1-\rho},
\end{aligned}
$$

where (a) follows from (16.2) and (b) follows from the derivative of geometric progression, i.e.,

$$
\sum_{x=1}^{\infty} nx^{n-1} = \frac{1}{(1-x)^2}, \quad \forall |x| < 1.
$$

From (15.28) in Theorem 15.2, we have

$$
\begin{aligned}
E[N_Q] &= E[N] - \rho \\
&\overset{(a)}{=} \frac{\rho}{1-\rho} - \rho \\
&= \frac{\rho^2}{1-\rho},
\end{aligned}
$$

where (a) follows from (16.3). □

Theorem 16.3. *The average response time and waiting time in $M/M/1$ queue are, respectively, given by*

$$
E[T_R] = \frac{1}{\mu - \lambda}, \tag{16.5}
$$

$$
E[T_W] = \frac{\rho}{\mu - \lambda}. \tag{16.6}
$$

Proof. From the Little's law in Section 15.4.5, we have

$$
\begin{aligned}
E[T_R] &= \frac{E[N]}{\lambda} \\
&\overset{(a)}{=} \frac{(\rho/(1-\rho))}{\lambda} \\
&= \frac{1}{\mu - \lambda}.
\end{aligned}
$$

where (a) follows from (16.3).

We also have

$$\begin{aligned} E[T_W] &= E[T_R] - E[T_S] \\ &= \frac{1}{\mu - \lambda} - \frac{1}{\mu} \\ &= \frac{\rho}{\mu - \lambda}. \end{aligned}$$

□

Theorem 16.4. *Server utilization and the average busy period of the server in $M/M/1$ queue are, respectively, given by*

$$U_S = \rho, \tag{16.7}$$

$$E[T_B] = \frac{1}{\mu - \lambda}. \tag{16.8}$$

Proof. From (15.21), we have

$$\begin{aligned} U_S &= 1 - P_0 \\ &= 1 - (1 - \rho) \\ &= \rho, \end{aligned}$$

where the second equal sign follows from (16.2) in Theorem 16.1 with $n = 0$, i.e., $P_0 = 1 - \rho$.

From (15.22) with a notice that the average idle time of the server, i.e., $E[T_0] = 1/\lambda$, we have

$$\begin{aligned} P_0 &= \frac{E[T_0]}{E[T_B] + E[T_0]} \\ &= \frac{1/\lambda}{E[T_B] + (1/\lambda)}. \end{aligned}$$

Therefore, we can obtain $E[T_B]$ as

$$\begin{aligned} E[T_B] &= \frac{1/\lambda}{1 - \rho} - \frac{1}{\lambda} \\ &= \frac{(\rho/(1 - \rho))}{\lambda} \\ &= \frac{1}{\mu - \lambda}. \end{aligned}$$

□

16.2 *M/M/1/K* Queue

An $M/M/1/K$ queue is an extended $M/M/1$ queue with a finite capacity of K packets, i.e., maximum K packets, in the system (cf. Figure 16.2). Both interarrival time, i.e., T_I, and service time, i.e., T_S, are also exponentially distributed with $E[T_I] = 1/\lambda$ and $E[T_S] = 1/\mu$.

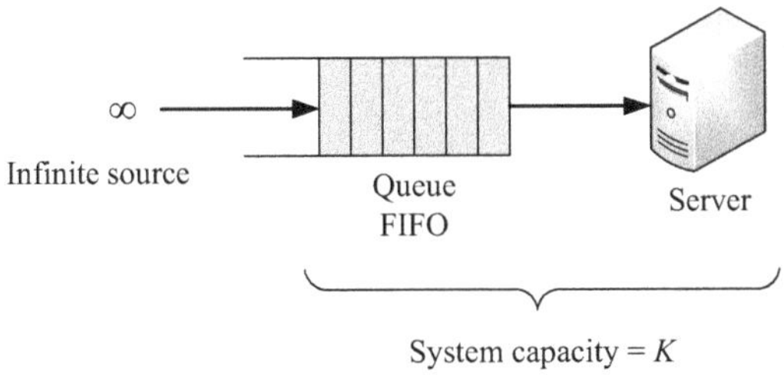

Figure 16.2 $M/M/1/K$ *queue*

Theorem 16.5. *The steady-state probability that there are n packets,* $n = 0, 1, 2, \ldots, K$, *in* $M/M/1/K$ *queue is given by*

$$P_n = \frac{\rho^n}{\sum_{k=0}^{K} \rho^k} = \begin{cases} \dfrac{1}{K+1} & \text{if } \rho = 1, \\ \dfrac{1-\rho}{1-\rho^{K+1}}\rho^n & \text{if } 0 \le \rho < 1. \end{cases} \tag{16.9}$$

Proof. From Theorem 15.1, we have

$$\begin{aligned} P_0 &= \left(1 + \sum_{n=1}^{K} \frac{\prod_{k=0}^{n-1} \lambda_k}{\prod_{k=1}^{n} \mu_k}\right)^{-1} \\ &\overset{(a)}{=} \left(1 + \sum_{n=1}^{K} \left(\frac{\lambda}{\mu}\right)^n\right)^{-1} \\ &= \left(\sum_{n=0}^{K} \rho^n\right)^{-1} \\ &\overset{(b)}{=} \begin{cases} \dfrac{1}{K+1} & \text{if } \rho = 1, \\ \dfrac{1-\rho}{1-\rho^{K+1}} & \text{if } 0 \le \rho < 1, \end{cases} \end{aligned}$$

where (a) follows from the assumption that $\lambda_k = \lambda$ and $\mu_k = \mu \ \forall k$ and (b) follows from the infinite sum of geometric series, i.e.,

$$\sum_{x=0}^{K} q^x = \frac{1 - q^{K+1}}{1 - q}, \quad \forall |q| < 1.$$

Substituting P_0 into (15.23), we obtain

$$\begin{aligned} P_n &= P_0 \prod_{k=0}^{n-1} \frac{\lambda}{\mu} \\ &= \frac{\rho^n}{\sum_{k=0}^{K} \rho^k} \\ &= \begin{cases} \dfrac{1}{K+1} & \text{if } \rho = 1, \\ \dfrac{1-\rho}{1-\rho^{K+1}} \rho^n & \text{if } \rho < 1. \end{cases} \end{aligned}$$

□

Theorem 16.6. *The average number of packets in the system and in the queue of the $M/M/1/K$ system is, respectively, given by*

$$E[N] = \sum_{k=1}^{K} \frac{k\rho^k}{\sum_{n=0}^{K} \rho^n}, \tag{16.10}$$

$$E[N_Q] = \sum_{k=1}^{K} \frac{k\rho^k}{\sum_{n=0}^{K} \rho^n} + \left(\sum_{n=0}^{K} \rho^n \right)^{-1} - 1. \tag{16.11}$$

Proof. We have

$$\begin{aligned} E[N] &= \sum_{k=0}^{K} k P_k \\ &\overset{(a)}{=} \sum_{k=0}^{K} \frac{k\rho^k}{\sum_{n=0}^{K} \rho^n} \\ &= \sum_{k=1}^{K} \frac{k\rho^k}{\sum_{n=0}^{K} \rho^n}, \end{aligned}$$

where (a) follows from (16.9).

We also have

$$\begin{aligned}
E[N_Q] &= \sum_{k=1}^{K}(k-1)P_k \\
&= \sum_{k=1}^{K} kP_k - \sum_{k=1}^{K} P_k \\
&\overset{(a)}{=} E[N] - (1 - P_0) \\
&\overset{(b)}{=} \sum_{k=1}^{K} \frac{k\rho^k}{\sum_{n=0}^{K}\rho^n} + \left(\sum_{n=0}^{K}\rho^n\right)^{-1} - 1
\end{aligned}$$

where (a) follows from $\sum_{k=0}^{K} P_k = 1$ and (b) follows from (16.10) and (16.9). □

Remark 16.2. *In an $M/M/1/K$ queue, due to its limited capacity, the system is blocked when there are K packets in the system. In other words, the $(K+1)$th packet will be dropped. The blocking probability of the queuing system can be understood as the probability that the system is full, which is given by*

$$P_K = \frac{\rho^K}{\sum_{k=0}^{K}\rho^k}. \tag{16.12}$$

It is also noticed that the packet arrival only occurs when the system is unblocked. Let $\overline{\lambda}$ denote the average arrival rate into the system. We then have

$$\begin{aligned}
\overline{\lambda} &= \lambda(1 - P_K) \\
&= \lambda\left(1 - \frac{\rho^K}{\sum_{k=0}^{K}\rho^k}\right)
\end{aligned} \tag{16.13}$$

Remark 16.3. *Another observation from* (16.10), (16.11), *and* (16.13) *is that*

$$E[N] = E[N_Q] + \rho(1 - P_K) \tag{16.14}$$

Theorem 16.7. *The average response time and waiting time in $M/M/1/K$ queue are, respectively, given by*

$$E[T_R] = \frac{\sum_{k=1}^{K}(k\rho^k/\sum_{n=0}^{K}\rho^n)}{\lambda\left(1 - (\rho^K/\sum_{k=0}^{K}\rho^k)\right)}, \tag{16.15}$$

$$E[T_W] = \frac{\sum_{k=1}^{K}(k\rho^k/\sum_{n=0}^{K}\rho^n) + \left(\sum_{n=0}^{K}\rho^n\right)^{-1} - 1}{\lambda\left(1 - (\rho^K/\sum_{k=0}^{K}\rho^k)\right)}. \tag{16.16}$$

Proof. The proof follows from the Little's law in Section 15.4.5 with

$$E[T_R] = \frac{E[N]}{\bar{\lambda}},$$

$$E[T_W] = \frac{E[N_Q]}{\bar{\lambda}},$$

where $\bar{\lambda}$ is given by (16.13), and $E[N]$ and $E[N_Q]$ are given by (16.10) and (16.11), respectively. □

Theorem 16.8. *Server utilization and the average busy period of the server in* $M/M/1/K$ *queue are, respectively, given by*

$$U_S = 1 - \left(\sum_{n=0}^{K} \rho^n\right)^{-1}, \tag{16.17}$$

$$E[T_B] = \frac{1}{\lambda}\left(\sum_{n=0}^{K} \rho^n - 1\right). \tag{16.18}$$

Proof. The proof follows from (15.21) and (15.20), i.e.,

$$U_S = 1 - P_0,$$

$$E[T_B] = \frac{U_S}{1 - U_S} E[T_0],$$

where P_0 is given by (16.9) and $E[T_0] = 1/\lambda$. □

16.3 M/M/1/K/K Queue

An $M/M/1/K/K$ queue is a generalized version of $M/M/1/K$ queue with a finite source having population size of K (cf. Figure 16.3). All the other parameters are similar to those in the $M/M/1/K$ queuing system with the system capacity K, exponentially distributed interarrival time, i.e., T_I, with $E[T_I] = 1/\lambda$, and exponentially distributed service time, i.e., T_S, with $E[T_S] = 1/\mu$.

Remark 16.4. *With a finite source, it can be noticed that the arrival rate at the nth stage is reduced as the birth rate in the birth–death process with*

$$\lambda_n = \begin{cases} (K - n)\lambda & \text{if } 0 \le n \le K, \\ 0 & \text{if } n > K, \end{cases} \tag{16.19}$$

while the service rate is a constant death rate due to the fact that there is only a single server in the system, i.e.,

$$\mu_n = \mu, \quad 1 \le n \le K. \tag{16.20}$$

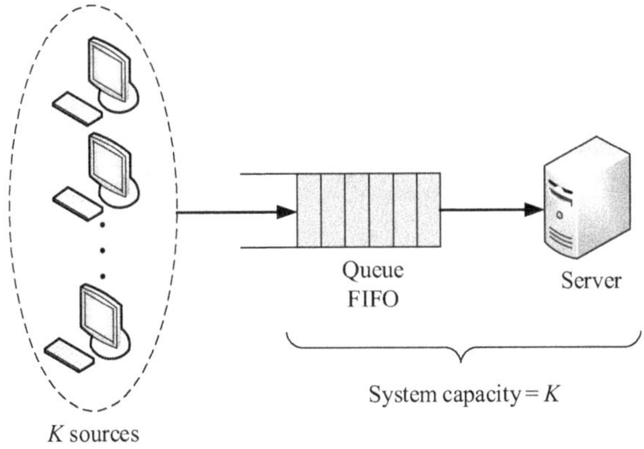

Figure 16.3 $M/M/1/K/K$ queue

Theorem 16.9. *The steady-state probability that there are n packets, $n = 0, 1, 2, \ldots, K$, in $M/M/1/K/K$ queue is given by*

$$P_n = \frac{(K!/(K-n)!)\rho^n}{\sum_{k=0}^{K}(K!/(K-k)!)\rho^k}. \tag{16.21}$$

Proof. From Theorem 15.1, we have

$$\begin{aligned} P_0 &= \left(1 + \sum_{n=1}^{K} \frac{\prod_{k=0}^{n-1}\lambda_k}{\prod_{k=1}^{n}\mu_k}\right)^{-1} \\ &\overset{(a)}{=} \left(1 + \sum_{n=1}^{K}\prod_{k=0}^{n-1}(K-k)\left(\frac{\lambda}{\mu}\right)^n\right)^{-1} \\ &\overset{(b)}{=} \left(1 + \sum_{n=1}^{K}\frac{K!}{(K-n)!}\rho^n\right)^{-1} \\ &= \left(\sum_{n=0}^{K}\frac{K!}{(K-n)!}\rho^n\right)^{-1}, \end{aligned}$$

where (a) follows from (16.19), (16.20), and $\rho = \lambda/\mu$, and (b) follows from the fact that

$$\prod_{k=0}^{n-1}(K-k) = \frac{K!}{(K-n)!}.$$

Substituting P_0 into (15.23), we obtain

$$P_n = P_0 \prod_{k=0}^{n-1} (K-k) \left(\frac{\lambda}{\mu}\right)^n = \frac{(K!/(K-n)!)\rho^n}{\sum_{k=0}^{K} (K!/(K-k)!)\rho^k}.$$

□

Theorem 16.10. *The average number of packets in the system and in the queue of the $M/M/1/K/K$ system is, respectively, given by*

$$E[N] = K - \frac{1}{\rho}\left(1 - \left(\sum_{n=0}^{K} \frac{K!}{(K-n)!}\rho^n\right)^{-1}\right), \tag{16.22}$$

$$E[N_Q] = K - \left(1 + \frac{1}{\rho}\right)\left(1 - \left(\sum_{n=0}^{K} \frac{K!}{(K-n)!}\rho^n\right)^{-1}\right). \tag{16.23}$$

Proof. We have

$$\begin{aligned} E[N] &= \sum_{k=0}^{K} kP_k \\ &= \sum_{k=0}^{K} (KP_k - (K-k)P_k) \\ &\overset{(a)}{=} K - \sum_{k=0}^{K} (K-k)P_k \\ &\overset{(b)}{=} K - \frac{1}{\rho}\sum_{k=0}^{K-1} P_{k+1} \\ &\overset{(c)}{=} K - \frac{1}{\rho}(1-P_0) \\ &\overset{(d)}{=} K - \frac{1}{\rho}\left(1 - \left(\sum_{n=0}^{K} \frac{K!}{(K-n)!}\rho^n\right)^{-1}\right), \end{aligned}$$

where (a) and (c) follow from $\sum_{k=0}^{K} P_k = 1$, (b) follows from the fact that $P_{k+1} = \rho(K-k)P_k$, and (d) follows from (16.21).

We also have

$$\begin{aligned}
E[N_Q] &= \sum_{k=1}^{K}(k-1)P_k \\
&= \sum_{k=1}^{K} kP_k - \sum_{k=1}^{K} P_k \\
&\overset{(a)}{=} E[N] - (1-P_0) \\
&\overset{(b)}{=} K - \frac{1}{\rho}(1-P_0) - (1-P_0) \\
&= K - \left(1+\frac{1}{\rho}\right)(1-P_0) \\
&\overset{(c)}{=} K - \left(1+\frac{1}{\rho}\right)\left(1-\left(\sum_{n=0}^{K}\frac{K!}{(K-n)!}\rho^n\right)^{-1}\right)
\end{aligned}$$

where (a) follows from $\sum_{k=0}^{K} P_k = 1$, (b) follows from the derivation of (16.22), and (c) follows from (16.21). □

Remark 16.5. *Notice that in an $M/M/1/K/K$ queue, apart from the packets in the system, there are packets waiting in the source. Let N_S denote the number of packets in the source. The average number of packets in the source can be given by*

$$\begin{aligned}
E[N_S] &= K - E[N] \\
&= \frac{1}{\rho}\left(1-\left(\sum_{n=0}^{K}\frac{K!}{(K-n)!}\rho^n\right)^{-1}\right). \qquad (16.24)
\end{aligned}$$

The utilization of the source, denoted by U_{Source}, is thus given by

$$\begin{aligned}
U_{Source} &= \frac{E[N_S]}{K} \\
&= \frac{1}{K\rho}\left(1-\left(\sum_{n=0}^{K}\frac{K!}{(K-n)!}\rho^n\right)^{-1}\right). \qquad (16.25)
\end{aligned}$$

Theorem 16.11. *The average response time and waiting time in $M/M/1/K/K$ queue are, respectively, given by*

$$E[T_R] = \frac{K-(1/\rho)\left(1-\left(\sum_{n=0}^{K}(K!/(K-n)!)\rho^n\right)^{-1}\right)}{\mu\left(1-\left(\sum_{n=0}^{K}(K!/(K-n)!)\rho^n\right)^{-1}\right)}, \qquad (16.26)$$

$$E[T_W] = \frac{K - (1 + (1/\rho))\left(1 - \left(\sum_{n=0}^{K}(K!/(K-n)!)\rho^n\right)^{-1}\right)}{\mu\left(1 - \left(\sum_{n=0}^{K}(K!/(K-n)!)\rho^n\right)^{-1}\right)}. \tag{16.27}$$

Proof. The proof follows from the Little's law in Section 15.4.5 with

$$E[T_R] = \frac{E[N]}{\lambda E[N_S]},$$

$$E[T_W] = \frac{E[N_Q]}{\lambda E[N_S]},$$

where $E[N_S]$, $E[N]$, and $E[N_Q]$ are given by (16.24), (16.22), and (16.23), respectively. □

Theorem 16.12. *Server utilization and the average busy period of the server in $M/M/1/K/K$ queue are, respectively, given by*

$$U_S = 1 - \left(\sum_{n=0}^{K}\frac{K!}{(K-n)!}\rho^n\right)^{-1}, \tag{16.28}$$

$$E[T_B] = \frac{1}{K\lambda}\left(\left(\sum_{n=0}^{K}\frac{K!}{(K-n)!}\rho^n\right) - 1\right). \tag{16.29}$$

Proof. The proof is similar to that for Theorem 16.8 by using (15.21) and (15.20), i.e.,

$$U_S = 1 - P_0,$$

$$E[T_B] = \frac{U_S}{1 - U_S}E[T_0],$$

with P_0 given by (16.21) and the average idle time of the server $E[T_0] = 1/(K\lambda)$. □

16.4 Related works

For more information, the readers are suggested to refer to the following selected references: Takacs in 1962 and 1963 [268,280], Chang in 1970 [281], Kleinrock in 1975 [270], Gopinath and Morrison in 1977 [282], Eckberg in 1979 [283], Wang and Mandayam in 2004 [284], Stewart in 2009 [246], Steiner and Shamai in 2010 [285], Cohen in 2012 [286], and Zhao *et al.* in 2018 [287].

16.5 Review questions

[Q16.1] Describe the following single-server queuing systems: $M/M/1$, $M/M/1/K$ and $M/M/1/K/K$. Provide an illustrative example for each of them.

[Q16.2] What is the stability condition for a single-server queue?

[Q16.3] Derive the steady-state probability that there are n packets in $M/M/1$, $M/M/1/K$ and $M/M/1/K/K$ queues.

[Q16.4] Derive the average number of packets in the system and in the queue of $M/M/1$, $M/M/1/K$ and $M/M/1/K/K$ queues.

[Q16.5] Derive the average response time and waiting time in $M/M/1$, $M/M/1/K$, and $M/M/1/K/K$ queues.

[Q16.6] Derive server utilization and the average busy period of the server in $M/M/1$, $M/M/1/K$, and $M/M/1/K/K$ queues.

[Q16.7] When is a queuing system blocked?

[Q16.8] Derive the blocking probability of an $M/M/1/K$ queue.

[Q16.9] Derive the average number of packets in the source and the utilization of the source in an $M/M/1/K/K$ queue.

16.6 Problems

[P16.1] Consider a server in a network. Packets arrive on an average of 5 s and it takes an average of 4 s for processing at the server. Both the inter-arrival time and service time are exponentially distributed. The queue is assumed to be infinitely large (unbounded) and the serving discipline is FIFO.

- Give the Kendall's notation describing this system and check the condition for stability of this system.
- Find the probability that the server is busy.
- Find the probability that there are less than three packets in the system.
- Find the average number of packets waiting to be processed and the average number of packets in the system.
- Find the average time the packet spends in the queue and totally spends in the system.

[P16.2] An M/M/1 queue has a service rate of two customers per hour.

- Find the maximum arrival rate such that the average queuing interval is not longer than 30 min.
- Using the obtained maximum arrival rate,
  - Find the average number of customers in the queue and in the system.
  - Find the probability that the server is busy.
  - Find the probability that there are no more than two customers in the system.

[P16.3] The data arrivals at a server are assumed to follow Poisson point process. The time to process data at the server is exponentially distributed with a mean of 1.5 s. The average waiting time of the packet in the queue is 3 s.

- Find the arrival rate of the packets.
- Find the server utilization and the average busy period of the server.
- Find the average number of packets in the queue and in the system.
- Find the probability that there is at least a packet in the system.

[P16.4] At a database server, it takes an average of 25 s to store a particular data packet. The packet arrival and storing time at the server follow exponential distribution.

- Find the average interarrival time of the data packets given that the average queuing length is five data packets.
- Find the average response and waiting time in the system.

[P16.5] A single-server queue has a finite capacity of five packets. Both the inter-arrival and service time are exponentially distributed with the average interarrival of 10 s and the average service time of 8 s.

- Give the Kendall's notation describing this system.
- Find the probability that the server is idle and the blocking probability of the queuing system.
- Find the average number of packets in the system and in the queue.
- Find the average response time and waiting time of the data packets.
- Find the server utilization and the average busy period of the server.

[P16.6] Considering a finite source in a single-server queuing system with a limited capacity of five packets and a population size of five packets. Both the packet arrival and service follow Poisson point process with an arrival rate of 5 packets/s and a service rate of 8 packets/s.

- Give the Kendall's notation describing this system.
- Find the probability that the server is busy.
- Find the average number of packets in the system, in the queue and in the source.
- Find the average response time and waiting time of the data packets.
- Find the server utilization, source utilization and the average busy period of the server.

Chapter 17

Multi-server queues—network behaviors and analysis

Considering queuing systems with multiple servers, this chapter is devoted to present multi-server queues in which packets are served by multiple packets in parallel. There exist various models for multi-server queues depending on the system capacity and population size of the sources.

In this chapter, I will first introduce $M/M/c$ queue with infinite system capacity and infinite population in Section 17.1. Its variant models with finite system capacity and finite population will be sequentially discussed in Sections 17.2 and 17.3. For every queuing model, its behaviors will be presented along with detailed analyses of relevant performance measures.

17.1 $M/M/c$ Queue

In $M/M/c$ queue, there are c servers. By default, the capacity and population without stated means infinite (cf. Figure 17.1). Both interarrival time, i.e., T_I, and service time, i.e., T_S, are exponentially distributed with $E[T_I] = 1/\lambda$ and $E[T_S] = 1/\mu$, respectively.

Remark 17.1. *With multiple servers to serve in parallel, it can be noticed that the service rate at the nth stage increases as the death rate in the birth–death process with*

$$\mu_n = \begin{cases} n\mu & \text{if } 0 \leq n \leq c, \\ c\mu & \text{if } n > c, \end{cases} \tag{17.1}$$

where the second case is due to the maximum service rate with c servers is $c\mu$. The arrival rate is however a constant birth rate, i.e.,

$$\lambda_n = \lambda, \quad n \geq 0. \tag{17.2}$$

Remark 17.2. *In multi-server queues, in order to prevent from the overload of the queue, it is required that the arrival rate should be less than the service rate, i.e., $\lambda < c\mu$. In other terms, the traffic intensity, i.e., $\rho = \lambda/\mu$, should be*

$$\rho < c. \tag{17.3}$$

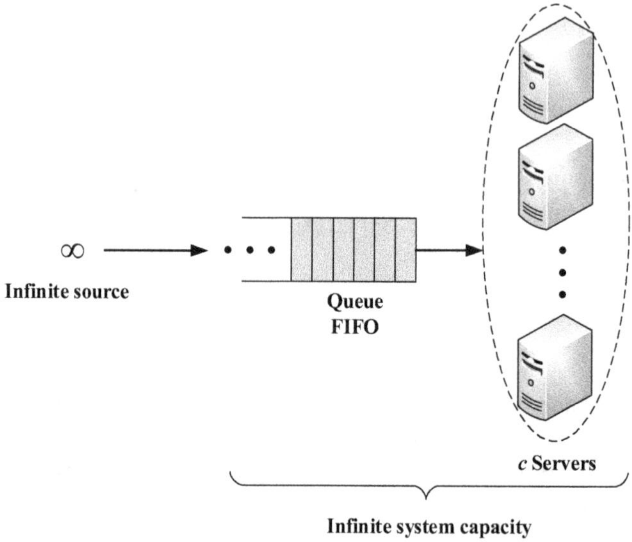

Figure 17.1 M/M/c Queue

For simplicity, we define an auxiliary parameter

$$a \triangleq \frac{\rho}{c}. \tag{17.4}$$

Theorem 17.1. *The steady-state probability that there are n packets, $n = 0, 1, 2, \ldots,$ in the $M/M/c$ queue is given by*

$$P_n = \begin{cases} \dfrac{\rho^n}{n!}\left(\sum\limits_{k=0}^{c-1}\dfrac{\rho^k}{k!} + \dfrac{\rho^c}{c!(1-a)}\right)^{-1} & \text{if } 0 \le n \le c, \\ \dfrac{a^n c^c}{c!}\left(\sum\limits_{k=0}^{c-1}\dfrac{\rho^k}{k!} + \dfrac{\rho^c}{c!(1-a)}\right)^{-1} & \text{if } n > c. \end{cases} \tag{17.5}$$

Proof. From Theorem 15.5, we have

$$\begin{aligned} P_0 &= \left(1 + \sum_{n=1}^{\infty}\frac{\prod_{k=0}^{n-1}\lambda_k}{\prod_{k=1}^{n}\mu_k}\right)^{-1} \\ &\overset{(a)}{=} \left(1 + \sum_{n=1}^{c-1}\frac{\lambda^n}{n!\mu^n} + \sum_{n=c}^{\infty}\frac{\lambda^n}{c!c^{n-c}(\mu)^n}\right)^{-1} \\ &= \left(\sum_{n=0}^{c-1}\frac{\rho^n}{n!} + \frac{\rho^c}{c!}\sum_{n=c}^{\infty}\frac{\rho^{n-c}}{c^{n-c}}\right)^{-1} \end{aligned}$$

$$\overset{(b)}{=} \left(\sum_{n=0}^{c-1} \frac{\rho^n}{n!} + \frac{\rho^c}{c!} \sum_{n=0}^{\infty} a^n \right)^{-1}$$
$$\overset{(c)}{=} \left(\sum_{n=0}^{c-1} \frac{\rho^n}{n!} + \frac{\rho^c}{c!(1-a)} \right)^{-1},$$

where (a) follows from (17.1) and (17.2), (b) is due to $a = \rho/c$, and (c) follows from the infinite sum of geometric series, i.e.,

$$\sum_{x=0}^{\infty} q^x = \frac{1}{1-q}, \quad \forall |q| < 1.$$

Substituting P_0 into (15.23), we obtain

$$\begin{aligned} P_n &= P_0 \frac{\prod_{k=0}^{n-1} \lambda_k}{\prod_{k=1}^{n} \mu_k} \\ &= \begin{cases} P_0 \dfrac{\lambda^n}{n!\mu^n} & \text{if } 0 \leq n \leq c, \\ P_0 \dfrac{\lambda^n}{c!c^{n-c}\mu^n} & \text{if } n > c \end{cases} \\ &= \begin{cases} P_0 \dfrac{\rho^n}{n!} & \text{if } 0 \leq n \leq c, \\ P_0 \dfrac{a^n c^c}{c!} & \text{if } n > c \end{cases} \\ &= \begin{cases} \dfrac{\rho^n}{n!} \left(\sum_{k=0}^{c-1} \dfrac{\rho^k}{k!} + \dfrac{\rho^c}{c!(1-a)} \right)^{-1} & \text{if } 0 \leq n \leq c, \\ \dfrac{a^n c^c}{c!} \left(\sum_{k=0}^{c-1} \dfrac{\rho^k}{k!} + \dfrac{\rho^c}{c!(1-a)} \right)^{-1} & \text{if } n > c. \end{cases} \end{aligned}$$

□

Remark 17.3. *Notice that, if the number of arriving packets is larger than the number of servers, then only c packets can be served in parallel with c servers, while the remaining packets have to wait.*

Theorem 17.2. *The waiting probability of a packet, i.e., P_W, in $M/M/c$ queue is given by*

$$P_W = \sum_{n=c}^{\infty} \frac{a^n c^c}{c!} \left(\sum_{k=0}^{c-1} \frac{\rho^k}{k!} + \frac{\rho^c}{c!(1-a)} \right)^{-1}. \tag{17.6}$$

Proof. By definition, the waiting probability of a packet is corresponding to the probability that the system already has at least c packets, i.e.,

$$\begin{aligned} P_W &= P_{n\geq c} \\ &= \sum_{n=c}^{\infty} P_n \\ &= \sum_{n=c}^{\infty} \frac{a^n c^c}{c!} \left(\sum_{k=0}^{c-1} \frac{\rho^k}{k!} + \frac{\rho^c}{c!(1-a)} \right)^{-1}, \end{aligned}$$

where the last equal sign follows from (17.5). □

Theorem 17.3. *The average number of packets in the system and in the queue of $M/M/c$ system are, respectively, given by*

$$E[N] = \rho + \sum_{n=c+1}^{\infty} (n-c) \frac{a^n c^c}{c!} \left(\sum_{k=0}^{c-1} \frac{\rho^k}{k!} + \frac{\rho^c}{c!(1-a)} \right)^{-1}, \tag{17.7}$$

$$E[N_Q] = \sum_{n=c+1}^{\infty} (n-c) \frac{a^n c^c}{c!} \left(\sum_{k=0}^{c-1} \frac{\rho^k}{k!} + \frac{\rho^c}{c!(1-a)} \right)^{-1}. \tag{17.8}$$

Proof. Let us first derive $E[N_Q]$. We have

$$\begin{aligned} E[N_Q] &= \sum_{n=c+1}^{\infty} (n-c) P_n \\ &\overset{(a)}{=} \sum_{n=c+1}^{\infty} (n-c) \frac{a^n c^c}{c!} \left(\sum_{k=0}^{c-1} \frac{\rho^k}{k!} + \frac{\rho^c}{c!(1-a)} \right)^{-1}, \end{aligned}$$

where (a) follows from (17.5).

From (15.28) in Theorem 15.2, we have

$$\begin{aligned} E[N] &= E[N_Q] + \rho \\ &= \rho + \sum_{n=c+1}^{\infty} (n-c) \frac{a^n c^c}{c!} \left(\sum_{k=0}^{c-1} \frac{\rho^k}{k!} + \frac{\rho^c}{c!(1-a)} \right)^{-1}. \end{aligned}$$

□

Theorem 17.4. *The average response time and waiting time in $M/M/c$ queue are, respectively, given by*

$$E[T_R] = \frac{1}{\lambda} \left[\rho + \sum_{n=c}^{\infty} (n-c) \frac{a^n c^c}{c!} \left(\sum_{k=0}^{c-1} \frac{\rho^k}{k!} + \frac{\rho^c}{c!(1-a)} \right)^{-1} \right], \tag{17.9}$$

$$E[T_W] = \frac{1}{\lambda}\sum_{n=c}^{\infty}(n-c)\frac{a^n c^c}{c!}\left(\sum_{k=0}^{c-1}\frac{\rho^k}{k!} + \frac{\rho^c}{c!(1-a)}\right)^{-1}. \tag{17.10}$$

Proof. From the Little's law in Section 15.4.5, we have

$$\begin{aligned} E[T_R] &= \frac{E[N]}{\lambda} \\ &\overset{(a)}{=} \frac{1}{\lambda}\left[\rho + \sum_{n=c}^{\infty}(n-c)\frac{a^n c^c}{c!}\left(\sum_{k=0}^{c-1}\frac{\rho^k}{k!} + \frac{\rho^c}{c!(1-a)}\right)^{-1}\right]. \end{aligned}$$

where (a) follows from (17.7).

We also have

$$\begin{aligned} E[T_W] &= \frac{E[N_Q]}{\lambda} \\ &\overset{(a)}{=} \frac{1}{\lambda}\sum_{n=c}^{\infty}(n-c)\frac{a^n c^c}{c!}\left(\sum_{k=0}^{c-1}\frac{\rho^k}{k!} + \frac{\rho^c}{c!(1-a)}\right)^{-1}, \end{aligned}$$

where (a) follows from (17.8). □

Remark 17.4. *Note that the traffic intensity in multi-server queue is in fact the utilization of all servers. Let U_A denote the utilization of all servers. We then have*

$$U_A = \rho. \tag{17.11}$$

This overall utilization also means the average number of busy servers in the system. Let c_B and c_I denote the number of busy servers and the number of idle servers, respectively. We can arrive at

$$E[c_B] = \rho, \tag{17.12}$$

$$E[c_I] = c - \rho. \tag{17.13}$$

Remark 17.5. *Considering the whole system, the system utilization in the multi-server queue, denoted by U_{System}, is similarly defined as in (15.21), i.e.,*

$$\begin{aligned} U_{System} &= 1 - P_0 \\ &= 1 - \left(\sum_{n=0}^{c-1}\frac{\rho^n}{n!} + \frac{\rho^c}{c!(1-a)}\right)^{-1}, \end{aligned} \tag{17.14}$$

where P_0 is given by (17.5).

Theorem 17.5. *Server utilization and the average busy period of the server in $M/M/c$ queue are, respectively, given by*

$$U_S = a = \frac{\rho}{c}, \tag{17.15}$$

$$E[T_B] = \frac{1}{\lambda}\left(\left(\sum_{n=0}^{c-1}\frac{\rho^n}{n!} + \frac{\rho^c}{c!(1-a)}\right) - 1\right). \tag{17.16}$$

Proof. We have

$$U_A = cU_S,$$

and thus

$$\begin{aligned} U_S &= \frac{U_A}{c} \\ &= \frac{\rho}{c} = a. \end{aligned}$$

From (15.22) with a notice that the average idle time of the server, i.e., $E[T_0] = 1/\lambda$, we have

$$\begin{aligned} P_0 &= \frac{E[T_0]}{E[T_B] + E[T_0]} \\ &= \frac{1/\lambda}{E[T_B] + (1/\lambda)}. \end{aligned}$$

Therefore, we can obtain $E[T_B]$ as

$$\begin{aligned} E[T_B] &= \frac{1 - P_0}{\lambda P_0} \\ &\overset{(a)}{=} \frac{1}{\lambda}\left(\left(\sum_{n=0}^{c-1}\frac{\rho^n}{n!} + \frac{\rho^c}{c!(1-a)}\right) - 1\right), \end{aligned}$$

where (a) follows from (17.5) with $n = 0$. □

17.2 *M/M/c/K* Queue

An $M/M/c/K$ queue is an extended version of $M/M/c$ queue for the case of finite capacity with a maximum of K packets in the system (cf. Figure 17.2). Both interarrival time, i.e., T_I, and service time, i.e., T_S, are also exponentially distributed with $E[T_I] = 1/\lambda$ and $E[T_S] = 1/\mu$.

Theorem 17.6. *The steady-state probability that there are n packets, $n = 0, 1, 2, \ldots, K$, in the $M/M/c/K$ queue is given by*

$$P_n = \begin{cases} \dfrac{\rho^n}{n!}\left(\displaystyle\sum_{k=0}^{c-1}\frac{\rho^k}{k!} + \frac{\rho^c(1-a^{K-c+1})}{c!(1-a)}\right)^{-1} & \text{if } 0 \le n \le c, \\ \dfrac{a^n c^c}{c!}\left(\displaystyle\sum_{k=0}^{c-1}\frac{\rho^k}{k!} + \frac{\rho^c(1-a^{K-c+1})}{c!(1-a)}\right)^{-1} & \text{if } c < n \le K, \end{cases} \tag{17.17}$$

where $a \triangleq (\rho/c) < 1$.

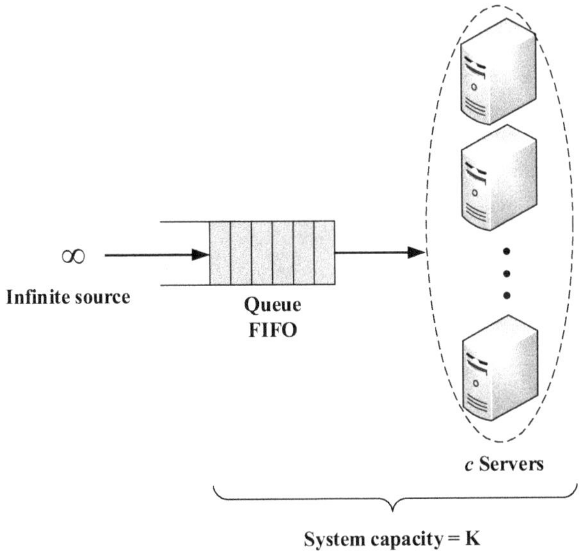

Figure 17.2 $M/M/c/K$ *Queue*

Proof. From Theorem 15.1, we have

$$\begin{aligned}
P_0 &= \left(1+\sum_{n=1}^{K}\frac{\prod_{k=0}^{n-1}\lambda_k}{\prod_{k=1}^{n}\mu_k}\right)^{-1}\\
&\overset{(a)}{=} \left(1+\sum_{n=1}^{c-1}\frac{\lambda^n}{n!\mu^n}+\sum_{n=c}^{K}\frac{\lambda^n}{c!c^{n-c}(\mu)^n}\right)^{-1}\\
&= \left(\sum_{n=0}^{c-1}\frac{\rho^n}{n!}+\frac{\rho^c}{c!}\sum_{n=c}^{K}\frac{\rho^{n-c}}{c^{n-c}}\right)^{-1}\\
&= \left(\sum_{n=0}^{c-1}\frac{\rho^n}{n!}+\frac{\rho^c}{c!}\sum_{n=0}^{K-c}a^n\right)^{-1}\\
&\overset{(b)}{=} \left(\sum_{n=0}^{c-1}\frac{\rho^n}{n!}+\frac{\rho^c(1-a^{K-c+1})}{c!(1-a)}\right)^{-1},
\end{aligned}$$

where (a) follows from (17.1) and (17.2), and (b) follows from

$$\sum_{k=c}^{K}a^{k-c}=\frac{1-a^{K-c+1}}{1-a}\ \text{if } a\neq 1. \tag{17.18}$$

Substituting P_0 into (15.23), we obtain

$$
\begin{aligned}
P_n &= P_0 \frac{\prod_{k=0}^{n-1} \lambda_k}{\prod_{k=1}^{n} \mu_k} \\
&= \begin{cases} P_0 \dfrac{\lambda^n}{n!\mu^n} & \text{if } 0 \le n \le c, \\ P_0 \dfrac{\lambda^n}{c!c^{n-c}\mu^n} & \text{if } c < n \le K \end{cases} \\
&= \begin{cases} P_0 \dfrac{\rho^n}{n!} & \text{if } 0 \le n \le c, \\ P_0 \dfrac{a^n c^c}{c!} & \text{if } c < n \le K \end{cases} \\
&= \begin{cases} \dfrac{\rho^n}{n!} \left(\sum_{k=0}^{c-1} \dfrac{\rho^k}{k!} + \dfrac{\rho^c(1 - a^{K-c+1})}{c!(1-a)} \right)^{-1} & \text{if } 0 \le n \le c, \\ \dfrac{a^n c^c}{c!} \left(\sum_{k=0}^{c-1} \dfrac{\rho^k}{k!} + \dfrac{\rho^c(1 - a^{K-c+1})}{c!(1-a)} \right)^{-1} & \text{if } c < n \le K. \end{cases}
\end{aligned}
$$

□

Theorem 17.7. *The waiting probability of a packet in $M/M/c/K$ queue is given by*

$$
P_W = \sum_{n=c}^{K} \frac{a^n c^c}{c!} \left(\sum_{k=0}^{c-1} \frac{\rho^k}{k!} + \frac{\rho^c(1 - a^{K-c+1})}{c!(1-a)} \right)^{-1}. \tag{17.19}
$$

Proof. The waiting probability can be similarly derived as in Theorem 17.2 with a notice that the number of packets allowed in the system is limited by K, i.e.,

$$
\begin{aligned}
P_W &= P_{c \le n \le K} \\
&= \sum_{n=c}^{K} P_n \\
&= \sum_{n=c}^{K} \frac{a^n c^c}{c!} \left(\sum_{k=0}^{c-1} \frac{\rho^k}{k!} + \frac{\rho^c(1 - a^{K-c+1})}{c!(1-a)} \right)^{-1},
\end{aligned}
$$

where the last equal sign follows from (17.17). □

Theorem 17.8. *The average number of packets in the system and in the queue of $M/M/c/K$ system is, respectively, given by*

$$
\begin{aligned}
E[N] &= \rho \left(1 - \frac{a^K c^c}{c!} \left(\sum_{k=0}^{c-1} \frac{\rho^k}{k!} + \frac{\rho^c(1 - a^{K-c+1})}{c!(1-a)} \right)^{-1} \right) \\
&\quad + \sum_{n=c+1}^{K} (n-c) \frac{a^n c^c}{c!} \left(\sum_{k=0}^{c-1} \frac{\rho^k}{k!} + \frac{\rho^c(1 - a^{K-c+1})}{c!(1-a)} \right)^{-1},
\end{aligned} \tag{17.20}
$$

$$E[N_Q] = \sum_{n=c+1}^{K} (n-c)\frac{a^n c^c}{c!}\left(\sum_{k=0}^{c-1}\frac{\rho^k}{k!} + \frac{\rho^c(1-a^{K-c+1})}{c!(1-a)}\right)^{-1}. \quad (17.21)$$

Proof. Let us first derive $E[N_Q]$. We have

$$\begin{aligned} E[N_Q] &= \sum_{n=c+1}^{K} (n-c)P_n \\ &\stackrel{(a)}{=} \sum_{n=c+1}^{K} (n-c)\frac{a^n c^c}{c!}\left(\sum_{k=0}^{c-1}\frac{\rho^k}{k!} + \frac{\rho^c(1-a^{K-c+1})}{c!(1-a)}\right)^{-1}, \end{aligned}$$

where (a) follows from (17.17).

Similar to $M/M/1/K$, from (16.14), we have

$$\begin{aligned} E[N] &= E[N_Q] + \rho(1-P_K) \\ &= \rho\left(1 - \frac{a^K c^c}{c!}\left(\sum_{k=0}^{c-1}\frac{\rho^k}{k!} + \frac{\rho^c(1-a^{K-c+1})}{c!(1-a)}\right)^{-1}\right) \\ &\quad + \sum_{n=c+1}^{K} (n-c)\frac{a^n c^c}{c!}\left(\sum_{k=0}^{c-1}\frac{\rho^k}{k!} + \frac{\rho^c(1-a^{K-c+1})}{c!(1-a)}\right)^{-1}. \end{aligned}$$

□

Theorem 17.9. *The average response time and waiting time in $M/M/c/K$ queue are, respectively, given by*

$$E[T_R] = \frac{1}{\mu} + \frac{\sum_{n=c+1}^{K}(n-c)(a^n c^c/c!)\left(\sum_{k=0}^{c-1}(\rho^k/k!) + ((\rho^c(1-a^{K-c+1}))/(c!(1-a)))\right)^{-1}}{\lambda\left(1-(a^K c^c/c!)\left(\sum_{k=0}^{c-1}(\rho^k/k!) + ((\rho^c(1-a^{K-c+1}))/(c!(1-a))\right)^{-1}\right)}, \quad (17.22)$$

$$E[T_W] = \frac{\sum_{n=c+1}^{K}(n-c)(a^n c^c/c!)\left(\sum_{k=0}^{c-1}(\rho^k/k!) + ((\rho^c(1-a^{K-c+1}))/(c!(1-a)))\right)^{-1}}{\lambda\left(1-(a^K c^c/c!)\left(\sum_{k=0}^{c-1}(\rho^k/k!) + ((\rho^c(1-a^{K-c+1}))/(c!(1-a))\right)^{-1}\right)}. \quad (17.23)$$

Proof. Let us first derive $E[T_W]$. Similar to the proof of Theorem 16.7, from the Little's law, we have

$$E[T_W] = \frac{E[N_Q]}{\lambda(1-P_K)}$$
$$\overset{(a)}{=} \frac{\sum_{n=c+1}^{K}(n-c)(a^n c^c/c!)\left(\sum_{k=0}^{c-1}(\rho^k/k!)+((\rho^c(1-a^{K-c+1}))/(c!(1-a)))\right)^{-1}}{\lambda\left(1-(a^K c^c/c!)\left(\sum_{k=0}^{c-1}(\rho^k/k!)+((\rho^c(1-a^{K-c+1}))/(c!(1-a))\right)^{-1}\right)},$$

where (a) follows from (17.21) and (17.17).

We also have from the Little's law that

$$\begin{aligned} E[T_R] &= \frac{E[N]}{\lambda(1-P_K)} \\ &\overset{(a)}{=} \frac{E[N_Q]+\rho(1-P_K)}{\lambda(1-P_K)} \\ &= E[T_W]+\frac{1}{\mu} \\ &\overset{(b)}{=} \frac{1}{\mu} \\ &+ \frac{\sum_{n=c+1}^{K}(n-c)(a^n c^c/c!)\left(\sum_{k=0}^{c-1}(\rho^k/k!)+((\rho^c(1-a^{K-c+1}))/(c!(1-a)))\right)^{-1}}{\lambda\left(1-(a^K c^c/c!)\left(\sum_{k=0}^{c-1}(\rho^k/k!)+((\rho^c(1-a^{K-c+1}))/(c!(1-a))\right)^{-1}\right)}. \end{aligned}$$

where (a) follows from (16.14) and (b) follows from (17.23). □

Remark 17.6. *We can similarly obtain the following measures of the $M/M/c/K$ queue:*

- *Utilization of all servers:*

$$\begin{aligned} U_A &= \rho(1-P_K) \\ &= \rho\left(1-\frac{a^K c^c}{c!}\left(\sum_{k=0}^{c-1}\frac{\rho^k}{k!}+\frac{\rho^c(1-a^{K-c+1})}{c!(1-a)}\right)^{-1}\right). \end{aligned} \tag{17.24}$$

- *System utilization:*

$$\begin{aligned} U_{System} &= 1-P_0 \\ &= 1-\left(\sum_{n=0}^{c-1}\frac{\rho^n}{n!}+\frac{\rho^c(1-a^{K-c+1})}{c!(1-a)}\right)^{-1}. \end{aligned} \tag{17.25}$$

- *The average number of busy servers:*

$$E[c_B] = U_A = \rho\left(1 - \frac{a^K c^c}{c!}\left(\sum_{k=0}^{c-1}\frac{\rho^k}{k!} + \frac{\rho^c(1-a^{K-c+1})}{c!(1-a)}\right)^{-1}\right). \tag{17.26}$$

- *The average number of idle servers:*

$$E[c_I] = c - \rho\left(1 - \frac{a^K c^c}{c!}\left(\sum_{k=0}^{c-1}\frac{\rho^k}{k!} + \frac{\rho^c(1-a^{K-c+1})}{c!(1-a)}\right)^{-1}\right). \tag{17.27}$$

Theorem 17.10. *Server utilization and the average busy period of the server in $M/M/c/K$ queue are, respectively, given by*

$$U_S = a\left(1 - \frac{a^K c^c}{c!}\left(\sum_{k=0}^{c-1}\frac{\rho^k}{k!} + \frac{\rho^c(1-a^{K-c+1})}{c!(1-a)}\right)^{-1}\right), \tag{17.28}$$

$$E[T_B] = \frac{1}{\lambda}\left(\left(\sum_{n=0}^{c-1}\frac{\rho^n}{n!} + \frac{\rho^c(1-a^{K-c+1})}{c!(1-a)}\right) - 1\right). \tag{17.29}$$

Proof. We have

$$U_A = cU_S,$$

and thus

$$\begin{aligned} U_S &= \frac{U_A}{c} \\ &= a\left(1 - \frac{a^K c^c}{c!}\left(\sum_{k=0}^{c-1}\frac{\rho^k}{k!} + \frac{\rho^c(1-a^{K-c+1})}{c!(1-a)}\right)\right). \end{aligned}$$

From (15.22) with a notice that the average idle time of the server, i.e., $E[T_0] = 1/\lambda$, we can obtain $E[T_B]$ as

$$\begin{aligned} E[T_B] &= \frac{1-P_0}{\lambda P_0} \\ &= \frac{1}{\lambda}\left(\left(\sum_{n=0}^{c-1}\frac{\rho^n}{n!} + \frac{\rho^c(1-a^{K-c+1})}{c!(1-a)}\right) - 1\right). \end{aligned}$$

□

17.3 *M/M/c/K/K* Queue

An $M/M/c/K/K$ queue is a generalized version of $M/M/c/K$ queue considering a finite source with population size K (cf. Figure 17.3). All the other parameters are similar to those in the $M/M/c/K$ queuing system with c servers, system capacity K, exponentially distributed interarrival time, i.e., T_I, with $E[T_I] = 1/\lambda$ and exponentially distributed service time, i.e., T_S, with $E[T_S] = 1/\mu$.

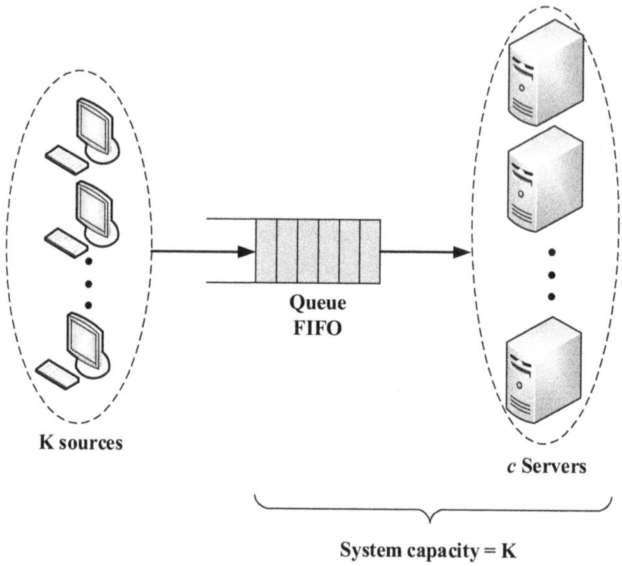

Figure 17.3 $M/M/c/K/K$ Queue

Remark 17.7. *With a finite source, it can be noticed that the arrival rate at the nth stage is reduced as the birth rate in the birth–death process with*

$$\lambda_n = \begin{cases} (K-n)\lambda & \text{if } 0 \le n \le K, \\ 0 & \text{if } n > K, \end{cases} \tag{17.30}$$

while the service rate at the nth stage increases as follows:

$$\mu_n = \begin{cases} n\mu & \text{if } 1 \le n \le c, \\ c\mu & \text{if } c \le n \le K. \end{cases} \tag{17.31}$$

Theorem 17.11. *The steady-state probability that there are n packets, $n = 0, 1, 2, \ldots, K$, in the $M/M/c/K/K$ queue is given by*

$$P_n = \begin{cases} \binom{K}{n}\rho^n \left(\sum\limits_{n=0}^{c-1} \binom{K}{n}\rho^n + \dfrac{\rho^c}{c!} \sum\limits_{n=c}^{K} \dfrac{K!}{(K-n)!} a^{n-c} \right)^{-1} & \text{if } 0 \le n \le c, \\ \dfrac{K! a^n c^c}{(K-n)! c!} \left(\sum\limits_{n=0}^{c-1} \binom{K}{n}\rho^n + \dfrac{\rho^c}{c!} \sum\limits_{n=c}^{K} \dfrac{K!}{(K-n)!} a^{n-c} \right)^{-1} & \text{if } c < n \le K, \end{cases} \tag{17.32}$$

where $a \triangleq \rho/c$.

Proof. From Theorem 15.1, we have

$$
\begin{aligned}
P_0 &= \left(1+\sum_{n=1}^{K}\frac{\prod_{k=0}^{n-1}\lambda_k}{\prod_{k=1}^{n}\mu_k}\right)^{-1}\\
&\overset{(a)}{=} \left(1+\sum_{n=1}^{c-1}\frac{K!\lambda^n}{(K-n)!n!\mu^n}+\sum_{n=c}^{K}\frac{K!\lambda^n}{(K-n)!c!c^{n-c}\mu^n}\right)^{-1}\\
&\overset{(b)}{=} \left(\sum_{n=0}^{c-1}\binom{K}{n}\rho^n+\frac{\rho^c}{c!}\sum_{n=c}^{K}\frac{K!}{(K-n)!}a^{n-c}\right)^{-1},
\end{aligned}
$$

where (a) follows from (17.30), (17.31), and $\prod_{k=0}^{n-1}(K-n)=(K!/(K-n)!)$ and (b) is due to binomial coefficient $\binom{K}{n}\triangleq K!/(n!(K-n)!)$ and $a=\rho/c$.

Substituting P_0 into (15.23), we obtain

$$
\begin{aligned}
P_n &= P_0\frac{\prod_{k=0}^{n-1}\lambda_k}{\prod_{k=1}^{n}\mu_k}\\
&= \begin{cases} P_0\dfrac{K!\lambda^n}{(K-n)!n!\mu^n} & \text{if } 0\le n\le c,\\ P_0\dfrac{K!\lambda^n}{(K-n)!c!c^{n-c}\mu^n} & \text{if } c<n\le K\end{cases}\\
&= \begin{cases} P_0\binom{K}{n}\rho^n & \text{if } 0\le n\le c,\\ P_0\dfrac{K!a^nc^c}{(K-n)!c!} & \text{if } c<n\le K\end{cases}\\
&= \begin{cases} \binom{K}{n}\rho^n\left(\sum_{n=0}^{c-1}\binom{K}{n}\rho^n+\dfrac{\rho^c}{c!}\sum_{n=c}^{K}\dfrac{K!}{(K-n)!}a^{n-c}\right)^{-1} & \text{if } 0\le n\le c,\\ \dfrac{K!a^nc^c}{(K-n)!c!}\left(\sum_{n=0}^{c-1}\binom{K}{n}\rho^n+\dfrac{\rho^c}{c!}\sum_{n=c}^{K}\dfrac{K!}{(K-n)!}a^{n-c}\right)^{-1} & \text{if } c<n\le K,\end{cases}
\end{aligned}
$$

□

Theorem 17.12. *The waiting probability of a packet in $M/M/c/K/K$ queue is given by*

$$
P_W=\sum_{n=c}^{K}\frac{K!a^nc^c}{(K-n)!c!}\left(\sum_{n=0}^{c-1}\binom{K}{n}\rho^n+\frac{\rho^c}{c!}\sum_{n=c}^{K}\frac{K!}{(K-n)!}a^{n-c}\right)^{-1}. \tag{17.33}
$$

Proof. The proof is similar to that of Theorem 17.7. □

Theorem 17.13. *The average number of packets in the system and in the queue of $M/M/c/K/K$ system is, respectively, given by*

$$E[N] = \sum_{n=0}^{c} n\binom{K}{n}\rho^n \left(\sum_{n=0}^{c-1}\binom{K}{n}\rho^n + \frac{\rho^c}{c!}\sum_{n=c}^{K}\frac{K!}{(K-n)!}a^{n-c}\right)^{-1} + \sum_{n=c+1}^{K} n\frac{K!a^n c^c}{(K-n)!c!}\left(\sum_{n=0}^{c-1}\binom{K}{n}\rho^n + \frac{\rho^c}{c!}\sum_{n=c}^{K}\frac{K!}{(K-n)!}a^{n-c}\right)^{-1}, \tag{17.34}$$

$$E[N_Q] = \sum_{n=c+1}^{K}(n-c)\frac{K!a^n c^c}{(K-n)!c!}\left(\sum_{n=0}^{c-1}\binom{K}{n}\rho^n + \frac{\rho^c}{c!}\sum_{n=c}^{K}\frac{K!}{(K-n)!}a^{n-c}\right)^{-1}. \tag{17.35}$$

Proof. We have

$$\begin{aligned} E[N] &= \sum_{n=0}^{K} nP_n \\ &\overset{(a)}{=} \sum_{n=0}^{c} n\binom{K}{n}\rho^n \left(\sum_{n=0}^{c-1}\binom{K}{n}\rho^n + \frac{\rho^c}{c!}\sum_{n=c}^{K}\frac{K!}{(K-n)!}a^{n-c}\right)^{-1} \\ &\quad + \sum_{n=c+1}^{K} n\frac{K!a^n c^c}{(K-n)!c!}\left(\sum_{n=0}^{c-1}\binom{K}{n}\rho^n + \frac{\rho^c}{c!}\sum_{n=c}^{K}\frac{K!}{(K-n)!}a^{n-c}\right)^{-1}, \end{aligned}$$

where (a) follows from (17.32).

We also have

$$\begin{aligned} E[N_Q] &= \sum_{n=c+1}^{K}(n-c)P_n \\ &= \sum_{n=c+1}^{K}(n-c)\frac{K!a^n c^c}{(K-n)!c!}\left(\sum_{n=0}^{c-1}\binom{K}{n}\rho^n + \frac{\rho^c}{c!}\sum_{n=c}^{K}\frac{K!}{(K-n)!}a^{n-c}\right)^{-1}. \end{aligned}$$

□

Remark 17.8. *Notice that in an $M/M/c/K/K$ queue, apart from the packets in the system, there are packets waiting in the source. Let N_S denote the number of packets in the source. The average number of packets in the source can be determined by*

$$E[N_S] = K - E[N], \tag{17.36}$$

where $E[N]$ is given by (17.34).

The utilization of the sources, denoted by U_{Source}, *can be thus given by*

$$\begin{aligned} U_{Source} &= \frac{E[N_S]}{K} \\ &= 1 - \frac{E[N]}{K}. \end{aligned} \tag{17.37}$$

Theorem 17.14. *The average response time and waiting time in* $M/M/c/K/K$ *queue are, respectively, given by*

$$E[T_R] = \frac{1}{\lambda}\frac{E[N]}{K - E[N]}, \tag{17.38}$$

$$E[T_W] = \frac{1}{\lambda}\frac{E[N_Q]}{K - E[N]}, \tag{17.39}$$

where $E[N]$ *and* $E[N_Q]$ *are given by* (17.34) *and* (17.35), *respectively.*

Proof. The proof follows from the Little's law in Section 15.4.5 with

$$E[T_R] = \frac{E[N]}{\lambda E[N_S]},$$

$$E[T_W] = \frac{E[N_Q]}{\lambda E[N_S]},$$

where $E[N_S]$ is given by (17.36). □

Remark 17.9. *We can similarly obtain the following measures of the* $M/M/c/K/K$ *queue:*

- *Utilization of all servers:*

$$\begin{aligned} U_A &= \rho E[N_S] \\ &= \rho(K - E[N]). \end{aligned} \tag{17.40}$$

- *System utilization:*

$$\begin{aligned} U_{System} &= 1 - P_0 \\ &= 1 - \left(\sum_{n=0}^{c-1} \binom{K}{n} \rho^n + \frac{\rho^c}{c!} \sum_{n=c}^{K} \frac{K!}{(K-n)!} a^{n-c} \right)^{-1}. \end{aligned} \tag{17.41}$$

- *The average number of busy servers:*

$$E[c_B] = U_A = \rho(K - E[N]). \tag{17.42}$$

- *The average number of idle servers:*

$$E[c_I] = c - \rho(K - E[N]). \tag{17.43}$$

Theorem 17.15. *Server utilization and the average busy period of the server in $M/M/c/K/K$ queue are, respectively, given by*

$$U_S = a(K - E[N]), \tag{17.44}$$

$$E[T_B] = \frac{1}{K\lambda}\left(\left(\sum_{n=0}^{c-1}\binom{K}{n}\rho^n + \frac{\rho^c}{c!}\sum_{n=c}^{K}\frac{K!}{(K-n)!}a^{n-c}\right) - 1\right). \tag{17.45}$$

Proof. We have

$$U_A = cU_S,$$

and thus

$$\begin{aligned} U_S &= \frac{U_A}{c} \\ &= a(K - E[N]). \end{aligned}$$

From (15.22) with a notice that the average idle time of the server, i.e., $E[T_0] = 1/(K\lambda)$, we can obtain $E[T_B]$ as

$$\begin{aligned} E[T_B] &= \frac{1 - P_0}{K\lambda P_0} \\ &= \frac{1}{K\lambda}\left(\left(\sum_{n=0}^{c-1}\binom{K}{n}\rho^n + \frac{\rho^c}{c!}\sum_{n=c}^{K}\frac{K!}{(K-n)!}a^{n-c}\right) - 1\right). \end{aligned}$$

□

17.4 Related works

For more information, the readers are suggested to refer to the following selected references: Takacs in 1962 [268], Kleinrock in 1975 [270], Daley and Servi in 1998 [288], Artalejo and Lopez-Herrero in 2001 [289], Elhanany and Kahane in 2005 [290], Dudin and Kim in 2017 [291], Fidler *et al.* in 2018 [292], and Jiang *et al.* in 2018 [293].

17.5 Review questions

[Q17.1] Describe the following multi-server queuing systems: $M/M/c$, $M/M/c/K$, and $M/M/c/K/K$. Provide an illustrative example for each of them.

[Q17.2] What is the stability condition for a multi-server queue?

[Q17.3] Derive the steady-state probability that there are n packets in $M/M/c$, $M/M/c/K$, and $M/M/c/K/K$ queues.

[Q17.4] Derive the average number of packets in the system and in the queue of $M/M/c$, $M/M/c/K$, and $M/M/c/K/K$ queues.

[Q17.5] Derive the average response time and waiting time in $M/M/c$, $M/M/c/K$, and $M/M/c/K/K$ queues.

[Q17.6] Derive server utilization and system utilization of $M/M/c$, $M/M/c/K$, and $M/M/c/K/K$ queues.

[Q17.7] Derive the average busy period of the server in $M/M/c$, $M/M/c/K$ and $M/M/c/K/K$ queues.

[Q17.8] When is a multi-server queue blocked?

[Q17.9] Derive the blocking probability of an $M/M/c/K$ queue.

[Q17.10] Derive the average number of packets in the source and the utilization of the source in an $M/M/c/K/K$ queue.

17.6 Problems

[P17.1] Derive the waiting probability of data packets in $M/M/c$, $M/M/c/K$, and $M/M/c/K/K$ queuing systems.

[P17.2] At a company, two mail servers are working in parallel. E-mails arrive on an average of every 1 min throughout the day. It takes an average of 2 s to process an e-mail at the servers. Both interarrival time and service time are exponentially distributed. The queue is assumed to be infinitely large and the serving discipline is FIFO.

- Give the Kendall's notation describing this system.
- Find the probability that there is no e-mail to be processed at the servers.
- Find the average number of e-mails waiting in the queue and in the system.
- Find the average response time of the servers, the average waiting time of the e-mails, and the average busy time of the servers.
- Find the server and system utilizations.
- Discuss the impacts on performance measures of the queue when a third e-mail server working at the same speed as the first two servers is added to the system.

[P17.3] A queuing system consists of three servers with a finite capacity of six packets. Both the interarrival and service time are exponentially distributed with the average interarrival of 6 s and the average service time of 9 s.

- Give the Kendall's notation describing this system.
- Find the probability that all the servers are idle, the blocking probability of the queuing system, and the waiting probability of the packets.
- Find the average number of packets in the system and in the queue.
- Find the average response time and waiting time of the data packets.
- Find the server utilization, system utilization, and the average busy period of the servers.

[P17.4] Considering a finite source in a four-server queuing system with a limited capacity of eight packets and the same population size of eight packets. Both the packet arrival and service follow Poisson point process with an arrival rate of 20 packets/s and a service rate at each server of 10 packets/s.

- Give the Kendall's notation describing this system.

- Find the probability that all the servers are busy and the waiting probability of the packets.
- Find the average number of packets in the system, in the queue, and in the source.
- Find the average response time and waiting time of the data packets.
- Find the server utilization, source utilization, and system utilization.
- Find the average idle period and busy period of the server.

References

[1] D. L. Spohn, T. Brown, and S. Grau, *Data Network Design*, 3rd ed. New York, NY, USA: McGraw-Hill, Inc., 2002.

[2] W. Stallings, *Data and Computer Communications*, 8th ed. Upper Saddle River, NJ, USA: Prentice-Hall, Inc., 2006.

[3] M. Dye, R. McDonald, and A. Rufi, *Network Fundamentals, CCNA Exploration Companion Guide*, 1st ed. Cisco Press, Indiana, USA, 2011.

[4] A. Fraser, "Towards a universal data transport system," *IEEE J. Sel. Areas Commun.*, vol. 1, no. 5, pp. 803–816, 1983.

[5] T.-S. Chu and M. Gans, "High speed infrared local wireless communication," *IEEE Commun. Mag.*, vol. 25, no. 8, pp. 4–10, 1987.

[6] K. Pahlavan, T. H. Probert, and M. E. Chase, "Trends in local wireless networks," *IEEE Commun. Mag.*, vol. 33, no. 3, pp. 88–95, 1995.

[7] B. P. Crow, I. Widjaja, J. G. Kim, and P. T. Sakai, "IEEE 802.11 wireless local area networks," *IEEE Commun. Mag.*, vol. 35, no. 9, pp. 116–126, 1997.

[8] J. Woodcock, *Step Up to Networking (IT Professional)*. Microsoft Press, USA, 1999.

[9] K. Clark and K. Hamilton, *CCIE Professional Development: Cisco LAN Switching*. Cisco Press, Indiana, USA, 1999.

[10] J. Bray and C. Sturman, *Bluetooth: Connect Without Cables*, 1st ed. Upper Saddle River, NJ, USA: Prentice Hall PTR, 2000.

[11] W. Stallings, *Wireless Communications and Networks*, 1st ed. Prentice Hall Professional Technical Reference, New Jersey, USA, 2001.

[12] P. Roshan and J. Leary, *802.11 Wireless LAN Fundamentals*. Cisco Press, Indiana, USA, 2003.

[13] M. Decina and E. Scace, "CCITT recommendations on the ISDN: A review," *IEEE J. Sel. Areas Commun.*, vol. 4, no. 3, pp. 320–325, 1986.

[14] J. W. Lechleider, "High bit rate digital subscriber lines: A review of HDSL progress," *IEEE J. Sel. Areas Commun.*, vol. 9, no. 6, pp. 769–784, 1991.

[15] N. M. Thorpe and D. Ross, *X.25 Made Easy*. Upper Saddle River, NJ, USA: Prentice-Hall, Inc., 1992.

[16] D. E. McDysan and D. L. Spohn, *ATM: Theory and Application*. New York, NY, USA: McGraw-Hill, Inc., 1994.

[17] C. Brown and A. Malis, "RFC 2427: Multiprotocol Interconnect over Frame Relay," IETF, Tech. Rep., 1998.

[18] D. Ginsburg, *ATM: Solutions for Enterprise Internetworking*, 2nd ed. Boston, MA, USA: Addison-Wesley Longman Publishing Co., Inc., 1999.

[19] J. C. Bellamy, *Digital Telephony*, 3rd ed. Wiley, New York, USA, 2000.

[20] ITU-T, "Information technology—open systems interconnection—basic reference model: The basic model," ITU, Recommendation X.200, 1994.
[21] V. G. Cerf and R. E. Khan, "A protocol for packet network intercommunication," *IEEE Trans. Commun.*, vol. 22, pp. 637–648, 1974.
[22] A. Rodriguez, J. Gatell, J. Karas, and R. Peschke, *TCP/IP Tutorial and Technical Overview*, 7th ed. Upper Saddle River, NJ, USA: Prentice-Hall, Inc., 2001.
[23] D. Boggs, J. Shoch, E. Taft, and R. Metcalfe, "Pup: An internetwork architecture," *IEEE Trans. Commun.*, vol. 28, no. 4, pp. 612–624, 1980.
[24] D. Teare, Ed., *Designing Cisco Networks*. Cisco Press, Indiana, USA, 1999.
[25] W. Stallings, *Computer Networking with Internet Protocols and Technology*. Pearson, New Jersey, USA, 2003.
[26] M. A. Sportack and J. Fairweather, *IP Routing Fundamentals*. Cisco Press, Indiana, USA, 1999.
[27] J. Doyle and J. Carroll, *Routing TCP/IP, Volume 1*, 2nd ed. Cisco Press, Indiana, USA, 2005.
[28] R. Graziani and A. Johnson, *Routing Protocols and Concepts, CCNA Exploration Companion Guide*. Cisco Press, Indiana, USA, 2007.
[29] L. R. J. Ford, *Network Flow Theory*. Santa Monica, CA, USA: RAND Corporation, 1956.
[30] L. R. J. Ford and D. R. Fulkerson, "Maximal flow through a network," *Can. J. Math.*, vol. 8, pp. 399–404, 1956.
[31] R. Bellman, "On a routing problem," *Q. Appl. Math.*, vol. 16, no. 1, pp. 87–90, 1958.
[32] C. Hedrick, "RFC 1058: Routing Information Protocol," IETF, Tech. Rep., 1988.
[33] G. Malkin and R. Minnear, "RFC 2080: RIPng for IPv6," IETF, Tech. Rep., 1997.
[34] G. Malkin, "RFC 2453: RIP Version 2," IETF, Tech. Rep., 1998.
[35] C. L. H. Rutgers, "An Introduction to IGRP (White Paper)," Cisco Systems, Inc., Tech. Rep. 26825, 1991.
[36] Cisco, "IGRP Metric (Troubleshooting TechNotes)," Cisco Systems, Inc., Tech. Rep. 13678, 2005.
[37] T. Clausen and P. Jacquet, "RFC 3626: Optimized Link State Routing Protocol (OLSR)," IETF, Tech. Rep., 2003.
[38] J. T. Moy, *OSPF: Anatomy of an Internet Routing Protocol*. Boston, MA, USA: Addison-Wesley Longman Publishing Co., Inc., 1998.
[39] J. Moy, "RFC 2328: OSPF Version 2," IETF, Tech. Rep., 1998.
[40] R. Coltun, D. Ferguson, J. Moy, and A. Lindem, "RFC 5340: OSPF for IPv6," IETF, Tech. Rep., 2008.
[41] E. W. Dijkstra, "A note on two problems in connexion with graphs," *Numer. Math.*, vol. 1, no. 1, pp. 269–271, 1959.
[42] J. J. Garcia-Lunes-Aceves, "Loop-free routing using diffusing computations," *IEEE/ACM Trans. Networking*, vol. 1, no. 1, pp. 130–141, 1993.
[43] I. Pepelnjak, *EIGRP Network Design Solutions*. Cisco Press, Indiana, USA, 1999.

[44] D. Savage, J. Ng, S. Moore, D. Slice, P. Paluch, and R. White, "RFC 7868: Cisco's Enhanced Interior Gateway Routing Protocol (EIGRP)," IETF, Tech. Rep., 2016.
[45] Y. Rekhter and P. Gross, "RFC 1772: Application of the Border Gateway Protocol in the Internet," IETF, Tech. Rep., 1995.
[46] Y. Rekhter, T. Li, and S. Hares, "RFC 4271: A Border Gateway Protocol 4 (BGP-4)," IETF, Tech. Rep., 2006.
[47] S. Halabi and D. McPherson, *Internet Routing Architectures*, 2nd ed. Cisco Press, Indiana, USA, 2000.
[48] R. White, D. McPherson, and S. Sangli, *Practical BGP*. Redwood City, CA, USA: Addison Wesley Longman Publishing Co., Inc., 2004.
[49] T. Bates, E. Chen, and R. Chandra, "RFC 4456: BGP Route Reflection: An Alternative to Full Mesh Internal BGP (iBGP)," IETF, Tech. Rep., 2006.
[50] P. Traina, D. McPherson, and J. Scudder, "RFC 5065: Autonomous System Confederations for BGP," IETF, Tech. Rep., 2007.
[51] J. Postel, "RFC 791: Internet Protocol—DARPA Internet Program Protocol Specification," IETF, Tech. Rep., 1981.
[52] P. Mockapetris, "RFC 1035: Domain Names—Implementation and Specification," IETF, Tech. Rep., 1987.
[53] R. Braden, "RFC 1122: Requirements for Internet Hosts—Communication Layers," IETF, Tech. Rep., 1989.
[54] K. Egevang and P. Francis, "RFC 1631: The IP Network Address Translator (NAT)," IETF, Tech. Rep., 1994.
[55] Y. Rekhter, B. Moskowitz, D. Karrenberg, G. J. de Groot, and E. Lear, "RFC 1918: Address Allocation for Private Internets," IETF, Tech. Rep., 1996.
[56] M. Cotton, L. Vegoda, R. Bonica, and B. Haberman, "RFC 6890: Special-Purpose IP Address Registries," IETF, Tech. Rep., 2013.
[57] C. Perkins, "RFC 2003: IP Encapsulation within IP," IETF, Tech. Rep., 1996.
[58] C. Perkins, "RFC 5944: IP Mobility Support for IPv4, Revised," IETF, Tech. Rep., 2010.
[59] J. Touch, "RFC 6864: Updated Specification of the IPv4 ID Field," IETF, Tech. Rep., 2013.
[60] J. Mogul, "RFC 917: Internet Subnets," IETF, Tech. Rep., 1984.
[61] P. Tsuchiya, "RFC 1219: On the Assignment of Subnet Numbers," IETF, Tech. Rep., 1991.
[62] T. Pummill and B. Manning, "RFC 1878: Variable Length Subnet Table For IPv4," IETF, Tech. Rep., 1995.
[63] Y. Rekhter and T. Li, "RFC 1518: An Architecture for IP Address Allocation with CIDR," IETF, Tech. Rep., 1993.
[64] Y. Rekhter, "RFC 1817: CIDR and Classful Routing," IETF, Tech. Rep., 1995.
[65] V. Fuller and T. Li, "RFC 4632: Classless Inter-domain Routing (CIDR): The Internet Address Assignment and Aggregation Plan," IETF, Tech. Rep., 2006.
[66] R. Hinden and B. Haberman, "RFC 4193: Unique Local IPv6 Unicast Addresses," IETF, Tech. Rep., 2005.

[67] R. Hinden and S. Deering, "RFC 4291: IP Version 6 Addressing Architecture," IETF, Tech. Rep., 2006.

[68] M. Boucadair and S. Venaas, "RFC 7371: Updates to the IPv6 Multicast Addressing Architecture," IETF, Tech. Rep., 2014.

[69] S. Deering and R. Hinden, "RFC 8200: Internet Protocol, Version 6 (IPv6) Specification," IETF, Tech. Rep., 2017.

[70] D. Minoli, *Broadband Network Analysis and Design*. Norwood, MA, USA: Artech House, Inc., 1993.

[71] C. Partridge, *Gigabit Networking*. Boston, MA, USA: Addison-Wesley Longman Publishing Co., Inc., 1994.

[72] M. Pióro and D. Medhi, *Routing, Flow, and Capacity Design in Communication and Computer Networks*. San Francisco, CA, USA: Morgan Kaufmann Publishers Inc., 2004.

[73] J. McCabe, Ed., *Network Analysis, Architecture, and Design*, 3rd ed. Morgan Kaufmann, Massachusetts, USA, 2007.

[74] K. Raza and M. Turner, *Large-scale IP Network Solutions*. Cisco Press, Indiana, USA, 1999.

[75] E. Ravasz and A.-L. Barabasi, "Hierarchical organization in complex networks," *Phys. Rev. E*, vol. 67, no. 2, p. 026112, 2003.

[76] Cisco Networking Academy, *Connecting Networks Companion Guide*. Cisco Press, Indiana, USA, 2014.

[77] P. Oppenheimer, *Top-Down Network Design*, 3rd ed. Cisco Press, Indiana, USA, 2010.

[78] R. White, A. Retana, and D. Slice, *Optimal Routing Design*, 1st ed. Cisco Press, Indiana, USA, 2010.

[79] D. E. Robling Denning, *Cryptography and Data Security*. Boston, MA, USA: Addison-Wesley Longman Publishing Co., Inc., 1982.

[80] W. Stallings, *Cryptography and Network Security: Principles and Practice*, 3rd ed. Pearson Education, New Jersey, USA, 2002.

[81] S. Kent and K. Seo, "RFC 4301: Security Architecture for the Internet Protocol," IETF, Tech. Rep., 2005.

[82] W. Jiang, H. Song, and Y. Dai, "Real-time intrusion detection for high-speed networks," *Comput. Secur.*, vol. 24, no. 4, pp. 287–294, 2005.

[83] Y. S. Shiu, S. Y. Chang, H. C. Wu, S. C. H. Huang, and H. H. Chen, "Physical layer security in wireless networks: A tutorial," *IEEE Wireless Commun.*, vol. 18, no. 2, pp. 66–74, 2011.

[84] H.-J. Liao, C.-H. R. Lin, Y.-C. Lin, and K.-Y. Tung, "Intrusion detection system: A comprehensive review," *J. Network Comput. Appl.*, vol. 36, no. 1, pp. 16–24, 2013.

[85] J. Pescatore, "DDoS attacks advancing and enduring: A SANS survey," Feb. 2014. [Online]. Available: https://www.sans.org/reading-room/whitepapers/analyst/ddos-attacks-advancing-enduring-survey-34700.

[86] R. K. Deka, K. P. Kalita, D. Bhattacharya, and J. K. Kalita, "Network defense: Approaches, methods and techniques," *J. Network Comput. Appl.*, vol. 57, pp. 71–84, 2015.

[87] T. A. Le, H. X. Nguyen, Q.-T. Vien, and M. Karamanoglu, "Secure information transmission and power transfer in cellular networks," *IEEE Commun. Lett.*, vol. 19, no. 9, pp. 1532–1535, 2015.

[88] T. A. Le, H. X. Nguyen, Q. T. Vien, and M. Karamanoglu, "Secure information transmission in the presence of energy-harvesting eavesdroppers in multi-cell networks," in *2015 IEEE Global Communications Conference (GLOBECOM)*, 2015, pp. 1–6.

[89] T. A. Le, Q. T. Vien, H. X. Nguyen, D. W. K. Ng, and R. Schober, "Robust optimization with probabilistic constraints for power-efficient and secure SWIPT," in *2016 IEEE Global Communications Conference (GLOBECOM)*, 2016, pp. 1–7.

[90] T. A. Le, Q.-T. Vien, H. X. Nguyen, D. W. K. Ng, and R. Schober, "Robust chance-constrained optimization for power-efficient and secure SWIPT systems," *IEEE Trans. Green Commun. Networking*, vol. 1, no. 3, pp. 333–346, 2017.

[91] Q. T. Vien, T. A. Le, H. X. Nguyen, and H. Phan, "A secure network coding based modify-and-forward scheme for cooperative wireless relay networks," in *2016 IEEE 83rd Vehicular Technology Conference (VTC Spring)*, 2016, pp. 1–5.

[92] Q. T. Vien, T. A. Le, and T. Q. Duong, "Opportunistic secure transmission for wireless relay networks with modify-and-forward protocol," in *2017 IEEE International Conference on Communications (ICC)*, 2017, pp. 1–6.

[93] Q. T. Vien, T. A. Le, X. S. Yang, and T. Q. Duong, "On the handover security key update and residence management in LTE networks," in *2017 IEEE Wireless Communications and Networking Conference (WCNC)*, 2017, pp. 1–6.

[94] I. Karim and Q.-T. Vien, *Snort-based Intrusion Detection System for Practical Computer Networks: Implementation and Comparative Study*. LAP LAMBERT Academic Publishing, Saarland, Germany, 2017.

[95] D. Bertsekas and R. Gallager, *Data Networks*, 2nd ed. Upper Saddle River, NJ, USA: Prentice-Hall, Inc., 1992.

[96] C. White, *Data Communications and Computer Networks: A Business User's Approach*, 6th ed. Boston, MA, USA: Course Technology Press, 2010.

[97] R. Braden, D. Clark, and S. Shenker, "RFC 1633: Integrated Services in the Internet Architecture: An Overview," IETF, Tech. Rep., 1994.

[98] G. Huston, "RFC 2990: Next Steps for the IP QoS Architecture," IETF, Tech. Rep., 2000.

[99] J. W. Evans and C. Filsfils, *Deploying IP and MPLS QoS for Multiservice Networks: Theory & Practice*. San Francisco, CA, USA: Morgan Kaufmann Publishers Inc., 2007.

[100] M. Marchese, *QoS Over Heterogeneous Networks*. Wiley Publishing, New York, USA, 2007.

[101] T. Braun, M. Diaz, J. Enrquez-Gabeiras, and T. Staub, *End-to-End Quality of Service Over Heterogeneous Networks*, 1st ed. Springer Publishing Company, Incorporated, New York, USA, 2008.

[102] D. Papadimitriou, "RFC 6003: Ethernet Traffic Parameters," IETF, Tech. Rep., 2010.

[103] F. Baker, J. Polk, and M. Dolly, "RFC 5865: A Differentiated Services Code Point (DSCP) for Capacity-Admitted Traffic," IETF, Tech. Rep., 2010.

[104] G. Rubino, *Network Reliability Evaluation*. Newark, NJ, USA: Gordon and Breach Science Publishers, Inc., 1999, pp. 275–302.

[105] J. G. Proakis and M. Salehi, *Communication Systems Engineering*, 2nd ed. Upper Saddle River, NJ, USA: Prentice-Hall, Inc., 2001.

[106] J. G. Proakis and M. Salehi, *Digital Communications*, 5th ed. McGraw-Hill, New York, USA, 2007.

[107] H. Nyquist, "Certain topics in telegraph transmission theory," *Trans. Am. Inst. Electr. Eng.*, vol. 47, no. 2, pp. 617–644, 1928.

[108] C. E. Shannon, "A mathematical theory of communication," *Bell Syst. Tech. J.*, vol. 27, no. 3, pp. 379–423, 1948.

[109] C. E. Shannon, "Communication in the presence of noise," *Proc. IRE*, vol. 37, no. 1, pp. 10–21, 1949.

[110] R. J. Marks, II, *Introduction to Shannon Sampling and Interpolation Theory*. Berlin, Heidelberg: Springer-Verlag, 1991.

[111] T. M. Cover and J. A. Thomas, *Elements of Information Theory (Wiley Series in Telecommunications and Signal Processing)*. New York, NY, USA: Wiley-Interscience, 2006.

[112] R. Ramaswamy, N. Weng, and T. Wolf, "Characterizing network processing delay," in *Global Telecommunications Conference, 2004. GLOBECOM '04. IEEE*, vol. 3, 2004, pp. 1629–1634.

[113] J. F. Kurose and K. W. Ross, *Computer Networking: A Top-Down Approach*, 6th ed. Pearson, New Jersey, USA, 2012.

[114] R. Jain, *The Art of Computer Systems Performance Analysis: Techniques for Experimental Design, Measurement, Simulation, and Modeling*, 1st ed. John Wiley & Sons, New York, USA, 1991.

[115] G. Kotonya and I. Sommerville, *Requirements Engineering: Processes and Techniques*, 1st ed. Wiley Publishing, New York, USA, 1998.

[116] D. C. Hay, *Requirements Analysis: From Business Views to Architecture*, 1st ed. Prentice Hall, New Jersey, USA, 2003.

[117] P. A. Laplante, *Requirements Engineering for Software and Systems*, 2nd ed. Boston, MA, USA: Auerbach Publications, 2013.

[118] D. Smith, *Reliability, Maintainability and Risk*, 8th ed. Butterworth-Heinemann, Oxford, UK, 2011.

[119] M. L. Ayers, *Telecommunications System Reliability Engineering, Theory, and Practice*, 1st ed. Wiley-IEEE Press, New York, USA, 2012.

[120] E. A. Elsayed, *Reliability Engineering*, 2nd ed. Wiley, New York, USA, 2012.

[121] J. Lienig and H. Bruemmer, *Reliability Analysis*. Cham: Springer International Publishing, 2017, pp. 45–73.

[122] C. G. Omidyar and G. Pujolle, "Guest editorial—Introduction to flow and congestion control," *IEEE Commun. Mag.*, vol. 34, no. 11, pp. 30, 1996.

[123] N. Brownlee, C. Mills, and G. Ruth, "RFC 2722: Traffic Flow Measurement: Architecture," IETF, Tech. Rep., 1999.

[124] S. Floyd and K. Fall, "Promoting the use of end-to-end congestion control in the Internet," *IEEE/ACM Trans. Networking*, vol. 7, no. 4, pp. 458–472, 1999.

[125] A. E. Gamal and Y.-H. Kim, *Network Information Theory*. New York, NY, USA: Cambridge University Press, 2012.

[126] P. L. Frana, "Telematics and the early history of international digital information flows," *IEEE Ann. Hist. Comput.*, vol. 40, no. 2, pp. 32–47, 2018.

[127] X. Shen, H. Yu, J. Buford, and M. Akon, *Handbook of Peer-to-Peer Networking*, 1st ed. Springer Publishing Company, Incorporated, New York, USA, 2009.

[128] F. H. P. Fitzek and H. Charaf, *Mobile Peer-to-Peer Networks: An Introduction to the Tutorial Guide*. Wiley Telecom, New York, USA, 2009.

[129] C. G. Gurler and M. Tekalp, "Peer-to-peer system design for adaptive 3D video streaming," *IEEE Commun. Mag.*, vol. 51, no. 5, pp. 108–114, 2013.

[130] N. Kumar and J. H. Lee, "Peer-to-peer cooperative caching for data dissemination in urban vehicular communications," *IEEE Syst. J.*, vol. 8, no. 4, pp. 1136–1144, 2014.

[131] Z. Yin, C. Wu, Z. Yang, and Y. Liu, "Peer-to-peer indoor navigation using smartphones," *IEEE J. Sel. Areas Commun.*, vol. 35, no. 5, pp. 1141–1153, 2017.

[132] W. Wu, H. Mao, Y. Wang, J. Wang, W. Wang, and C. Tian, "CoolConferencing: Enabling robust peer-to-peer multi-party video conferencing," *IEEE Access*, vol. 5, pp. 25474–25486, 2017.

[133] C. Jiang, L. Gao, L. Duan, and J. Huang, "Scalable mobile crowdsensing via peer-to-peer data sharing," *IEEE Trans. Mob. Comput.*, vol. 17, no. 4, pp. 898–912, 2018.

[134] A. Asadi, Q. Wang, and V. Mancuso, "A survey on device-to-device communication in cellular networks," *IEEE Commun. Surv. Tutorials*, vol. 16, no. 4, pp. 1801–1819, Fourthquarter 2014.

[135] R. Trestian, Q. T. Vien, H. X. Nguyen, and O. Gemikonakli, "ECO-M: Energy-efficient cluster-oriented multimedia delivery in a LTE D2D environment," in *2015 IEEE International Conference on Communications (ICC)*, 2015, pp. 55–61.

[136] Y. Wu, S. Wang, W. Liu, W. Guo, and X. Chu, "Iunius: A cross-layer peer-to-peer system with device-to-device communications," *IEEE Trans. Wireless Commun.*, vol. 15, no. 10, pp. 7005–7017, 2016.

[137] K. Ali, H. X. Nguyen, P. Shah, Q. T. Vien, and N. Bhuvanasundaram, "Architecture for public safety network using D2D communication," in *2016 IEEE Wireless Communications and Networking Conference*, 2016, pp. 1–6.

[138] K. Ali, H. X. Nguyen, P. Shah, Q. T. Vien, and E. Ever, "D2D multi-hop relaying services towards disaster communication system," in *2017 24th International Conference on Telecommunications (ICT)*, 2017, pp. 1–5.

[139] K. Ali, H. X. Nguyen, Q. T. Vien, P. Shah, and Z. Chu, "Disaster management using D2D communication with power transfer and clustering techniques," *IEEE Access*, vol. 6, pp. 14643–14654, 2018.

[140] A. Sinha, "Client-server computing," *Commun. ACM*, vol. 35, no. 7, pp. 77–98, 1992.

[141] S. Ramesh and H. G. Perros, "A multilayer client-server queueing network model with synchronous and asynchronous messages," *IEEE Trans. Softw. Eng.*, vol. 26, no. 11, pp. 1086–1100, 2000.

[142] V. Kanitkar and A. Delis, "Real-time processing in client-server databases," *IEEE Trans. Comput.*, vol. 51, no. 3, pp. 269–288, 2002.

[143] I. Abdul-Fatah and S. Majumdar, "Performance of CORBA-based client-server architectures," *IEEE Trans. Parallel Distrib. Syst.*, vol. 13, no. 2, pp. 111–127, 2002.

[144] Y. Zhu, W. Wu, and D. Li, "Efficient client assignment for client-server systems," *IEEE Trans. Netw. Serv. Manage.*, vol. 13, no. 4, pp. 835–847, 2016.

[145] J. M. Taylor, "Co-operative computing and control," *IEE Proc. E— Comput. Digital Tech.*, vol. 137, no. 1, pp. 1–16, 1990.

[146] T. Richardson, Q. Stafford-Fraser, K. R. Wood, and A. Hopper, "Virtual network computing," *IEEE Internet Comput.*, vol. 2, no. 1, pp. 33–38, 1998.

[147] Q. Wu, B. Qin, L. Zhang, J. Domingo-Ferrer, and J. A. Manjon, "Fast transmission to remote cooperative groups: A new key management paradigm," *IEEE/ACM Trans. Networking*, vol. 21, no. 2, pp. 621–633, 2013.

[148] C. You and K. Huang, "Exploiting non-causal CPU-state information for energy-efficient mobile cooperative computing," *IEEE Trans. Wireless Commun.*, vol. 17, no. 6, pp. 4104–4117, 2018.

[149] Z. Sheng, C. Mahapatra, V. C. M. Leung, M. Chen, and P. K. Sahu, "Energy efficient cooperative computing in mobile wireless sensor networks," *IEEE Trans. Cloud Comput.*, vol. 6, no. 1, pp. 114–126, 2018.

[150] T. L. Casavant and J. G. Kuhl, "A taxonomy of scheduling in general-purpose distributed computing systems," *IEEE Trans. Software Eng.*, vol. 14, no. 2, pp. 141–154, 1988.

[151] H. S. Morse, *Practical Parallel Computing*. San Diego, CA, USA: Academic Press Professional, Inc., 1994.

[152] G. Bianchi, "Performance analysis of the IEEE 802.11 distributed coordination function," *IEEE J. Sel. Areas Commun.*, vol. 18, no. 3, pp. 535–547, 2000.

[153] O. Younis and S. Fahmy, "HEED: A hybrid, energy-efficient, distributed clustering approach for ad hoc sensor networks," *IEEE Trans. Mob. Comput.*, vol. 3, no. 4, pp. 366–379, 2004.

[154] H. Nishida and T. Nguyen, "Optimal client-server assignment for internet distributed systems," *IEEE Trans. Parallel Distrib. Syst.*, vol. 24, no. 3, pp. 565–575, 2013.

[155] P. Kandaswamy, J. A. Flint, and V. A. Chouliaras, "System on fabrics architecture using distributed computing," *IEEE Sens. J.*, vol. 18, no. 14, pp. 5929–5936, 2018.

[156] E. D. Pascale, I. Macaluso, A. Nag, M. Kelly, and L. Doyle, "The network as a computer: A framework for distributed computing over IoT mesh networks," *IEEE Internet Things J.*, vol. 5, no. 3, pp. 2107–2119, 2018.

[157] D. Estrin, J. C. Mogul, and G. Tsudik, "Visa protocols for controlling interorganizational datagram flow," *IEEE J. Sel. Areas Commun.*, vol. 7, no. 4, pp. 486–498, 1989.

[158] B. Quoitin, C. Pelsser, L. Swinnen, O. Bonaventure, and S. Uhlig, "Interdomain traffic engineering with BGP," *IEEE Commun. Mag.*, vol. 41, no. 5, pp. 122–128, 2003.

[159] C. Partridge, "RFC 1363: A Proposed Flow Specification," IETF, Tech. Rep., 1992.

[160] P. Marques, N. Sheth, R. Raszuk, B. Greene, J. Mauch, and D. McPherson, "RFC 5575: Dissemination of Flow Specification Rules," IETF, Tech. Rep., 2009.

[161] R. L. Cruz, "A calculus for network delay. ii. Network analysis," *IEEE Trans. Inf. Theory*, vol. 37, no. 1, pp. 132–141, 1991.

[162] A. K. Parekh and R. G. Gallager, "A generalized processor sharing approach to flow control in integrated services networks: The single-node case," *IEEE/ACM Trans. Networking*, vol. 1, no. 3, pp. 344–357, 1993.

[163] C. C. Jane, J. S. Lin, and J. Yuan, "Reliability evaluation of a limited-flow network in terms of minimal cutsets," *IEEE Trans. Reliab.*, vol. 42, no. 3, pp. 354–361, 368, 1993.

[164] S. Sen and J. Wang, "Analyzing peer-to-peer traffic across large networks," *IEEE/ACM Trans. Networking*, vol. 12, no. 2, pp. 219–232, 2004.

[165] P. Popovski and H. Yomo, "Wireless network coding by amplify-and-forward for bi-directional traffic flows," *IEEE Commun. Lett.*, vol. 11, no. 1, pp. 16–18, 2007.

[166] S. Katti, H. Rahul, W. Hu, D. Katabi, M. Medard, and J. Crowcroft, "XORs in the air: Practical wireless network coding," *IEEE/ACM Trans. Networking*, vol. 16, no. 3, pp. 497–510, 2008.

[167] K. Sohn and D. Kim, "Dynamic origin-destination flow estimation using cellular communication system," *IEEE Trans. Veh. Technol.*, vol. 57, no. 5, pp. 2703–2713, 2008.

[168] Q. Liu and C. W. Chen, "Smart downlink scheduling for multimedia streaming over LTE networks with hard handoff," *IEEE Trans. Circuits Syst. Video Technol.*, vol. 25, no. 11, pp. 1815–1829, 2015.

[169] W. J. Dally, "Virtual-channel flow control," *IEEE Trans. Parallel Distrib. Syst.*, vol. 3, no. 2, pp. 194–205, 1992.

[170] N. A. Jagadeesan and B. Krishnamachari, "Software-defined networking paradigms in wireless networks: A survey," *ACM Comput. Surv.*, vol. 47, no. 2, pp. 27:1–27:11, 2014.

[171] A. Lara, A. Kolasani, and B. Ramamurthy, "Network innovation using OpenFlow: A survey," *IEEE Commun. Surv. Tutorials*, vol. 16, no. 1, pp. 493–512, First 2014.

[172] D. Kreutz, F. M. V. Ramos, P. E. Verissimo, C. E. Rothenberg, S. Azodolmolky, and S. Uhlig, "Software-defined networking: A comprehensive survey," *Proc. IEEE*, vol. 103, no. 1, pp. 14–76, 2015.

[173] M. A. Hassan, Q.-T. Vien, and M. Aiash, "Software defined networking for wireless sensor networks: A survey," *Adv. Wireless Commun. Networks*, vol. 3, no. 2, pp. 10–22, 2017.

[174] G. Li, H. Zhou, B. Feng, and G. Li, "Context-aware service function chaining and its cost-effective orchestration in multi-domain networks," *IEEE Access*, vol. 6, pp. 34976–34991, 2018.

[175] Y. Jia, C. Wu, Z. Li, F. Le, and A. Liu, "Online scaling of NFV service chains across Geo-distributed datacenters," *IEEE/ACM Trans. Networking*, vol. 26, no. 2, pp. 699–710, 2018.

[176] J. Gray, *Benchmark Handbook: For Database and Transaction Processing Systems*. San Francisco, CA, USA: Morgan Kaufmann Publishers Inc., 1992.

[177] R. Madhavan, E. Tunstel, and E. Messina, *Performance Evaluation and Benchmarking of Intelligent Systems*, 1st ed. Springer Publishing Company, Incorporated, New York, USA, 2009.

[178] M. Harchol-Balter, *Performance Modeling and Design of Computer Systems: Queueing Theory in Action*, 1st ed. New York, NY, USA: Cambridge University Press, 2013.

[179] R. Nambiar and M. Poess, *Performance Evaluation and Benchmarking: Traditional to Big Data to Internet of Things*. Springer International Publishing, New York, USA, 2016.

[180] Wolfram, "Mathematica." [Online]. Available: http://www.wolfram.com/mathematica/.

[181] MathWorks, "Matlab." [Online]. Available: https://uk.mathworks.com/products/matlab.html.

[182] M. L. Abell and J. P. Braselton, *Mathematica by Example*, 5th ed. Orlando, FL, USA: Academic Press, Inc., 2017.

[183] S. Wolfram, *An Elementary Introduction to the Wolfram Language*, 2nd ed. Wolfram Media, Inc., 2017.

[184] A. Varga, "Using the OMNeT++ discrete event simulation system in education," *IEEE Trans. Educ.*, vol. 42, no. 4, 11 pp, 1999.

[185] J. Banks, J. S. Carson, B. L. Nelson, and D. M. Nicol, *Discrete-Event System Simulation*, 5th ed. Pearson, New Jersey, USA, 2010.

[186] K. Wehrle, M. Gnes, and J. Gross, *Modeling and Tools for Network Simulation*, 1st ed. Springer Publishing Company, Incorporated, New York, USA, 2010.

[187] T. A. Davis, *MATLAB Primer*, 8th ed. Boca Raton, FL, USA: CRC Press, Inc., 2010.

[188] B. Rimoldi, *Principles of Digital Communication: A Top-Down Approach*. Cambridge University Press, Cambridge, UK, 2016.

[189] K. Choi and H. Liu, *Problem-Based Learning in Communication Systems Using MATLAB and Simulink.* John Wiley & Sons, Inc., New York, USA, 2016.

[190] S. Attaway, *MATLAB: A Practical Introduction to Programming and Problem Solving*, 4th ed. Elsevier Science, Amsterdam, Netherlands, 2017.

[191] NS-3 Consortium, "ns-3." [Online]. Available: https://www.nsnam.org.

[192] OpenSim Ltd., "Omnet++." [Online]. Available: https://omnetpp.org.

[193] MathWorks, "Simulink." [Online]. Available: https://uk.mathworks.com/products/simulink.html.

[194] J. Walrand, K. Bagchi, and G. W. Zobrist, Eds., *Network Performance Modeling and Simulation.* Newark, NJ, USA: Gordon and Breach Science Publishers, Inc., 1999.

[195] S. Robinson, *Simulation: The Practice of Model Development and Use.* USA: John Wiley & Sons, Inc., New York, USA, 2004.

[196] D. Murray-Smith, *Modelling and Simulation of Integrated Systems in Engineering: Issues of Methodology, Quality, Testing and Application.* Woodhead Publishing, Limited, Cambridge, UK, 2012.

[197] B. Kumar and E. S. Davidson, "Performance evaluation of highly concurrent computers by deterministic simulation," *Commun. ACM*, vol. 21, no. 11, pp. 904–913, 1978.

[198] D. Poole and A. E. Raftery, "Inference for deterministic simulation models: The Bayesian melding approach," *J. Am. Stat. Assoc.*, vol. 95, no. 452, pp. 1244–1255, 2000.

[199] J. P. C. Kleijnen, *Design and Analysis of Simulation Experiments*, 1st ed. Springer Publishing Company, Incorporated, New York, USA, 2007.

[200] H. C. Tijms, *Stochastic Modelling and Analysis: A Computational Approach.* New York, NY, USA: John Wiley & Sons, Inc., 1986.

[201] B. D. Ripley, *Stochastic Simulation.* New York, NY, USA: John Wiley & Sons, Inc., 1987.

[202] V. G. Kulkarni, *Modeling and Analysis of Stochastic Systems*. London, UK: Chapman & Hall, Ltd., 1995.

[203] M. A. Pinsky and S. Karlin, *An Introduction to Stochastic Modeling*. Orlando, FL, USA: Academic Press, Inc., 2011.

[204] G. S. Fishman, *Discrete-Event Simulation*. Berlin, Heidelberg: Springer-Verlag, 2001.

[205] B. K. Choi and D. Kang, *Modeling and Simulation of Discrete Event Systems*, 1st ed. Wiley Publishing, New York, USA, 2013.

[206] B. P. Zeigler, T. G. Kim, and H. Praehofer, *Theory of Modeling and Simulation*, 2nd ed. Orlando, FL, USA: Academic Press, Inc., 2000.

[207] F. E. Cellier and E. Kofman, *Continuous System Simulation*. Berlin, Heidelberg: Springer-Verlag, 2006.

[208] L. G. Birta and G. Arbez, *Modelling and Simulation: Exploring Dynamic System Behaviour*. Berlin, Heidelberg: Springer-Verlag, 2007.

[209] S. Bandyopadhyay and R. Bhattacharya, *Discrete and Continuous Simulation: Theory and Practice*, 1st ed. Boca Raton, FL, USA: CRC Press, Inc., 2014.

[210] A. C. Davison, *Statistical Models*. Cambridge University Press, Cambridge, UK, 2003.

[211] D. R. Cox, *Principles of Statistical Inference*. Cambridge University Press, Cambridge, UK, 2006.

[212] J. Rissanen, *Information and Complexity in Statistical Modeling*, 1st ed. Springer Publishing Company, Incorporated, New York, USA, 2007.

[213] L. Wasserman, *All of Statistics: A Concise Course in Statistical Inference*. Springer Publishing Company, Incorporated, New York, USA, 2010.

[214] D. P. Kroese and J. C. C. Chan, *Statistical Modeling and Computation*. Springer Publishing Company, Incorporated, New York, USA, 2013.

[215] A. Papoulis, *Probability, Random Variables, and Stochastic Processes*, 4th ed. McGraw Hill, New York, USA, 2002.

[216] C. Forbes, M. Evans, N. Hastings, and B. Peacock, *Statistical Distributions*, 4th ed. Wiley, New York, USA, 2010.

[217] D. C. Montgomery and G. C. Runger, *Applied Statistics and Probability for Engineers*, 6th ed. Wiley, New York, USA, 2013.

[218] S. M. Ross, *Introduction to Probability Models*, 11th ed. Orlando, FL, USA: Academic Press, Inc., 2014.

[219] S. M. Ross, *Stochastic Processes*, 2nd ed. Wiley, New York, USA, 1995.

[220] R. Nelson, *Probability, Stochastic Processes, and Queueing Theory: The Mathematics of Computer Performance Modeling*. Berlin, Heidelberg: Springer-Verlag, 1995.

[221] D. Daley and D. Vere-Jones, *An Introduction to the Theory of Point Processes: Volume I: Elementary Theory and Methods*, 2nd ed. New York, NY, USA: Springer-Verlag New York, 2003.

[222] D. Daley and D. Vere-Jones, *An Introduction to the Theory of Point Processes: Volume II: General Theory and Structure*, 2nd ed. New York, NY, USA: Springer-Verlag New York, 2008.

[223] J. Illian, A. Penttinen, H. Stoyan, and D. Stoyan, *Statistical Analysis and Modelling of Spatial Point Patterns*. John Wiley & Sons, New York, USA, 2008.

[224] R. F. Bass, *Stochastic Processes*, 1st ed. New York, NY, USA: Cambridge University Press, 2011.

[225] D. Snyder and M. Miller, *Random Point Processes in Time and Space*, 2nd ed. New York, NY, USA: Springer New York, 2012.

[226] M. S. Obaidat, F. Zarai, and P. Nicopolitidis, *Modeling and Simulation of Computer Networks and Systems: Methodologies and Applications*, 1st ed. San Francisco, CA, USA: Morgan Kaufmann Publishers Inc., 2015.

[227] A. van den Bos, *Parameter Estimation for Scientists and Engineers*. New York, NY, USA: Wiley-Interscience, 2007.

[228] B. C. Levy, *Principles of Signal Detection and Parameter Estimation*, 1st ed. Springer Publishing Company, Incorporated, New York, USA, 2008.

[229] R. C. Aster, B. Borchers, and C. H. Thurber, *Parameter Estimation and Inverse Problems*, 2nd ed. Academic Press, Massachusetts, USA, 2012.

[230] F. J. Massey, "The Kolmogorov–Smirnov test for goodness of fit," *J. Am. Stat. Assoc.*, vol. 46, no. 253, pp. 68–78, 1951.

[231] R. B. D'Agostino and M. A. Stephens, Eds., *Goodness-of-fit Techniques*. New York, NY, USA: Marcel Dekker, Inc., 1986.

[232] P. E. Greenwood and M. S. Nikulin, Eds., *A Guide to Chi-Squared Testing*. Wiley, New York, USA, 1996.

[233] Y. Fan, "Goodness-of-fit tests for a multivariate distribution by the empirical characteristic function," *J. Multivar. Anal.*, vol. 62, no. 1, pp. 36–63, 1997.

[234] A. Justel, D. Pena, and R. Zamar, "A multivariate Kolmogorov–Smirnov test of goodness of fit," *Stat. Probab. Lett.*, vol. 35, no. 3, pp. 251–259, 1997.

[235] W. Feller, *An Introduction to Probability Theory and Its Applications, Volume 1*, 3rd ed. Wiley, New York, USA, 1968.

[236] R. D. Yates and D. J. Goodman, *Probability and Stochastic Processes: A Friendly Introduction for Electrical and Computer Engineers*, 2nd ed. John Wiley & Sons, New York, USA, 2004.

[237] J. A. Gubner, *Probability and Random Processes for Electrical and Computer Engineers*. New York, NY, USA: Cambridge University Press, 2006.

[238] A. Leon-Garcia, *Probability and Random Processes For EE's*, 3rd ed. Upper Saddle River, NJ, USA: Prentice-Hall, Inc., 2007.

[239] R. Durrett, *Probability: Theory and Examples*, 4th ed. New York, NY, USA: Cambridge University Press, 2010.

[240] R. G. Gallager, *Stochastic Processes: Theory for Applications*. Cambridge University Press, Cambridge, UK, 2014.

[241] R. Steyer and W. Nagel, *Probability and Conditional Expectation: Fundamentals for the Empirical Sciences*. Wiley, New York, USA, 2017.

[242] M. Mitzenmacher and E. Upfal, *Probability and Computing: Randomization and Probabilistic Techniques in Algorithms and Data Analysis*, 2nd ed. New York, NY, USA: Cambridge University Press, 2017.

[243] Y. V. Prohorov and Y. A. Rozanov, *Probability Theory: Basic Concepts, Limit Theorems, Random Processes*, 1st ed. Springer-Verlag Berlin Heidelberg, Berlin, Germany, 1969.

[244] N. L. Johnson, S. Kotz, and N. Balakrishnan, *Continuous Univariate Distributions, Volume 1*, 2nd ed. Wiley, New York, USA, 1994.

[245] N. L. Johnson, A. W. Kemp, and S. Kotz, *Univariate Discrete Distributions*, 3rd ed. John Wiley & Sons, Inc., New York, USA, 2005.

[246] W. J. Stewart, *Probability, Markov Chains, Queues, and Simulation: The Mathematical Basis of Performance Modeling*. Princeton, NJ, USA: Princeton University Press, 2009.

[247] J. E. Gentle, *Computational Statistics*, 1st ed. Springer Publishing Company, New York, USA, Incorporated, 2009.

[248] J. Miller, *Table of Binomial Coefficients (Royal Society Mathematical Tables: Volume 3)*. Cambridge University Press, Cambridge, UK, 1954.

[249] R. Kaas and J. Buhrman, "Mean, median and mode in binomial distributions," *Stat. Neerl.*, vol. 34, no. 1, pp. 13–18, 1980.

[250] Y. H. Wang, "On the number of successes in independent trials," *Stat. Sin.*, vol. 3, no. 2, pp. 295–312, 1993.

[251] G. E. Andrews, "The geometric series in calculus," *Am. Math. Mon.*, vol. 105, no. 1, pp. 36–40, 1998.

[252] M. R. Spiegel, *Theory and Problems of probability and Statistics (Schaum's Outline Series)*. McGraw-Hill Education (India) Pvt Limited, 2003. [Online]. Available: https://books.google.co.uk/books?id=qXPAOq5qE5wC.

[253] H. S. Wilf, *Generatingfunctionology*. Natick, MA, USA: A. K. Peters, Ltd., 2006.

[254] J. Hilbe, *Negative Binomial Regression*. Cambridge University Press, Cambridge, UK, 2011.

[255] F. Haight, *Handbook of the Poisson distribution*, ser. Publications in Operations Research. Wiley, New York, USA, 1967.

[256] L. Knüsel, "Computation of the Chi square and Poisson distribution," *SIAM J. Sci. Stat. Comput.*, vol. 7, no. 3, pp. 1022–1036, 1986.

[257] N. Balakrishnan and A. P. Basu, *Exponential Distribution: Theory, Methods and Applications*. Taylor & Francis, Oxford, UK, 1996.

[258] M. V. Jambunathan, "Some properties of beta and gamma distributions," *Ann. Math. Stat.*, vol. 25, no. 2, pp. 401–405, 1954.

[259] J. Patel and C. Read, *Handbook of the Normal Distribution*, 2nd ed., ser. Statistics: A Series of Textbooks and Monographs. Taylor & Francis, Oxford, UK, 1996.

[260] J. Aitchison and J. Brown, *Lognormal Distribution*, ser. Department of Applied Economics Monographs. Cambridge University Press, Cambridge, UK, 1957.

[261] E. L. Crow and K. Shimizu, *Lognormal Distributions: Theory and Applications*, ser. Statistics: A Series of Textbooks and Monographs. CRC Press, Florida, USA, 1987.

[262] G. E. P. Box and M. E. Muller, "A note on the generation of random normal deviates," *Ann. Math. Stat.*, vol. 29, no. 2, pp. 610–611, 1958.

[263] L. Devroye, "On the computer generation of random variables with a given characteristic function," *Comput. Math. Appl.*, vol. 7, no. 6, pp. 547–552, 1981.

[264] L. Devroye, *Non-Uniform Random Variate Generation*, 1st ed. Springer-Verlag Berlin, Germany, 1986.

[265] L. Barabesi and L. Pratelli, "Universal methods for generating random variables with a given characteristic function," *J. Stat. Comput. Simul.*, vol. 85, no. 8, pp. 1679–1691, 2015.

[266] C. P. Robert and G. Casella, *Monte Carlo Statistical Methods (Springer Texts in Statistics)*. Berlin, Heidelberg: Springer-Verlag, 2005.

[267] R. Y. Rubinstein and D. P. Kroese, *Simulation and the Monte Carlo Method*, ser. Wiley Series in Probability and Statistics, 2nd ed. Wiley, New York, USA, 2007.

[268] L. Takács, *Introduction to the Theory of Queues*, ser. University Texts in the Mathematical Sciences. Oxford University Press, Oxford, UK, 1962.

[269] G. Newell, *Applications of Queueing Theory*, ser. Monographs on Applied Probability and Statistics. Chapman and Hall, London, UK, 1971.

[270] L. Kleinrock, *Queueing Systems: Volume 1: Theory*. New York, NY, USA: Wiley-Interscience, 1975.

[271] L. Kleinrock, *Queueing Systems: Volume 2: Computer Applications*. John Wiley & Sons, New York, USA, 1976.

[272] G. Giambene, *Queuing Theory and Telecommunications: Networks and Applications*. Berlin, Heidelberg: Springer-Verlag, 2005.

[273] A. S. Alfa, *Queueing Theory for Telecommunications: Discrete Time Modelling of a Single Node System*, 1st ed. Springer Publishing Company, Incorporated, New York, USA, 2010.

[274] R. W. Wolff, "Poisson arrivals see time averages," *Oper. Res.*, vol. 30, no. 2, pp. 223–231, 1982.

[275] E. A. van Doorn and G. Regterschot, "Conditional pasta," *Oper. Res. Lett.*, vol. 7, no. 5, pp. 229–232, 1988.

[276] J. D. C. Little, "A proof for the queuing formula: L = λw," *Oper. Res.*, vol. 9, no. 3, pp. 383–387, 1961.

[277] S. Stidham, "A last word on L = λw," *Oper. Res.*, vol. 22, no. 2, pp. 417–421, 1974.

[278] D. G. Kendall, "Stochastic processes occurring in the theory of queues and their analysis by the method of the imbedded Markov chain," *Ann. Math. Stat.*, vol. 24, no. 3, pp. 338–354, 1953.

[279] A. Lee, *Applied Queueing Theory*, ser. Studies in Management. Macmillan, London, UK, 1966.

[280] L. Takács, "A single-server queue with feedback," *Bell Syst. Tech. J.*, vol. 42, no. 2, pp. 505–519, 1963.

[281] W. Chang, "Single-server queuing processes in computing systems," *IBM Syst. J.*, vol. 9, no. 1, pp. 36–71, 1970.

[282] B. Gopinath and J. A. Morrison, "Discrete-time single server queues with correlated inputs," *Bell Syst. Tech. J.*, vol. 56, no. 9, pp. 1743–1768, 1977.

[283] A. Eckberg, "The single server queue with periodic arrival process and deterministic service times," *IEEE Trans. Commun.*, vol. 27, no. 3, pp. 556–562, 1979.

[284] H. Wang and N. B. Mandayam, "A simple packet-transmission scheme for wireless data over fading channels," *IEEE Trans. Commun.*, vol. 52, no. 7, pp. 1055–1059, 2004.

[285] A. Steiner and S. Shamai, "On queueing and multilayer coding," *IEEE Trans. Inf. Theory*, vol. 56, no. 5, pp. 2392–2415, 2010.

[286] J. Cohen, *The Single Server Queue*, 2nd ed., ser. North-Holland Series in Applied Mathematics and Mechanics. Elsevier Science, Amsterdam, Netherlands, 2012.

[287] L. Zhao, W. Sun, Y. Shi, and J. Liu, "Optimal placement of cloudlets for access delay minimization in SDN-based Internet of Things networks," *IEEE Internet Things J.*, vol. 5, no. 2, pp. 1334–1344, 2018.

[288] D. J. Daley and L. D. Servi, "Idle and busy periods in stable M/M/k queues," *J. Appl. Probab.*, vol. 35, no. 4, pp. 950–962, 1998.

[289] J. R. Artalejo and M. J. Lopez-Herrero, "Analysis of the busy period for the M/M/c queue: An algorithmic approach," *J. Appl. Probab.*, vol. 38, no. 1, pp. 209–222, 2001.

[290] I. Elhanany and M. Kahane, "Heterogeneous bursty traffic dispersion over multiple server clusters," *IEEE Commun. Lett.*, vol. 9, no. 3, pp. 261–263, 2005.

[291] S. Dudin and C. Kim, "Analysis of multi-server queue with spatial generation and location-dependent service rate of customers as a cell operation model," *IEEE Trans. Commun.*, vol. 65, no. 10, pp. 4325–4333, 2017.

[292] M. Fidler, B. Walker, and Y. Jiang, "Non-asymptotic delay bounds for multi-server systems with synchronization constraints," *IEEE Trans. Parallel Distrib. Syst.*, vol. 29, no. 7, pp. 1545–1559, 2018.

[293] Z. Jiang, B. Krishnamachari, S. Zhou, and Z. Niu, "Optimal sleeping mechanism for multiple servers with MMPP-based bursty traffic arrival," *IEEE Wireless Commun. Lett.*, vol. 7, no. 3, pp. 436–439, 2018.

Index

access layer 85–6
adaptive routing 42
Address Resolution Protocol (ARP) 32, 34
analytical modeling 131–2
anycast address 72
application-layer protocols 34–5
application-requirement analysis 100–1
asymmetric DSL (ADSL) 27
asynchronous transfer mode (ATM) 25–6
autonomous system (AS) 36
availability, defined 104

backbone flow 116, 122, 126
bandwidth, defined 5
banking system, components in 134
Basic Rate Interface (BRI) 24
Bayes' theorem 165–7
benchmarking 129–30
Bernoulli random variable 197–9, 202, 242–3
Bernoulli trial 198, 200
binomial distribution 153
binomial random variable 199–202, 248
birth–death process
  steady-state distribution of 261
  see also Poisson point process (PPP), for queuing systems
bit error rate (BER) 95–6
Border Gateway Protocols (BGPs) 47, 56–7
bridges/switches 38, 40
broadcast transmission 15
buffered repeaters 39
bus topology 7

C++ programming language 130
campus area network (CAN) 2–3
capacity 95, 100, 104
capacity planning 115–16
capital expenditures (CAPEX) 93
carrier equipment 26
carrier sense multiple access with collision avoidance (CSMA/CA) 13
carrier sense multiple access with collision detection (CSMA/CD) 12–13
Classes of Service (CoS) 95
classful IP addresses 63–6
classful routing protocol 50
classless inter-domain routing (CIDR) notation 62
client–server model 119, 126
closely coupled nodes 120
cloud computing applications 101
coarse granularity 120–1
collectively exhaustive set 160
Committed Information Rates (CIRs) 95
composite flow 116
computer and communication networks, queuing systems in 256–7
computers 39
computing cluster 121
computing nodes 120–1
conditional pmf 178
conditional probability 164–8
configuration of network 90

connectionless packet switching 21
connection-oriented packet switching 21
content-delivery network (CDN) 87
continuous distributions 146
continuous random variables 186–92, 217
  continuous uniform random variable 217–21
  Erlang random variable 226–9
  exponential random variable 222–6
  formulas of 218
  lognormal random variable 235–8
  normal random variable 230–4
  review of 217
continuous simulation models 140
continuous uniform distribution 153
continuous uniform random variable 187–8, 217, 246
  expected value and variance of 220–1
controlled-rate application 100–1
control token 13–14
cooperative computing model 119–20, 126
core layer 84
cost versus performance 82
count-to-infinity condition 52
count-to-infinity problem 50
covariance 184
cumulative distribution function (cdf) 178–80, 187–8
  of continuous uniform random variable 219–20
  of Erlang random variable 228–9
  of exponential random variable 222–4
  of lognormal random variable 237
  of normal random variable 231–3
customer premises equipment (CPE) 26

datagram packet switching 21
data sources and sinks 117
data transport applications 101
design approaches and evolution 86–7
design process 82
deterministic simulation models 139
device-to-device (D2D) communication systems 118
digital subscriber line (DSL) 26–8
Dijkstra's algorithm 55–6
directly connected networks 47
discrete distributions 146, 153
discrete random variables 176–86, 197
  Bernoulli random variable 197–9, 202
  binomial random variable 199–202
  discrete uniform random variable 211–14
  formulas of 198
  geometric random variable 202–5
  Pascal random variable 205–8
  Poisson random variable 208–11
  review of 197
discrete simulation models 139
discrete uniform random variable 211–14, 245
distance-vector routing algorithms 47
distance-vector routing protocols (DVRPs) 49
  Interior Gateway Routing Protocol (IGRP) 50–1
  Routing Information Protocol (RIP) 50
  routing loops 51–2
distributed computing model 120–1, 126
distribution fitness test 155
distribution layer 84–5
domain name system (DNS) 33, 35
dotted-quad format 63
dynamic routing algorithms 47
dynamic routing protocols 48–9
dynamic simulation models 139

empirical distribution 153
encapsulation 31
Enhanced Interior Gateway Routing Protocol (EIGRP) 45, 56
enterprise requirements 102

Erlang random variable 226–9, 250
  expected value and variance of 229
ES 36
event, defined 160–1
event notification via monitoring 90
exchange-termination equipment 23
expected value 180–2
  of continuous random variable 190
  of discrete random variables 180–2
experiment, defined 159
exponential distribution 145–6, 153
exponential random variable 222–6, 247
  expected value and variance of 224–5
exterior gateway protocol (EGP) 47–8

failure rate 104
fast Ethernet 11
File Transfer Protocol (FTP) 29, 35
financial requirements 102
fine granularity 120
finest grain set 160
flat routing algorithms 47
flow analysis model 125–6
flow boundaries 121–2
flow distributions 122–3
flow models 117
  client–server model 119
  cooperative computing model 119–20
  distributed computing model 120–1
  peer-to-peer (P2P) model 118
flows 115–16
flowspec 123–4
  developing 126
flow specification 123–5
flow types 116
frame relay 24–5
full-duplex system 117
full mesh topologies 19–20

Gamma and Erlang distribution 146
Gamma and Weibull distribution 146
gateways 40
Gaussian random variable: *see* normal random variable
generic computing devices 101
geometric random variable 202–5, 243–4

half-duplex systems 117
hierarchical design of network 49, 82–3
hierarchical layers of network 84
  access layer 85–6
  core layer 84
  distribution layer 84–5
high-bit-rate DSL 28
hold-down timers 52
home area network (HAN) 2
homogeneous Poisson point process 147–50
host-requirement analysis 101–2
hosts 93
hubs 5–6
100BaseTX and 100BaseFX fast Ethernet 11
HyperText Transfer Protocol (HTTP) 35

independence of events 168–70
individual flow 116
inhomogeneous Poisson point process 147
input modelling 152
  data collection 152–3
  distribution fitness test 155
  parameter estimation 154–5
  probability distribution identification 153–4
Integrated Services Digital Network (ISDN) 22–4
interarrival and service time in Poisson point process 259
interior gateway protocols (IGPs) 47, 49–56
Interior Gateway Routing Protocol (IGRP) 45, 50–1
intermediate system (IS) 36

Intermediate System to Intermediate System (IS–IS) 53
International Organization for Standardization (ISO) 28
Internet Control Message Protocol (ICMP) 32, 34
Internet Engineering Task Force (IETF) 94
Internet Protocol addressing 37, 61
  addressing mechanisms 63
  addressing types 63
  IP version 4 (IPv4) 63–4
    classful IP addresses 64–6
    private IP addresses 64
  IP version 6 71–3
  subnetting 66–8
  supernetting 70–1
  variable-length subnetting (VLSM) 69–70
Internet Protocols 32
  application-layer protocols 35
  ARP 34
  ICMP 34
  Internet Protocol 34
  TCP/IP suite 33
  Transmission Control Protocol (TCP) 34
  User datagram protocol (UDP) 34
Internet service providers (ISPs) 56
internetworking and network fundamentals 1–5
  Internet protocols 32–5
  internetworks 35
    addressing 36–7
    devices 38–40
    hierarchy 36
    standards organizations 38
  local area networks (LAN) 5
    access methods 12–14
    devices 5–7
    technologies 9–12
    topologies 7–8
    transmission methods 14–16
  open systems interconnection (OSI) reference model 28–9
    information exchange and information formats in OSI layers 31–2
    interaction between OSI layers in communications 30–1
    OSI layers 29–30
  wide area networks (WAN) 16
    devices 16–18
    technologies 20–8
    topologies 18–20
inverse transform technique 241
  Bernoulli random variable generation 242–3
  continuous uniform random variable generation 246
  discrete uniform random variable generation 245
  exponential random variable generation 247
  geometric random variable generation 243–4
IP version 4 (IPv4) 63–4
  classful IP addresses 64–6
  private IP addresses 64
IP version 6 71–3

Kendall’s notation for queuing systems 262–3
Kolmogorov–Smirnov (K–S) test 155
kurtosis 186

latency 95–6, 100, 104
layer-distributing network services 84
line-termination equipment 23
link-state advertisements (LSAs) 53
link-state routing protocols (LSRPs) 49, 52
  Dijkstra’s algorithm 55–6
  Open Shortest Path First (OSPF) 53–4
Little’s law 259
  for queuing systems 261–2
local addresses versus global addresses 63
local area network (LAN) 1–2

access methods 12
carrier sense multiple access with collision avoidance (CSMA/CA) 13
carrier sense multiple access with collision detection (CSMA/CD) 12–13
control token 13–14
devices 5
hubs 5–6
routers 6
switches 6
technologies 9
1000BaseT and 1000BaseX gigabit Ethernet 11
100BaseTX and 100BaseFX fast Ethernet 11
10BaseT Ethernet 10–11
10GBaseSR and 10GBaseLR 10Gbps Ethernet 11
token ring 11
wireless Ethernet 11
topologies 7
bus 7
extended star topology 9
ring 7–8
star 7–8
transmission methods 14
broadcast transmission 15
multicast transmission 15–16
unicast transmission 14–15
logical link control (LLC) sub-layer 30
lognormal random variable 235–8, 250–1
expected value and variance of 237–8
loosely coupled nodes 120

MAC addresses 36–7
Maclaurin series of exponential function 210
maintainability, defined 104
management and security of network 87
network design goals 88
network-management design 89–91
technology evaluation criteria 89
management requirements 102
Markov process 147
MATHEMATICA 131
mathematical modelling: *see* analytical modeling
MATLAB 131
maximum bit rate 95
mean time between failures (MTBF) 104–5
mean time to repair (MTTR) 104
media access control (MAC) sub-layer 30
metropolitan area network (MAN) 2
mission-critical applications 100
M/M/1/K/K queue 274–8
M/M/1/K queue 271–4
M/M/1 queue 267–70
M/M/c/K/K queue 291–6
M/M/c/K Queue 286–91
M/M/c queue 281–6
multicast address 72
multicast transmission 15–16
multipart flowspec 124–5
multi-server queues 256, 281
M/M/c/K/K queue 291–6
M/M/c/K Queue 286–91
M/M/c queue 281–6
mutually exclusive set 160

network analysis process 81
network architecture process 81–2
network interface card (NIC) 11
network layer addresses 37
network-management protocols 90
network mask 62
network-termination (NT) devices 23
non-real-time applications 100
nonstationary Poisson point process (NPPP) 147,150–1
normal distribution 145–6, 153
normal random variable 230–5
generation 250–1

number of packets in the queue and in the system 260
Nyquist's theorem 96–7

octet 37, 63, 71–2
OMNeT++ 130
one-part flowspec 123–4
Open Shortest Path First (OSPF) 45, 53–4
open systems interconnection (OSI) reference model 28–9
  information exchange and information formats in 31–2
  interaction between OSI layers in communications 30
  layer 1, physical layer 30
  layer 2, data-link layer 30
  layer 3, network layer 30
  layer 4, transport layer 29
  layer 5, session layer 29
  layer 6, presentation layer 29
  layer 7, application layer 29
operation, administration, and maintenance applications 101
OPEX 79, 82, 88, 93
optical fiber 9–10
outcome, defined 159

packet error rate (PER) 95–7
packet-switched networks (PSNs) 22, 25
packet switching 21, 46
parallel computing system 121
  distributed computing model with 126
partial mesh topologies 20
Pascal distribution 146
Pascal random variable 205–8
Pascal random variable generation 249
peer-to-peer (P2P) model 118, 126
performance evaluation 129
  analytical modeling 131–2
  benchmarking 129–30
  simulation 130–1
  systems and system environment 132–4
  via monitoring 90
performance measures in queuing systems 258
  interarrival and service time in Poisson point process 259
  Little's law for queuing systems 261–2
  number of packets in the queue and in the system 260
  traffic intensity and server utilization 260–1
  waiting and response time 259–60
personal area network (PAN) 2, 5
Poisson distribution 146–7, 151, 153
Poisson point process (PPP) 145, 147, 222
  interarrival and service time in 259
  nonstationary Poisson point process 150–1
  operations of 151
  for queuing systems 257–8
  stationary Poisson point process 148–50
Poisson random variable 208–11, 258
Primary Rate Interface (PRI) 24
private IP addresses 64
probabilities in performance analysis 159
  axioms and properties of 161–4
  basic concepts and terminology 159–61
  conditional probability 164–8
  independence of events 168–70
probability density function (pdf) 188–90, 259
  of continuous uniform random variable 218
  of Erlang random variable 226
  of exponential random variable 222
  of lognormal random variable 235
  of normal random variable 230
probability distribution identification 153–4

probability mass function (pmf) 177–8, 198, 200, 203
of Bernoulli random variable 198–9
of binomial random variable 200
of discrete uniform random variable 211
of geometric random variable 203
of Pascal random variable 206
of Poisson random variable 208
probability measure 161, 175
protocol data unit (PDU) 31–2
public data networks (PDNs) 26

Quality of Service (QoS) 95
quantile–quantile (Q–Q) plot 154
queuing system 141
simulation phases in 142
statistical models in 145–6
queuing theory 147, 255–6
computer and communication networks, queuing systems in 256–7
Kendall's notation for queuing systems 262–3
performance measures in queuing systems 258
interarrival and service time in Poisson point process 259
Little's law for queuing systems 261–2
number of packets in the queue and in the system 260
traffic intensity and server utilization 260–1
waiting and response time 259–60
Poisson point process (PPP) for queuing systems 257–8

random variable generation in network simulation 241
inverse transform technique 241
Bernoulli random variable generation 242–3
continuous uniform random variable generation 246
discrete uniform random variable generation 245
exponential random variable generation 247
geometric random variable generation 243–4
transformation techniques 248
binomial random variable generation 248
Erlang random variable generation 250
lognormal random variable generation 250–1
Pascal random variable generation 249
random variables 175–6
continuous random variables 186
cumulative distribution function 187–8
expected value 190
moments 191–2
probability density function 188–90
variance and standard deviation 190–1
discrete random variables 176
cumulative distribution function 178–80
expected value 180–2
moments 185–6
probability mass function 177–8
variance and standard deviation 182–5
real-time applications 100
reliability 95, 100
definition 104
reliability, maintainability, and availability analyses 104–9
remote networks 47
repeaters/hubs 38–40
requirement analysis 93, 102
application-requirement analysis 100–1
host-requirement analysis 101–2

network services and requirements 93–4
reliability, maintainability, and availability analyses 104–9
requirement-analysis model 103–4
requirement-analysis process 98–9
service characteristics 94–7
user-requirement analysis 99–100
requirement-analysis model 103–4
requirement-analysis process 98–9
response time, waiting and 259–60
ring topology 7–8
Riverbed Modeler 130
routers 6, 40
routing in computer networks 45
border gateway protocols (BGPs) 56–7
dynamic routing protocols 48–9
fundamentals 45
path determination 46
routing algorithms 46–8
interior gateway protocols (IGPs) 49
distance-vector routing protocols (DVRPs) 49–52
enhanced interior gateway routing protocol (EIGRP) 56
link-state routing protocols (LSRPs) 52–6
static routing protocols 48
Routing Information Protocol (RIP) 50
routing loops 51–2

sample space 159
security requirements 102
servers 101, 120, 256
server utilization 260, 270, 274, 296
service-based networking 93
service characteristics 94–7
service data unit (SDU) 31
service planning 115
set, defined 160
Shannon's theorem 97
shielded twisted-pair (STP) 9–10
Simple Mail Transfer Protocol (SMTP) 35
Simple Network Management Protocol (SNMP) 35, 89–91
simple repeaters 39
simplicity versus functionality 82
simulation, merits and demerits of 130–1
simulation model 137
deterministic 139
dynamic 139
for emulating and evaluating a system 137–8
example 140–1
static 139
steps/phases in 140
stochastic 139
of a system 139–40
single-path routing algorithms 47
single-server queues 256, 267
components in 134
M/M/1/K/K queue 274–8
M/M/1/K queue 271–4
M/M/1 queue 267–70
skewness 186
specialized equipment 101
split horizon rule 52
standards organizations 38
star topology 7–9
static routing algorithms 47
static routing protocols 48
static simulation models 139
stationary Poisson point process (SPPP) 147–50
statistical models 145
input modelling 152
data collection 152–3
distribution fitness test 155
parameter estimation 154–5
probability distribution identification 153–4
Poisson point process (PPP) 147
nonstationary Poisson point process 150–1
operations of 151
stationary Poisson point process 148–50

in queuing systems 145–6
statistical multiplexing 25
steady-state distribution of the birth–death process 261
steady-state probability 268
stochastic simulation models 139
subnet mask 62–3
subnetting 63, 66–8
variable-length subnetting 69–70
subscriber network interface (SNI) 26
supernetting 63, 70–1
switched multimegabit data service (SMDS) 26
switches 6
symmetric DSL (SDSL) 27–8
system, defined 93
system components 94
systems and system environment 132–4

telemetry control applications 101
teleservice applications 101
Telnet 35
temporary addresses versus persistent addresses 63
10BaseT Ethernet 10–11
10GBaseSR and 10GBaseLR 10Gbps Ethernet 11
terminal adapter (TA) 23
terminals 23
1000BaseT and 1000BaseX gigabit Ethernet 11
throughput, defined 6
time between failure (TBF) of a system 105
time to repair (TTR) of a system 105
token ring 9, 11
transformation techniques 248
binomial random variable generation 248
Erlang random variable generation 250
lognormal random variable generation 250–1
Pascal random variable generation 249
Transmission Control Protocol (TCP) 34
Transmission Control Protocol/Internet Protocol (TCP/IP) 1
TCP/IP suite 33–4
triangular distribution 153
Trivial FTP (TFTP) 35
troubleshooting 83, 90
truncated normal distribution 145
two-part flowspec 124
Types of Service (ToS) 95

unicast address 72
unicast transmission 14–15
unshielded twisted-pair (UTP) 9–10
user datagram protocol (UDP) 34
user-requirement analysis 99–100
users 93

variable-length packets 25
variable-length subnetting (VLSM) 63, 69–70
variance and standard deviation 182–5, 190–1
VC packet switching 21
very high-data-rate DSL 28
virtual circuits (VCs) 20
PermanentVC (PVC) 22
Switched VC (SVC) 22
visualization applications 101

waiting and response time 259–60
waiting time 259–60, 269
web applications 101
Weibull distribution 146, 153
wide area network (WAN) 1–2, 16, 122
devices 16
access server 18
channel service unit/digital service unit 17
modems 17
multiplexers 18
WAN switches 17
technologies 20

asynchronous transfer mode (ATM) 25–6
circuit switching 21
digital subscriber line (DSL) 26–8
frame relay 24–5
integrated services digital network (ISDN) 22–4
packet switching 21
switched multimegabit data service (SMDS) 26
virtual circuits 22
X.25 24
topologies 18
full mesh 19–20
partial mesh 20
wireless Ethernet 11

X.25 24–5

www.ingramcontent.com/pod-product-compliance
Lightning Source LLC
LaVergne TN
LVHW011910200926
848484LV00002B/584